GCSE physics

fourth edition

GCSE

physics

fourth edition

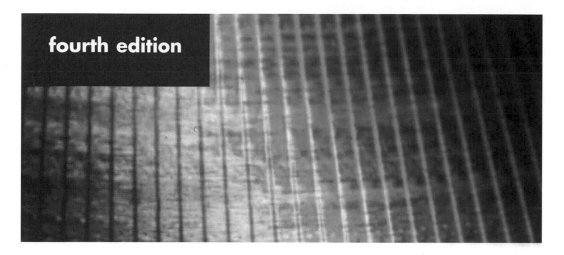

Tom Duncan Heather Kennett

HODDER
EDUCATION
PART OF HACHETTE LIVRE UK

To F.D.B.

© Tom Duncan 1986, 1987, 1995
© Tom Duncan and Heather Kennett 2001

First published in 1986
by Hodder Education,
part of Hachette Livre UK
338 Euston Road
London NW1 3BH

Second edition 1987
Reprinted 1987, 1988 (twice), 1990, 1991, 1992, 1993
Third edition 1995
Reprinted 1995, 1996, 1997, 1998, 1999
Fourth edition 2001
Reprinted 2002 (three times), 2003 (twice), 2004, 2005, 2006 (twice), 2008 (twice)

Layouts by Fiona Webb
Artwork by Wearset, Boldon, Tyne and Wear
Cover design by John Townson/Creation

Typeset in 11.5/13 pt Bembo by Wearset, Boldon, Tyne and Wear

Printed in Dubai

A catalogue entry for this title may be obtained from the British
Library

ISBN-13: 978-0-719-58614-9

Contents

Preface

Tom Duncan has been joined by Heather Kennett in the preparation of this new edition. A major revision has been undertaken to cover the core and extension content of the new GCSE courses and to meet the requirements of the revised National Curriculum. Material no longer in syllabuses has been removed and the following topics have been introduced, extended or brought up to date:

- **Ideas and evidence in science** – a section has been added with links to appropriate chapters.
- **Waves and sound** – details of a range of ultrasonic echo techniques such as medical ultrasound imaging have been included.
- **Motion and energy** – factors affecting stopping distances are analysed in greater detail and the sections on communication and monitoring satellites have been brought together and extended.
- **Heat and energy** – this section now includes a fuller comparison of different energy sources.
- **Electricity and electromagnetic effects** – the principle of inkjet printers has been included, potential divider circuits and applications are treated in more detail, and magnetic recording and metal detectors have been added.
- **Electrons and atoms** – extensions here include a review of sources of background radiation, radiation hazards, the uses of radioisotopes for dating and other applications, β^+ and β^- decay, nuclear stability, fundamental particles, optic fibre communications and digital signals.
- **Earth and space physics** – the method of location of earthquake epicentres has been introduced; the search for extra-terrestrial life is discussed and there is more on black holes and the big bang theory; material on the atmosphere and the weather, no longer required, has been removed.

- **Datalogging and computers** – investigations which are suitable for connection to a datalogger and computer have been identified throughout the text.
- **Questions** – older questions have been removed and many new questions from recent examination papers have been included. In the end-of-chapter questions, in the **Additional questions** (after each group of related topics, for homework) and in the **Revision questions** (at the end of the book, for quick, comprehensive revision before examinations), the more difficult 'higher' questions are marked with a stripe down the left-hand side.
- **Answers** – the answer section at the end of the book has been expanded to include all but long descriptive answers.

Many thanks are due to W. S. Tucker for his comprehensive analysis of the various GCSE syllabuses and for his detailed comments and helpful suggestions about where changes should be made to the text. Thanks again go to Keith Munnings, Neil Duncan and Brian and Malcolm Kennett for their helpful advice during the preparation of this or former editions. The authors are indebted to Jane Roth for her excellent editorial work in the preparation of this new edition.

T.D. and H.K.

Acknowledgement is made to the following examining boards (answers given being the sole responsibility of the authors):

AQA:
   NEAB (Northern Examinations and Assessment Board)
   SEG (Southern Examining Group)
EDEXCEL: *London* (London Examinations)
OCR (Oxford, Cambridge and RSA Examinations)

Photo acknowledgements

Cover Tom Carroll/Phototake NYC/robertharding.com;
p.ix *t* Philippe Plailly/Science Photo Library, *b* Space Telescope Science Institute/NASA/Science Photo Library; **p.x** *tl* NASA/Science Photo Library, *tr* courtesy Intelsat, *bl* Martin Bond/Science Photo Library, *br* MIT AI Lab/Surgical Planning Lab/Brigham & Women's Hospital/Science Photo Library; **p.xi** Christine Boyd; **p.1** George Post/Science Photo Library; **p.2** *t* Topham Picturepoint, *b* Alexander Tsiaras/Science Photo Library; **p.6** Corbis Stock Market; **p.8** Last Resort; **p.9** Last Resort; **p.11** *both* Andrew Lambert; **p.12** Last Resort; **p.13** Andrew Lambert; **p.15** Last Resort; **p.18** *tl* Last Resort, *tr* Roger Scruton, *b* CNRI/Science Photo Library; **p.20** *both* Last Resort; **p.27** *l* Alfred Pasieka/Science Photo Library, *r* Last Resort; **p.28** Last Resort; **p.29** Last Resort; **p.31** Kodak Limited; **p.33** Robin Scagell/Science Photo Library; **p.34** Stephen & Donna O'Meara/Science Photo Library; **p.35** *t* NASA, *b* François Gohier/Science Photo Library; **p.39** Adrienne Hart-Davis/Science Photo Library; **p.42** *tl, tr, cr, bl* Andrew Lambert, *br* HR Wallingford Ltd.; **p.43** © Peter Gould; **p.45** Last Resort; **p.46** *tl & bl* Andrew Lambert, *tr* Dr Jeremy Burgess/Science Photo Library, *br* Peter Aprahamian/Science Photo Library; **p.47** Pascal Goetgheluck/Science Photo Library; **p.48** *tl & bl* Last Resort, *tr* Ray Ellis/Science Photo Library, *br* API Foils, Advanced Holographic Laboratories, Loughborough University; **p.51** *l* US Geological Survey/Science Photo Library, *r* Corbis Stock Market; **p.52** Unilab (a Philip Harris Education brand); **p.53** Westinghouse; **p.54** Jonathan Watts/Science Photo Library; **p.57** Department of Clinical Radiology, Salisbury District Hospital/Science Photo Library; **p.59** Andrew Drysale/Rex Features; **p.60** © Steve Prezant/Corbis Stock Market; **p.61** *t* John Townson/Creation, *b* MSCUA, University of Washington Libraries; **p.65** Lawrence Livermore Laboratory/Science Photo Library; **p.66** *both* NASA/Science Photo Library; **p.69** Avery Berkel Salter Weigh-Tronix; **p.72** Israeli Government Tourist Board; **p.74** *both* Sporting Pictures; **p.77** Dr Linda Stannard, UCT/Science Photo Library; **p.79** *t* Claude Nuridsany & Marie Perennou/Science Photo Library, *b* Last Resort; **p.83** Rex Features; **p.87** Rex Features; **p.89** Kerstgens/SIPA Press/Rex Features; **p.90** *t* Silsoe Research Institute, *b* TransBus International; **p.93** © BP Amoco p.l.c. (1995); **p.96** Ray Fairall/Photoreporters/Rex Features; **p.97** *t* Alton Towers, *ct* Glen Dimplex Heating Ltd., *cb* Last Resort, *b* Scottish & Southern Energy plc; **p.103** JCB; **p.104** Last Resort; **p.107** *l* Robert Harding, *tr* Mark Burnett/Science Photo Library, *br* Ivor Walton; **p.110** Sporting Pictures; **p.111** *t* Pascal Rondeau/Allsport, *b* François Gohier/Ardea; **p.113** Sporting Pictures; **p.117** © David Stoeklein/Corbis Stock Market; **p.118** © Globus, Holway & Lobel/Corbis Stock Market; **p.120** Andrew Lambert; **p.126** Scala; **p.128** PSSC Physics © 1965, Education Development Center, Inc.; D.C. Heath & Company; **p.130** © Arbortech Industries Limited; **p.133** Agence DPPI/Rex Features; **p.137** *t* Coloursport, *bl* Patrick Eager, *br* Mike Powell/Allsport; **p.139** Charles Ommanney/Rex Features; **p.140** Last Resort; **p.141** Transport Research Laboratory; **p.143** Sporting Pictures; **p.144** Rex Features; **p.145** photo ESA; **p.151** © ALSTOM; **p.152** photo by Pete Mouginis-Mark; **p.155** *l* photo courtesy of BOC Limited, *r* Milepost 92 1/2; **p.157** John Townson/Creation; **p.161** Philippe Plailly/Eurelios/Science Photo Library; **p.169** John Townson/Creation; **p.174** *tl & bl* Rockwool Ltd, *r* John Townson/Creation; **p.175** *t* © Peter Gould, *b* John Townson/Creation; **p.176** Sky Systems Ltd.; **p.178** Gail Goodger/Hutchison; **p.180** James R. Sheppard; **p.182** *t* ETSU/AEAT Environment, *bl* Aurora Vehicle Association Inc., *br* © ALSTOM; **p.183** *tl & bl* Mark Edwards/Still Pictures, *r* © ALSTOM; **p.185** AP Photo/Ben Margot; **p.191** Richard R. Hansen/Science Photo Library; **p.192** Keith Kent/Science Photo Library; **p.193** Last Resort; **p.207** *both* RS Components; **p.212** *both* Andrew Lambert; **p.216** Sam Ogden/Science Photo Library; **p.217** Andrew Lambert; **p.223** RS Components; **p.224** Siemens Metering Limited; **p.232** Andrew Lambert; **p.236** Alex Bartel/Science Photo Library; **p.240** Elu Power Tools; **p.247** *both* © ALSTOM; **p.251** ALSTOM T & D Transformers Ltd.; **p.257** CERN/Science Photo Library; **p.261** Andrew Lambert; **p.266** *tl & bl* The Royal Society, Plate 16, Fig 1 from CTR Wilson, Proc. Roy. Soc. Lond. A104, pp1-24 (1923), *r* Lawrence Berkeley Laboratory/Science Photo Library; **p.268** *l* Lippke, *r* © University Museum of Cultural Heritage - University of Oslo, Norway (photo Eirik Irgens Johnsen); **p.275** courtesy of the Physics Department, University of Surrey; **p.276** CERN/Science Photo Library; **p.280** *t* Andrew Lambert, *b* Artema MEC; **p.281** *all* Andrew Lambert; **p.283** *both* RS Components; **p.284** John Townson/Creation; **p.287** Unilab (a Philip Harris Education brand); **p.290** *t* RS Components, *b* Unilab (a Philip Harris Education brand); **p.291** *t* Hugh Steeper Limited, *b* James King-Holmes/Science Photo Library; **p.292** Moviestore Collection; **p.297** *l* Courtesy of Cable & Wireless Archive, Porthcurno, *r* The Science Museum/Science & Society Picture Library; **p.298** *t* The Science Museum/Science & Society Picture Library, *b* Popperfoto; **p.302** John Townson/Creation; **p.303** courtesy Alcatel; **p.304** © Cable & Wireless; **p.313** Arc Science Simulations/Science Photo Library; **p.318** *l* Wendy Shattil/Bob Rozinski/Animals Animals/Oxford Scientific Films, *r* Peter MacDiarmid/Rex Features; **p.319** *l* J. Allan Cash, *tr* British Geological Survey © NERC. All rights reserved. *br* courtesy of the Northern Ireland Tourist Board; **p.320** *l* British Geological Survey © NERC. All rights reserved., *tr* © David W. Jones/Lakeland Life Picture Library, *br* © The Natural History Museum, London; **p.324** John Sanford/Science Photo Library; **p.325** NASA/Science Photo Library; **p.326** *both* NASA/Science Photo Library; **p.327** NASA/Science Photo Library; **p.328** Royal Greenwich Observatory/Science Photo Library; **p.330** *t* Mallan Morton/Science Photo Library, *b* NASA; **p.331** Pekka Parviainen/Science Photo Library; **p.332** Space Telescope Science Institute/Science Photo Library; **p.334** NASA/Science Photo Library.

t = top, *b* = bottom, *l* = left, *r* = right, *c* = centre.

Every effort has been made to contact copyright holders, and the publishers apologise for any omissions which they will be pleased to rectify at the earliest opportunity.

Physics and technology

Physicists explore the Universe. Their investigations range from particles that are smaller than atoms to stars that are millions and millions of kilometres away, Figures 1a, b.

As well as having to find the **facts** by observation and experiment, physicists also must try to discover the **laws** that summarize (often as mathematical equations) these facts. Sense has then to be made of the laws by thinking up and testing **theories** (thought-models) to explain the laws. The reward, apart from a satisfied curiosity, is a better understanding of the physical world. Engineers and technologists use physics to solve **practical problems** for the benefit of people, though in solving them social, environmental and other problems may arise.

In this book we will study the behaviour of **matter** (the stuff things are made of) and the different kinds of **energy** (such as light, sound, heat, electricity). We will also consider the applications of physics in the home, in transport, medicine, research, industry, energy production, meteorology, communications and electronics. Figures 2a, b, c and d show some examples.

Mathematics is an essential tool of physics and a 'reference section' of some of the basic mathematics is given at the end of the book along with a suggested procedure for solving physics problems.

Figure 1a This image, produced by a scanning tunnelling microscope, shows an aggregate of gold just three atoms thick on a graphite substrate. Individual graphite (carbon) atoms are shown as green

Figure 1b The many millions of stars in the Universe, of which the Sun is just one, are grouped in huge galaxies. This photograph of two interacting spiral galaxies was taken with the Hubble Space Telescope (see Figure 11.6). This orbiting telescope is enabling astronomers to tackle one of the most fundamental questions in science, i.e. the age and scale of the Universe, by giving much more detailed information about individual stars than is possible with ground-based telescopes

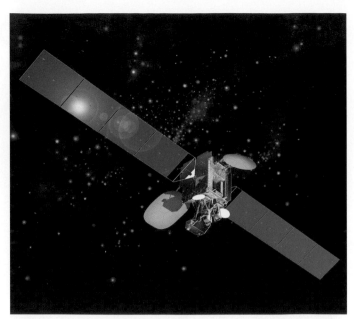

Figure 2c Communications satellites like the INTELSAT IX shown above are in geostationary orbit 36 000 km above the equator where they circle the Earth in 24 hours and so appear to be at rest. Microwave signals are sent to the satellite and received from it by Earth stations with large dish aerials (like those in Figure 61.19), enabling digital voice, data, internet and video transmissions across the globe

Figure 2a The manned exploration of space is such an expensive operation that international co-operation is seen as the way forward. This is the International Space Station, being built module-by-module in orbit around the Earth. It is operated as a joint venture by the USA and Russia

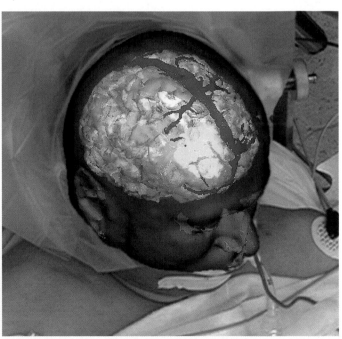

Figure 2b In the search for alternative energy sources, 'wind farms' of 20 to 100 wind turbines have been set up in suitable locations, such as this in North Wales, to generate at least enough electricity for the local community

Figure 2d An effect called 'virtual reality' can be produced using computer graphics. In this photograph a virtual reality image of a patient's brain, with tumour and blood vessels, is superimposed on to the patient's head. This enables the surgeon to locate the tumour before an incision is made, resulting in minimally invasive surgery avoiding the vital blood vessels

Scientific enquiry

During your course you will have to carry out a few scientific investigations aimed at encouraging you to develop some of the **skills** and **abilities** that scientists use to solve real-life problems.

Investigations may arise from the topic you are currently studying in class, or your teacher may provide you with suggestions to choose from, or you may have your own ideas. However an investigation arises, it will probably require at least one hour of laboratory time, but often longer and will involve you in the following four aspects.

1 **Planning** how you are going to set about finding answers to the questions the problem poses. Making predictions and hypotheses (informed guesses) may help you to focus on what is required at this stage.
2 **Obtaining** the necessary experimental data safely. You will have to decide on the equipment needed, what observations and measurements have to be made and what variable quantities need to be manipulated.
3 **Presenting** and **interpreting** the evidence in a way that enables any relationships between quantities to be established.
4 **Considering** and **evaluating** the evidence by drawing conclusions, assessing the reliability of data gathered and making comparisons with what was expected.

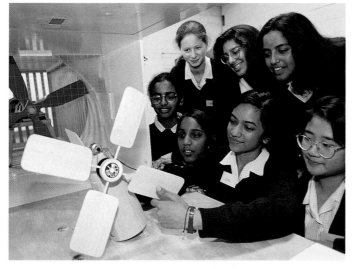

Figure 3 Girls from Copthall School, Mill Hill, London, with their winning entry for a contest to investigate, design and build the most efficient, elegant and cost-effective windmill

A **written report** of the investigation would normally be made, indicating the aim of the work, giving records of procedures, observations and measurements, and stating and evaluating the conclusions based on the evidence gathered.

Suggestions for investigations

1 Vibrating of a long steel strip clamped at one end (Chapter 16).
2 Resonance of an air column (Chapter 16).
3 Pitch of note from a vibrating wire (Chapter 16).
4 Stretching of rubber bands (Chapter 19).
5 Stretching of copper wires – **wear safety glasses** (Chapter 19).
6 Bending of 'beams' (strips or sheets) of different materials (Chapter 21).
7 Friction – factors affecting (Chapter 23).
8 Energy values from burning fuel, e.g. firelighter (Chapter 24).
9 Fall of ball-bearings in a liquid (Chapter 27).
10 Flow of liquid through tubes (Chapter 27).
11 Viscosity of different liquids (Chapter 27).
12 Speed of a bicycle and its stopping distance (Chapter 33).
13 Circular motion using a bung on a string (Chapter 34).
14 Heat loss using different insulating materials (Chapter 40).
15 Model wind turbine design (Chapter 42).
16 Resistance of a thermistor and temperature (Chapter 46).
17 Heating effect of an electric current (Chapter 49).
18 Strength of an electromagnet (Chapter 52).
19 Efficiency of an electric motor (Chapter 53).

Ideas and evidence in science

You may find that in some of the investigations you perform in the school laboratory you do not interpret your data in the same way as your friends do; perhaps you will argue with them as to the best way to explain your results and try to convince them that your interpretation is right. Scientific controversy frequently arises through people interpreting evidence differently.

Observations of the heavens led the ancient Greek philosophers to believe that the Earth was at the centre of the planetary system, but a complex system of rotation was needed to match observations of the apparent movement of the planets across the sky. In 1543 Nicolaus Copernicus made the radical suggestion that all the planets revolved not around the Earth but around the Sun. (His book 'On the Revolutions of the Celestial Spheres' gave us the modern usage of the word 'revolution'.) It took time for his ideas to gain acceptance. The careful astronomical observations of planetary motion documented by Tycho Brahe were studied by Johannes Kepler, who realised that the data could be explained if the planets moved in elliptical paths (not circular) with the Sun at one focus. Galileo's observations of the moons of Jupiter with the newly invented telescope led him to support this 'Copernican view' and to be imprisoned by the Catholic Church in 1633 for disseminating heretical views. About 50 years later, Isaac Newton introduced the idea of gravity and was able to explain the motion of all bodies, whether on Earth or in the heavens (Chapter 31), which led to full acceptance of the Copernican model. Newton's mechanics were refined further at the beginning of the 20th century when Einstein developed his theories of relativity; even today, data from the Hubble Space Telescope is providing new evidence which confirms Einstein's ideas.

Many other scientific theories have had to wait for new data, technological inventions, or time and the right social and intellectual climate for them to become accepted. In the field of health and medicine, for example, because cancer takes a long time to develop it took several years before people recognised that X-rays and radioactive materials could be dangerous (Chapter 58).

Wegener's theory of continental drift in the 1920s explained a wide range of different geological information, but a suitable mechanism for the movement of the continents had not been proposed. The ideas of plate tectonics (Chapter 62) developed in the 1960s. Studies of the magnetism of ancient rocks supported the idea that Europe and North America had once been joined and indicated that the magnetic field of the Earth reversed from time to time. The discovery of patterns of magnetic stripes on both sides of mid-ocean ridges could be explained by the creation of new rock which acquired magnetism based on the Earth's field at the time the material solidified. Continents could thus move by introducing new material between them.

At the beginning of the 20th century scientists were trying to reconcile the wave theory and the particle theory of light (Chapter 57) by means of the new ideas of quantum mechanics.

Today we are collecting evidence on possible health risks from microwaves used in mobile phone networks. The cheapness and popularity of mobile phones may make the public and the manufacturers reluctant to accept adverse findings, even if risks are made widely known in the press and on television. Although scientists can provide evidence and evaluation of that evidence, there may still be room for controversy and a reluctance to accept scientific findings, particularly if there are vested social or economic interests to contend with. This is most clearly shown today in the issue of global warming.

Light and sight

1 Light rays

■ Sources of light

You can see an object only if light from it enters your eyes. Some objects such as the Sun, electric lamps and candles make their own light. We call these **luminous** sources.

Most things you see do not make their own light but reflect it from a luminous source. They are **non-luminous** objects. This page, you and the Moon are examples. Figure 1.1 shows some others.

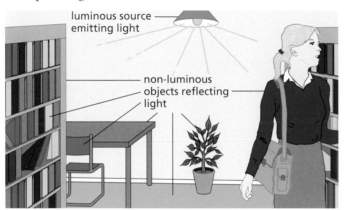

Figure 1.1 Luminous and non-luminous objects

Luminous sources radiate light when their atoms become 'excited' as a result of receiving energy. In a light bulb, for example, the energy comes from electricity. The 'excited' atoms give off their light haphazardly in most luminous sources.

A light source that works differently is the **laser**, invented in 1960. In it the 'excited' atoms act together and emit a narrow, very bright beam of light. The laser has a host of applications. It is used in industry to cut through plate metal, in scanners to read the bar code at shop and library check-outs, in CD players, in optical fibre telecommunication systems, in delicate medical operations on the eye or inner ear, for example, Figure 1.2, in printing, and in surveying and range-finding.

Figure 1.2 Laser surgery in the inner ear

■ Rays and beams

Sunbeams streaming through trees, Figure 1.3, and light from a cinema projector on its way to the screen both suggest that **light travels in straight lines**. The beams are visible because dust particles in the air reflect light into our eyes.

The direction of the path in which light is travelling is called a **ray** and is represented in diagrams by a straight line with an arrow on it. A **beam** is a stream of light and is shown by a number of rays; it may be parallel, diverging (spreading out) or converging (getting narrower), Figure 1.4.

Figure 1.3 Light travels in straight lines

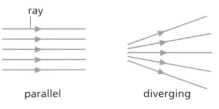

Figure 1.4 Beams of light

Practical work

The pinhole camera

A simple pinhole camera is shown in Figure 1.5a. Make a small pinhole in the centre of the black paper. Half-darken the room. Hold the box at arm's length so that the pinhole end is nearest to and about 1 metre from a luminous object, e.g. a carbon filament lamp or a candle. Look at the **image** on the screen (an image is a likeness of an object and need not be an exact copy).

Can you see *three* ways in which the image differs from the object? What is the effect of moving the camera closer to the object?

Make the pinhole larger. What happens to the

(i) brightness,
(ii) sharpness,
(iii) size of the image?

Make several small pinholes round the large hole, Figure 1.5b, and view the image again.

The forming of an image is shown in Figure 1.6.

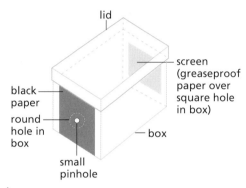

a A pinhole camera

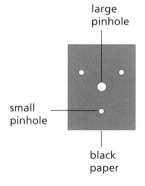

b

Figure 1.5

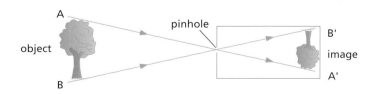

Figure 1.6 Forming an image in a pinhole camera

■ Shadows

Shadows are formed for two reasons. First, because some objects, which are said to be **opaque**, do not allow light to pass through them. Second, because light travels in straight lines. The sharpness of the shadow depends on the size of the light source. A very small source of light, called a **point** source, gives a sharp shadow which is equally dark all over. This may be shown as in Figure 1.7a where the small hole in the card acts as a point source.

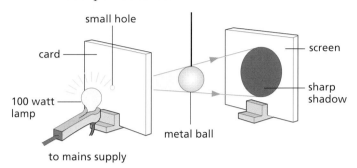

a With a point source

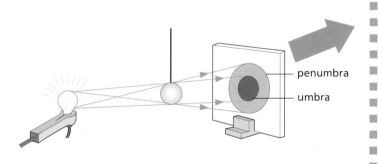

b With an extended source

Figure 1.7 Forming a shadow

If the card is removed the lamp acts as a large or **extended** source, Figure 1.7b. The shadow is then larger and has a central dark region, the **umbra**, surrounded by a ring of partial shadow, the **penumbra**. You can see by the rays that some light reaches the penumbra but none reaches the umbra.

■ Speed of light

Proof that light travels very much faster than sound is provided by a thunderstorm. The flash of lightning is seen before the thunder is heard. The length of the time lapse is greater the further the observer is from the storm.

The speed of light has a definite value; light does not travel instantaneously from one point to another but takes a certain, very small time. Its speed is about 1 million times greater than that of sound.

Questions

1 How would the size and brightness of the image formed by a pinhole camera change if the camera were made longer?

2 What changes would occur in the image if the single pinhole in a camera were replaced by
   a four pinholes close together,
   b a hole 1 cm wide?

3 In Figure 1.8 the completely dark region on the screen is

   A PQ   **B** PR   **C** QR   **D** QS   **E** RS

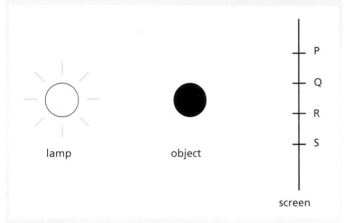

lamp   object

P
Q
R
S

screen

Figure 1.8

4 When watching a distant firework display do you see the cascade of lights before or after hearing the associated bang? Explain your answer.

■ *Checklist*

After studying this chapter you should be able to

■ give examples of effects which show that light travels in a straight line,

■ explain the operation of a pinhole camera and draw ray diagrams to show the result of varying the object distance or the length of the camera,

■ draw diagrams to show how shadows are formed using point and extended sources, and use the terms **umbra** and **penumbra**,

■ recall that light travels much faster than sound.

2 *Reflection of light*

Laws of reflection
Periscope
Regular and diffuse reflection

Practical work
Reflection by a plane mirror.

If we know how light behaves when it is reflected we can use a mirror to change the direction in which the light is travelling. This happens when a mirror is placed at the entrance of a concealed drive to give warning of approaching traffic.

An ordinary mirror is made by depositing a thin layer of silver on one side of a piece of glass and protecting it with paint. The silver – at the *back* of the glass – acts as the reflecting surface.

■ *Laws of reflection*

Terms used in connection with reflection are shown in Figure 2.1. The perpendicular to the mirror at the point where the incident ray strikes it is called the **normal**. Note that the angle of incidence i is the angle between the incident ray and the normal; similarly for the angle of reflection r.

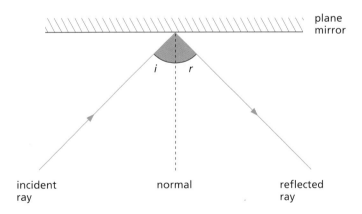

incident ray   normal   reflected ray

Figure 2.1 Reflection of light by a plane mirror

There are two laws of reflection.

1 The angle of incidence equals the angle of reflection.
2 The incident ray, the reflected ray and the normal all lie in the same plane. (*This means that they can all be drawn on a flat sheet of paper.*)

Practical work

Reflection by a plane mirror

Draw a line AOB on a sheet of paper and using a protractor mark angles on it. Measure them from the perpendicular ON, which is at right angles to AOB. Set up a plane (flat) mirror with its reflecting surface on AOB.

Shine a narrow ray of light along say the 30° line onto the mirror, Figure 2.2.

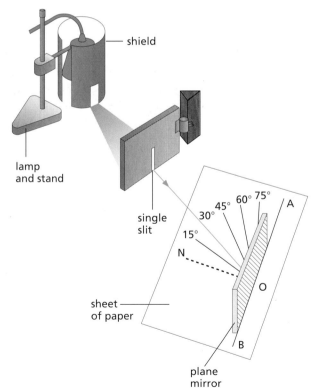

Figure 2.2

Mark the position of the reflected ray, remove the mirror and measure the angle between the reflected ray and ON. Repeat for rays at other angles. What can you conclude?

■ *Periscope*

A simple periscope consists of a tube containing two plane mirrors, fixed parallel to and facing one another. Each makes an angle of 45° with the line joining them, Figure 2.3. Light from the object is turned through 90° at each reflection and an observer is able to see over a crowd, for example, Figure 2.4, or over the top of an obstacle.

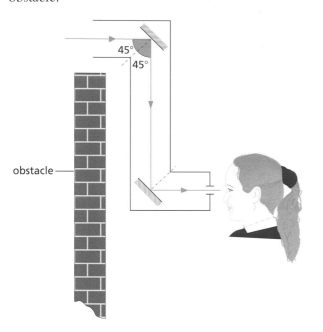

Figure 2.3 Action of a simple periscope

Figure 2.4 Periscopes being used by people in a crowd

In more elaborate periscopes like those used in submarines, prisms replace mirrors (see Chapter 6).

Make your own periscope from a long, narrow cardboard box measuring about 40 cm × 5 cm × 5 cm (e.g. one in which aluminium cooking foil or cling film is sold), two plane mirrors (7.5 cm × 5 cm) and sticky tape. When you have got it to work, make modifications that turn it into a 'See-back-o-scope' which lets you see what is behind you.

■ *Regular and diffuse reflection*

If a parallel beam of light falls on a plane mirror it is reflected as a parallel beam, Figure 2.5a, and **regular** reflection occurs. Most surfaces, however, reflect light irregularly and the rays in an incident parallel beam are reflected in many directions, Figure 2.5b.

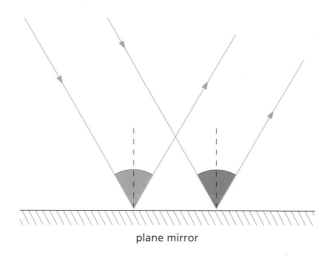

a Regular reflection

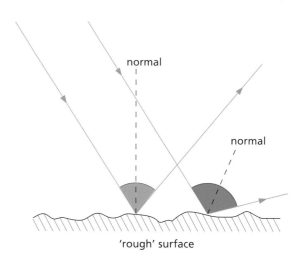

b Diffuse reflection

Figure 2.5

Irregular or **diffuse** reflection is due to the surface of the object not being perfectly smooth like a mirror. At each point on the surface the laws of reflection are obeyed but the angle of incidence and so the angle of reflection varies from point to point. The reflected rays are scattered haphazardly. Most objects, being rough, are seen by diffuse reflection.

Questions

1 Figure 2.6 shows a ray of light PQ striking a mirror AB. The mirror AB and the mirror CD are at right angles to each other. QN is a normal to the mirror AB.

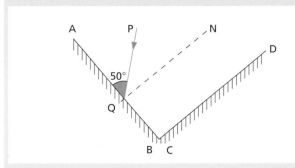

Figure 2.6

   a What is the value of the angle of incidence of the ray PQ on the mirror AB?
   b Copy the diagram and continue the ray PQ to show the path it takes after reflection at both mirrors.
   c What are the values of the angle of reflection at AB, the angle of incidence at CD and the angle of reflection at CD?
   d What do you notice about the path of the ray PQ and the final reflected ray?

2 A ray of light strikes a plane mirror at an angle of incidence of 60°, is reflected from the mirror and then strikes a second plane mirror placed so that the angle between the mirrors is 45°. The angle of reflection at the second mirror, in degrees, is

   A 15   **B** 25   **C** 45   **D** 65   **E** 75

3 A person stands in front of a mirror, Figure 2.7. How much of the mirror is used to see from eye to toes?

Figure 2.7

Checklist

After studying this chapter you should be able to

■ describe an experiment to show that the angle of incidence equals the angle of reflection,
■ state the laws of reflection and use them to solve problems,
■ draw a ray diagram to show how a periscope works.

3 Plane mirrors

Real and virtual images
Lateral inversion

Properties of the image
Kaleidoscope

Practical work
Position of the image.

When you look into a plane mirror on the wall of a room you see an image of the room behind the mirror; it is as if there were another room. Restaurants sometimes have a large mirror on one wall just to make them look larger. You may be able to say how much larger after the next experiment.

The position of the image formed by a mirror depends on the position of the object.

Practical work

Position of the image

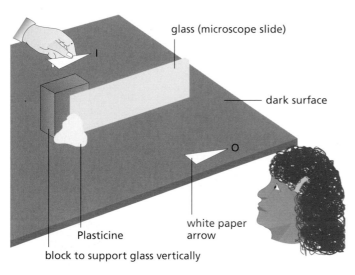

Figure 3.1

Support a piece of thin glass on the bench, as in Figure 3.1. It must be *vertical* (at 90° to the bench). Place a small paper arrow, O, about 10 cm from the glass. The glass acts as a poor mirror and an image of O will be seen in it; the darker the bench top the brighter the image.

Lay another identical arrow, I, on the bench behind the glass; move it until it coincides with the image of O. How do the sizes of O and its image compare? Imagine a line joining them. What can you say about it? Measure the distances of the points of O and I from the glass along the line joining them. How do they compare? Try O at other distances.

Real and virtual images

A **real** image is one which can be produced on a screen (as in a pinhole camera) and is formed by rays that actually pass through it.

A **virtual** image cannot be formed on a screen and is produced by rays which seem to come from it but do not pass through it. The image in a plane mirror is virtual. Rays from a point on an object are reflected at the mirror and appear to come from the point behind the mirror where the eye imagines the rays intersect when produced backwards, Figure 3.2. IA and IB are construction lines and are shown by broken lines.

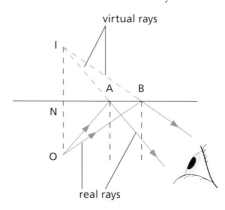

Figure 3.2 A plane mirror forms a virtual image

Lateral inversion

If you close your left eye your image in a plane mirror seems to close the right eye. In a mirror image, left and right are interchanged and the image is said to be **laterally inverted**. The effect occurs whenever an image is formed by one reflection and is very evident if print is viewed in a mirror, Figure 3.3. What happens if two reflections occur, as in a periscope?

Figure 3.3 The image in a plane mirror is laterally inverted

8

Properties of the image

The image in a plane mirror is

(i) as far behind the mirror as the object is in front and the line joining the object and image is perpendicular to the mirror,
(ii) the same size as the object,
(iii) virtual,
(iv) laterally inverted.

Kaleidoscope

Figure 3.4 A kaleidoscope produces hundreds of ever-changing coloured designs using the images formed by two plane mirrors

To see how a kaleidoscope works, draw on a sheet of paper two lines at right angles to one another. Using different coloured pens or pencils draw a design between them, Figure 3.5a. Place a small mirror along each line and look into the mirrors, Figure 3.5b. You will see three reflections which join up to give a circular pattern. If you make the angle between the mirrors smaller, more reflections appear but you always get a complete design.

In a kaleidoscope the two mirrors are usually kept at the same angle (about 60°) and different designs are made by hundreds of tiny coloured beads which can be moved around between the mirrors.

Now make a kaleidoscope using a cardboard tube (e.g. from half a kitchen roll), some thin card, grease-proof paper, clear sticky tape, small pieces of different coloured cellophane and two mirrors (10 cm × 3 cm) or a single plastic mirror (10 cm × 6 cm) bent to form two mirrors at 60° to each other.

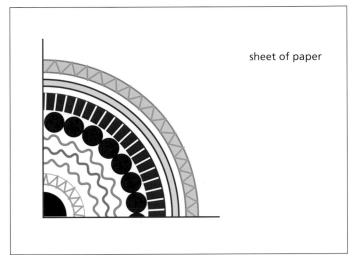

sheet of paper

a

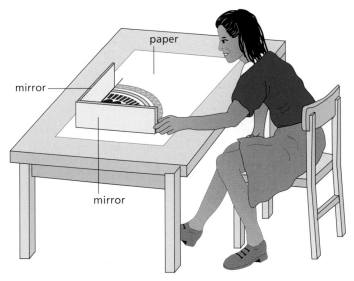

paper

mirror

mirror

b

Figure 3.5

Questions

1 In Figure 3.6 at which of the points **A** to **E** will the observer see the image of the object in the plane mirror?

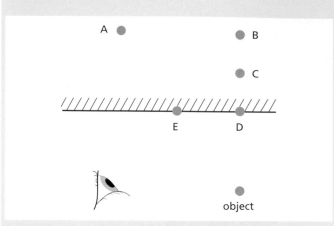

Figure 3.6

2 Figure 3.7 shows the image in a plane mirror of a clock. The correct time is

A 2.25   **B** 2.35   **C** 6.45   **D** 9.25

Figure 3.7

3 A girl stands 5 m away from a large plane mirror. How far must she walk to be 2 m away from her image?

■ *Checklist*

After studying this chapter you should be able to

- ■ describe an experiment to show that the image in a plane mirror is as far behind the mirror as the object is in front, and that the line joining the object and image is at right angles to the mirror,
- ■ draw a diagram to explain the formation of a virtual image by a plane mirror,
- ■ explain the term **lateral inversion**,
- ■ explain how a kaleidoscope works.

4 Curved mirrors

Concave and convex mirrors
Principal focus
Ray diagrams

Images in a concave mirror
Uses of curved mirrors

For some purposes curved mirrors are more useful than plane mirrors. They can make things look larger or smaller depending on which way they curve and how far the object is from them. You can see the images they give by looking into both sides of a large, well-polished spoon, Figure 4.1.

There are two main types of curved mirror.

Figure 4.1 Curved mirrors change the appearance of an object

■ *Concave and convex mirrors*

A **concave** mirror curves inwards like a cave, Figure 4.2a; a **convex** one curves outwards, Figure 4.2b. Many curved mirrors have spherical surfaces and the **principal axis** of such a mirror is the line joining the **pole** P or centre of the mirror to the **centre of curvature** C. C is the centre of the sphere of which the mirror is a part; it is in front of a concave mirror and behind a convex one. The **radius of curvature** r is the distance CP.

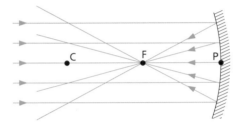

a Concave

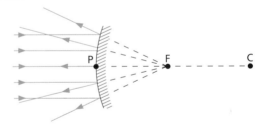

b Convex

Figure 4.2 Action of curved mirrors on a parallel beam of light

■ *Principal focus*

When a beam of light parallel to the principal axis is reflected from a concave mirror, the laws of reflection hold and it converges to a point on the axis called the **principal focus** F. Since light actually passes through it, it is a **real** focus and can be obtained on a screen. A convex mirror has a **virtual** principal focus behind the mirror, from which the reflected beam appears to diverge.

The **focal length** f is the distance FP and experiment and theory show that FP = CP/2 or $f = r/2$, i.e.

focal length = half the radius of curvature

These statements apply only to small mirrors, or to large mirrors if the beam is close to the axis.

■ **11**

Ray diagrams

Facts about the images formed by spherical mirrors can be found by drawing *two* of the following rays.

1 A ray parallel to the principal axis which is reflected through the principal focus F.
2 A ray through the centre of curvature C which hits the mirror normally and is reflected back along its own path (the radius of a sphere is perpendicular to the surface where it meets the surface).
3 A ray through the principal focus F which is reflected parallel to the principal axis.

In diagrams a curved mirror is represented by a straight line. A good-sized object can then be shown and rays from it regarded as forming a point focus, i.e. the mirror behaves as a small one. In numerical questions horizontal distances have sometimes to be scaled down.

Images in a concave mirror

The ray diagrams in Figure 4.3 show the images for four object positions. In each case two rays are drawn from the top A of an object OA and where they intersect after reflection gives the top B of the image IB. The foot I of each image is on the axis since ray OP hits the mirror normally and is reflected back along the axis. In part d the broken rays and the image are virtual (not real).

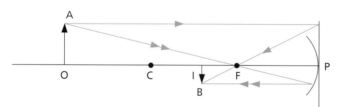

Image is between C and F, real, inverted, smaller

a Object beyond C

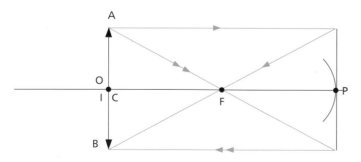

Image is at C, real, inverted, same size

c Object at C

Figure 4.3 Ray diagrams for a concave mirror

Uses of curved mirrors

a) Make-up, shaving and dental mirrors

A concave mirror forms a **magnified**, **upright** image of an object **inside** its focus, Figure 4.4. This accounts for these uses. The image appears to be behind the mirror and is virtual.

Figure 4.4 A concave mirror forms a magnified image of a close object

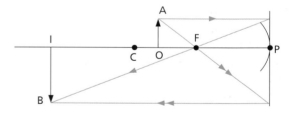

Image is beyond C, real, inverted, larger

b Object between C and F

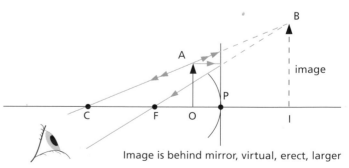

Image is behind mirror, virtual, erect, larger

d Object between F and P

b) Reflectors

Concave mirrors are used as reflectors in, for example, car headlamps and torches, Figure 4.5a, because a small lamp at their focus gives a **parallel** reflected beam. This is only strictly true if the mirror has a parabolic shape (rather than spherical) like that in Figure 4.5b.

Figure 4.5a A curved mirror is used in a torch to produce a beam of light

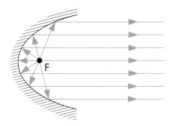

Figure 4.5b A parabolic mirror produces a parallel beam

c) Driving mirrors

A convex mirror gives a wider field of view than a plane mirror of the same size, Figures 4.6a, b. For this reason and because it always gives an upright (but smaller) image, it is used as a car driving mirror and on dangerous bends of roads. However it does give the driver a false idea of distance.

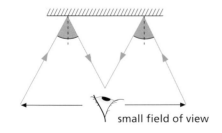

a

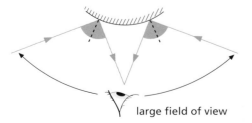

b

Figure 4.6 A convex mirror gives a wider field of view

Questions

1 Draw diagrams to represent
  a a convex mirror,
  b a concave mirror.
  Mark on each the principal axis, the principal focus F and the centre of curvature C if the focal length of **a** is 3 cm and of **b** 4 cm.

2 A concave mirror has a focal length of 4 cm and a real object 2 cm tall is placed 9 cm away from it. By means of an accurate full-size diagram find where the image would be and measure its length.

3 A communications satellite in orbit sends a parallel beam of signals down to Earth. If they obey the same laws of reflection as light and are to be focused on to a small receiving aerial, what is the best shape for the metal 'dish' used to collect them?

4 Account for the use of a convex mirror on the stairs of a double-decker bus.

■ *Checklist*

After studying this chapter you should be able to

■ draw diagrams to show the action of concave and convex mirrors on a parallel beam of light, and use the terms **principal focus**, **focal length**, **radius** and **centre of curvature**,
■ explain why concave mirrors are used as reflectors in car headlamps,
■ explain why concave mirrors are used as make-up, shaving and dental mirrors,
■ explain why convex mirrors are used as driving mirrors.

5 Refraction of light

Facts about refraction
Real and apparent depth
Refractive index

Refraction by a prism
Practical work
Refraction in glass.

If you place a coin in an empty mug and move back until you *just* cannot see it, the result is surprising if someone *gently* pours in water. Try it.

Although light travels in straight lines in one transparent material, such as air, if it passes into a different material, such as water, it changes direction at the boundary between the two, i.e. it is bent. The **bending of light** when it passes from one material (called a medium) to another is called **refraction**. It causes effects like the coin trick.

■ Facts about refraction

(i) A ray of light is bent **towards** the normal when it enters an optically denser medium at an angle (e.g. from air to glass), i.e. the angle of refraction r is less than the angle of incidence i, Figure 5.1a.

(ii) A ray of light is bent **away from** the normal when it enters an optically less dense medium (e.g. from glass to air).

(iii) A ray emerging from a parallel-sided block is **parallel** to the ray entering, but is displaced sideways.

(iv) A ray travelling along the normal is **not refracted**, Figure 5.1b.

Note 'Optically denser' means having a greater refraction effect; the actual density may or may not be greater.

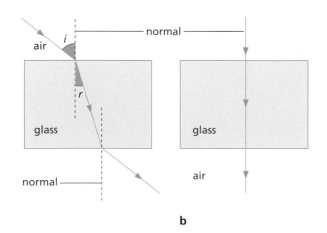

a   b

Figure 5.1 Refraction of light in glass

Practical work

Refraction in glass

Shine a ray of light at an angle on to a glass block (which has its lower face painted white or frosted), Figure 5.2. Draw the outline ABCD of the block on the sheet of paper under it. Mark the positions of the various rays in air and in glass.

Remove the block and draw the normals on the paper at the points where the ray enters AB (see Figure 5.2) and where it leaves CD.

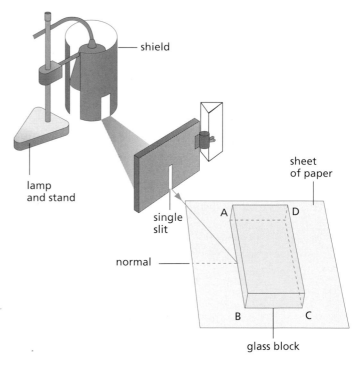

Figure 5.2

What *two* things happen to the light falling on AB? When the ray enters the glass at AB is it bent towards or away from the part of the normal in the block? How is it bent at CD? What can you say about the direction of the ray falling on AB and the direction of the ray leaving CD?

What happens if the ray hits AB at right angles?

Real and apparent depth

Rays of light from a point O on the bottom of a pool are refracted away from the normal at the water surface since they are passing into an optically less dense medium, i.e. air, Figure 5.3. On entering the eye they appear to come from a point I *above* O; I is the virtual image of O formed by refraction. The apparent depth of the pool is less than its real depth.

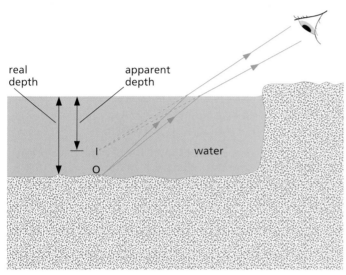

Figure 5.3 A pool of water appears shallower than it is

Figure 5.4 A straight stick seems bent in water. Why?

Refractive index

Light is refracted because its speed changes when it enters another medium. An analogy helps to explain why.

Suppose three people A, B, C are marching in line, with hands linked, on a good road surface. If they approach marshy ground at an angle, Figure 5.5a, A is slowed down first, followed by B and then C. This causes the whole line to swing round and change its direction of motion.

In air (and a vacuum) light travels at 300 000 km/s (3×10^8 m/s), in glass its speed falls to 200 000 km/s (2×10^8 m/s), Figure 5.5b. The **refractive index** n of the medium, i.e. glass, is defined by the equation

$$\text{refractive index } n = \frac{\text{speed of light in air (or a vacuum)}}{\text{speed of light in medium}}$$

$$\therefore \quad n = \frac{300\,000\,\text{km/s}}{200\,000\,\text{km/s}} = \frac{3}{2}$$

Experiments also show that

$$n = \frac{\text{sine of angle of incidence}}{\text{sine of angle of refraction}}$$

$$= \frac{\sin i}{\sin r} \quad \text{(see Figure 5.1a)}$$

The more light is slowed down when it enters a medium from air, the greater is the refractive index of the medium and the more it is bent.

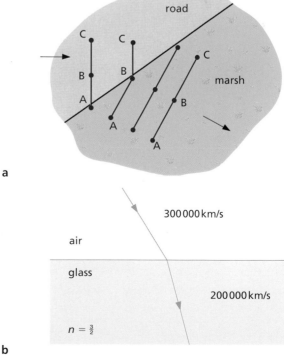

a

b

Figure 5.5

■ *Refraction by a prism*

In a triangular glass prism, Figure 5.6a, the deviation (bending) of a ray due to refraction at the first surface is added to the deviation at the second surface, Figure 5.6b. The deviations do not cancel out as in a parallel-sided block where the emergent ray, although displaced, is parallel to the incident ray.

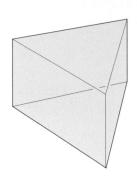

a

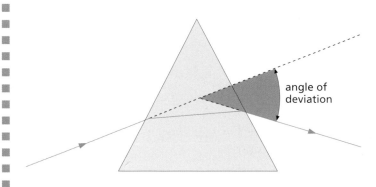

angle of deviation

b

Figure 5.6

Questions

1 Figure 5.7 shows a ray of light entering a rectangular block of glass.
  a Copy the diagram and draw the normal at the point of entry.
  b Sketch the approximate path of the ray through the block and out of the other side.

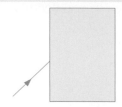

Figure 5.7

2 Draw two rays from a point on a fish in a stream to show where someone on the bank will see the fish. Where must the person aim to spear the fish?

3 What is the speed of light in a medium of refractive index 6/5 if its speed in air is 300 000 km/s?

4 Figure 5.8 shows a ray of light OP striking a glass prism and then passing through it. Which of the rays **A** to **D** is the correct representation of the emerging ray?

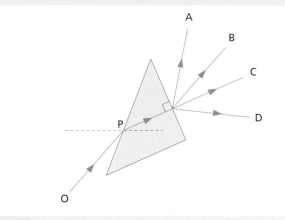

Figure 5.8

■ *Checklist*

After studying this chapter you should be able to

■ state what the term **refraction** means,
■ give examples of effects that show light can be refracted,
■ describe an experiment to study refraction,
■ draw diagrams of the passage of light rays through rectangular blocks and recall that lateral displacement occurs for a parallel-sided block,
■ recall that light is refracted because it changes speed when it enters another medium,
■ recall the definition of refractive index as $n =$ speed in air/speed in medium,
■ draw a diagram for the passage of a light ray through a prism.

6 Total internal reflection

■ *Critical angle*

When light passes at small angles of incidence from an optically denser to a less dense medium, e.g. from glass to air, there is a strong refracted ray and a weak ray reflected back into the denser medium, Figure 6.1a. Increasing the angle of incidence increases the angle of refraction.

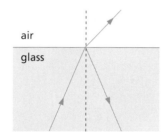

a

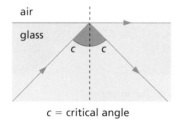

c = critical angle

b

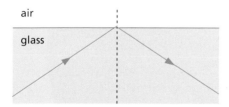

c

Figure 6.1

At a certain angle of incidence, called the **critical angle** c, the angle of refraction is 90°, Figure 6.1b. For angles of incidence greater than c, the refracted ray disappears and *all* the incident light is reflected inside the denser medium, Figure 6.1c. The light does not cross the boundary and is said to undergo **total internal reflection**.

Practical work

Critical angle of glass

Place a semicircular glass block on a sheet of paper, Figure 6.2, and draw the outline LOMN where O is the centre and ON the normal at O to LOM. Direct a narrow ray (at an angle of about 30°) *along a radius towards* O. The ray is not refracted at the curved surface. Why? Note the refracted ray in the air beyond LOM and also the weak internally reflected ray in the glass.

Slowly rotate the paper so that the angle of incidence increases until total internal reflection *just* occurs. Mark the incident ray. Measure the angle of incidence; it equals the critical angle.

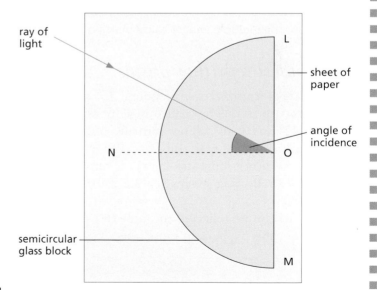

Figure 6.2

Multiple images in a mirror

An ordinary mirror silvered at the back forms several images of one object, due to multiple reflection inside the glass, Figures 6.3a, b. These blur the main image I (which is formed by one reflection at the silvering), especially if the glass is thick. The trouble is absent in front-silvered mirrors but they are easily damaged.

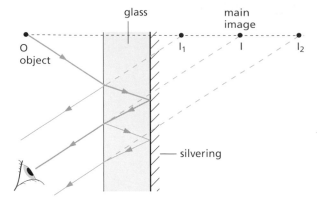

Figure 6.3a Multiple reflections in a mirror

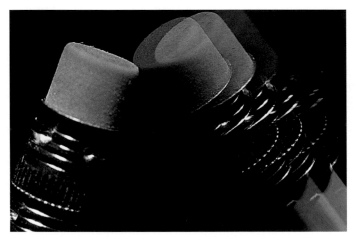

Figure 6.3b The multiple images cause blurring

Totally reflecting prisms

The defects of mirrors are overcome if 45° right-angled glass prisms are used. The critical angle of ordinary glass is about 42° and a ray falling normally on face PQ of such a prism, Figure 6.4a, hits face PR at 45°. Total internal reflection occurs and the ray is turned through 90°. Totally reflecting prisms replace mirrors in good periscopes.

Light can also be reflected through 180° by a prism, Figure 6.4b; this happens in binoculars.

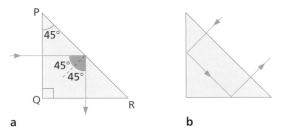

a　　　　　　**b**

18 ■ **Figure 6.4** Reflection of light by a prism

Light pipes and optical fibres

Light can be trapped by total internal reflection inside a bent glass rod and 'piped' along a curved path, Figure 6.5. A single, very thin glass fibre behaves in the same way. If several thousand such fibres are taped together a flexible light pipe is obtained that can be used, for example, by doctors as an 'endoscope', Figure 6.6a, to obtain an image of an internal organ in the body, Figure 6.6b, or by engineers to light up some awkward spot for inspection. The latest telephone 'cables' are optical (very pure glass) fibres carrying information as pulses of laser light.

Figure 6.5 Light travels through a curved glass rod or fibre by total internal reflection

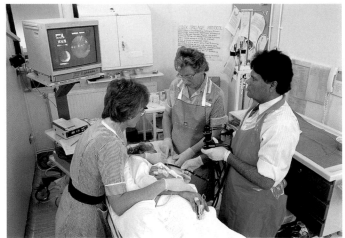

Figure 6.6a Endoscope in use

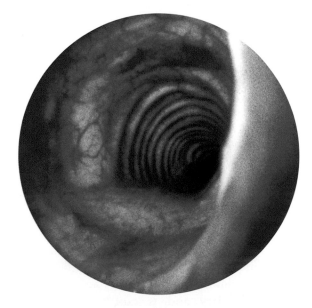

Figure 6.6b Trachea (windpipe) viewed by an endoscope

Questions

1 Figure 6.7 shows rays of light in a semicircular glass block.

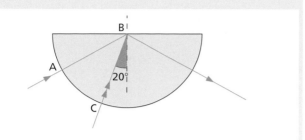

Figure 6.7

   a Explain why the ray entering the glass at A is not bent.
   b Explain why the ray AB is reflected at B and not refracted.
   c Ray CB does not stop at B. Copy the diagram and draw its approximate path after it leaves B.

2 Copy Figures 6.8a and b and complete the paths of the rays through the glass prisms.

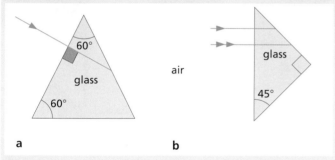

a   b

Figure 6.8

3 Name two instruments which use prisms to reflect light.

■ *Checklist*

After studying this chapter you should be able to

■ explain with the aid of diagrams what is meant by **critical angle** and **total internal reflection**,
■ describe an experiment to find the critical angle of glass or Perspex,
■ draw diagrams to show the action of totally reflecting prisms in periscopes and binoculars,
■ explain the action and state some uses of optical fibres.

7 Lenses

Convex and concave lenses

Lenses are used in many optical instruments (Chapters 10 and 11); they often have spherical surfaces and there are two types. A **convex** lens is thickest in the centre and is also called a **converging** lens because it bends light inwards, Figure 7.1a. You may have used one as a magnifying glass, Figure 7.2a, or as a burning glass. A **concave** or **diverging** lens is thinnest in the centre and spreads light out, Figure 7.1b; it always gives a diminished image, Figure 7.2b.

The centre of a lens is its **optical centre** C; the line through C at right angles to the lens is the **principal axis**.

The action of a lens can be understood by treating it as a number of prisms (most cut-off), each of which bends the ray towards its base, as in Figure 7.1c, d. The centre acts as a parallel-sided block.

Figure 7.2a A convex lens forms a magnified image of a close object

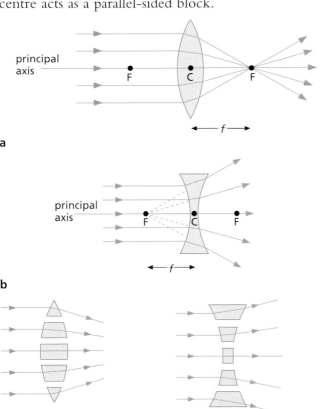

a

b

c   **d**

20 **Figure 7.1** Action of lenses on a parallel beam of light

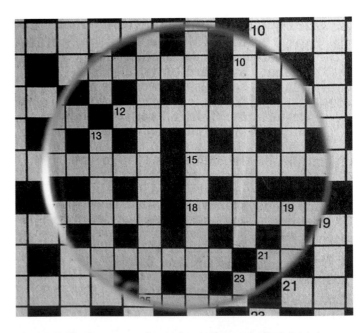

Figure 7.2b A concave lens always forms a diminished image

■ *Principal focus*

When a beam of light parallel to the principal axis passes through a convex lens it is refracted so as to converge to a point on the axis called the **principal focus** F. It is a real focus. A concave lens has a virtual principal focus behind the lens, from which the refracted beam seems to diverge.

Since light can fall on both faces of a lens it has two principal foci, one on each side, equidistant from C. The distance CF is the **focal length** f of the lens; it is an important property of a lens. The more curved the lens faces are, the smaller is f and the more powerful is the lens.

Practical work

f of a convex lens

We use the fact that rays from a **point** on a very distant object, i.e. at infinity, are nearly parallel, Figure 7.3a.

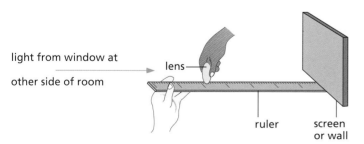

a

b

Figure 7.3

Move the lens, arranged as in Figure 7.3b, until a **sharp** image of a window at the other side of the room is obtained on the screen. The distance between the lens and the screen is f, roughly. Why?

Practical work

Images formed by a convex lens

In the formation of images by lenses, two important points on the principal axis are F and 2F; 2F is at a distance of twice the focal length from C.

First find the focal length of the lens by the 'distant object method', then fix the lens upright with Plasticine at the centre of a metre rule. Place small pieces of Plasticine at the points F and 2F on both sides of the lens, as in Figure 7.4.

Place a small light source, e.g. a torch bulb, as the object supported on the rule beyond 2F and move a white card, on the other side of the lens from the light, until a sharp image is obtained on the card.

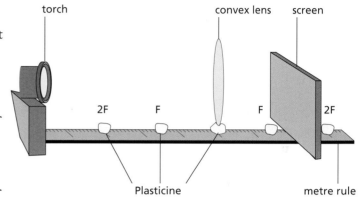

Figure 7.4

Note and record, in a table like the one below, the image position as 'beyond 2F', 'between 2F and F' or 'between F and lens'. Also note whether the image is 'larger' or 'smaller' than the actual bulb or 'same size' and if it is 'upright' or 'inverted'. Now repeat with the light at 2F, then between 2F and F.

Object position	Image position	Larger, smaller or same size?	Upright or inverted?
Beyond 2F			
At 2F			
Between 2F and F			
Between F and lens			

So far all the images have been real since they can be obtained on a screen. When the light is between F and the lens, the image is **virtual** and is seen by *looking through the lens* at the light. Do this. Is the virtual image larger or smaller than the object? Is it upright or inverted? Record your findings in the table.

■ *Ray diagrams*

Information about the images formed by a lens can be obtained by drawing *two* of the following rays.

> **1** A ray parallel to the principal axis which is refracted through the principal focus F.
> **2** A ray through the optical centre C which is undeviated for a thin lens.
> **3** A ray through the principal focus F which is refracted parallel to the principal axis.

In diagrams a thin lens is represented by a straight line at which all the refraction is considered to occur.

In each ray diagram in Figure 7.5 two rays are drawn from the top A of an object OA and where they intersect after refraction gives the top B of the image IB. The foot I of each image is on the axis since ray OC passes through the lens undeviated. In part d the broken rays, and the image, are virtual.

■ *Magnification*

The **linear magnification** *m* is given by

$$m = \frac{\text{height of image}}{\text{height of object}}$$

It can be shown that in all cases

$$m = \frac{\text{distance of image from lens}}{\text{distance of object from lens}}$$

■ *Power of a lens*

The shorter the focal length of a lens, the stronger it is, i.e. the more it converges or diverges a beam of light. We define the **power** of a lens *P* to be 1/focal length of the lens in metres:

$$P = \frac{1}{f}$$

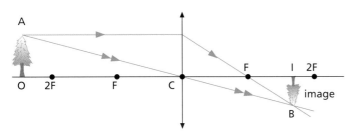

Image is between F and 2F, real, inverted, smaller

a Object beyond 2F

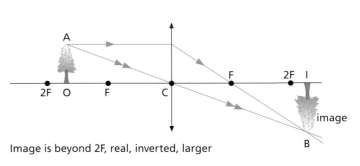

Image is beyond 2F, real, inverted, larger

b Object between 2F and F

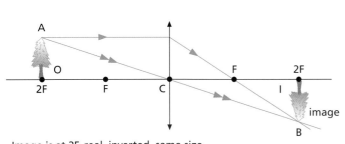

Image is at 2F, real, inverted, same size

c Object at 2F

Image is behind object, virtual, erect, larger

d Object between F and C

Figure 7.5 Ray diagrams for a convex lens

22

Questions

1 A small torch bulb is placed at the focal point of a converging lens. When the bulb is switched on does the lens produce a convergent, divergent or parallel beam of light?

2 a What kind of lens is shown in Figure 7.6?

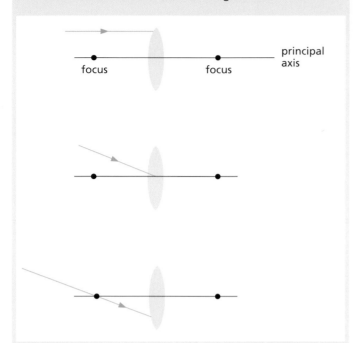

Figure 7.6

b Copy the diagrams and complete them to show the path of the light after passing through the lens.
c Figure 7.7 shows an object AB 6 cm high placed 18 cm in front of a lens of focal length 6 cm. Draw the diagram to scale and by tracing the paths of rays from A find the position and size of the image formed.

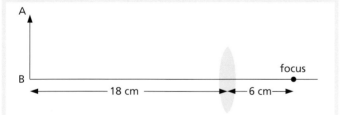

Figure 7.7

3 Where must the object be placed for the image formed by a convex lens to be
a real, inverted and smaller than the object,
b real, inverted and same size as the object,
c real, inverted and larger than the object,
d virtual, upright and larger than the object?

Checklist

After studying this chapter you should be able to

■ explain the action of a lens in terms of refraction by a number of small prisms,
■ draw diagrams showing the effects of convex and concave lenses on a beam of parallel rays,
■ recall the meaning of **optical centre**, **principal axis**, **principal focus** and **focal length**,
■ describe an experiment to measure the focal length of a convex lens,
■ draw ray diagrams to show image formation by a convex lens,
■ draw scale diagrams to solve problems on convex lenses,
■ recall the meaning of the term **linear magnification**,
■ recall the meaning of the term **power** of a lens.

8 The eye

Structure of the eye
Image formation
Defects of vision

Persistence of vision
Practical work
The eyes: blind spot, binocular vision.

■ Structure of the eye

An image is formed on the **retina** of the eye, Figure 8.1, by successive refraction at the **cornea**, the **aqueous humour**, the **lens** and the **vitreous humour**. Electrical signals then travel along the **optic nerve** to the brain to be interpreted. In good light, the **yellow spot** (**fovea**) is most sensitive to detail and the image is automatically formed there.

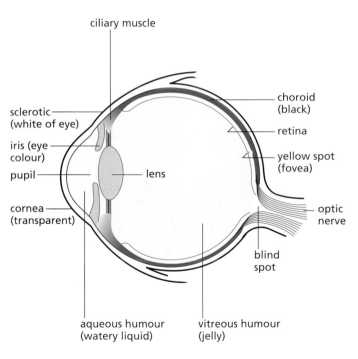

Figure 8.1 The structure of the eye

Objects at different distances are focused by the **ciliary muscles** changing the shape and so the focal length of the lens – a process called **accommodation**. The lens fattens to view near objects. The **iris** has a central hole, the **pupil**, whose size it decreases in bright light and increases in dim light. Excessive brightness can damage the eye.

Practical work

The eyes

a) Blind spot

This is the small area of the retina where the optic nerve leaves the eye. It has no light-sensitive nerve endings and in each eye it is closer to the nose than the yellow spot.

To show its existence, close your left eye and look at the cross in Figure 8.2. You will also see the black dot.

Slowly bring the book towards you. At a certain distance the dot disappears; its image has fallen on the blind spot of your right eye.

Figure 8.2

b) Binocular vision

Your two eyes see an object from slightly different angles, giving two slightly different images which the brain combines to give a three-dimensional impression. This also helps you to judge distances.

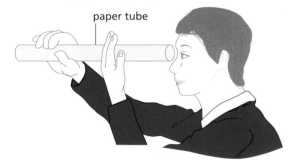

Figure 8.3

Roll a sheet of paper into a tube and hold it to your right eye with your right hand, Figure 8.3. Close your left eye and place your left hand halfway along the tube. Open your left eye. Is there a 'hole' in your hand to 'see' through? Explain.

24

Image formation

From the ray diagrams shown in Figure 7.5 (page 22) we would expect that the lens in the eye will form a real **inverted** image on the retina as shown in Figure 8.4. Since an object normally appears upright, the brain must invert the image.

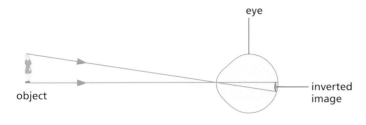

Figure 8.4 Inverted image on the retina

Defects of vision

The average adult eye can focus objects comfortably from about 25 cm (the **near point**) to infinity (the **far point**). Your near point may be less than 25 cm; it gets farther away with age.

a) Short sight

A short-sighted person sees near objects clearly but the far point is closer than infinity. The image of a distant object is formed in front of the retina because the eyeball is too long or because the eye lens cannot be made thin enough, Figure 8.5a. The defect is corrected by a concave spectacle lens (or contact lens) which diverges the light before it enters the eye, to give an image on the retina, Figure 8.5b.

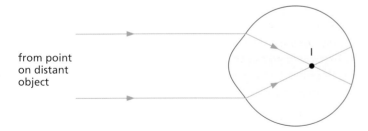

a

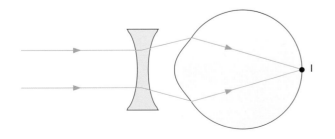

b

Figure 8.5 Short sight and its correction by a concave lens

b) Long sight

A long-sighted person sees distant objects clearly but the near point is beyond 25 cm. The image of a near object is focused behind the retina because the eyeball is too short or because the eye lens cannot be made thick enough, Figure 8.6a. A convex spectacle lens (or contact lens) corrects the defect, Figure 8.6b.

a

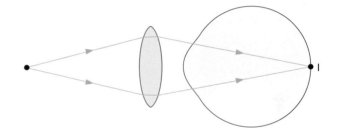

b

Figure 8.6 Long sight and its correction by a convex lens

c) Astigmatism

A person having this defect will see one set of lines in Figure 8.7 more sharply than the others. It is caused by the cornea being more curved in some directions than others. Correction may be achieved with a non-spherical spectacle lens whose curvature has the effect of increasing or decreasing that of the cornea in the required directions.

Figure 8.7

Persistence of vision

An image lasts on the retina for about one-tenth of a second after the object has disappeared, as can be shown by spinning a card like that in Figure 8.8. The effect makes possible the production of motion pictures. In a TV receiver 25 complete pictures are produced every second.

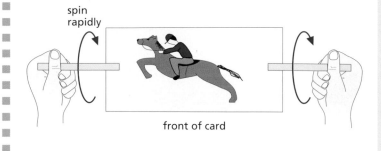

spin rapidly

front of card

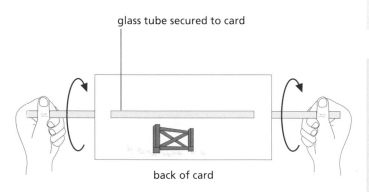

glass tube secured to card

back of card

Figure 8.8 Spin the card rapidly – what do you see?

Questions

1 Name the part of the eye
  a which controls how much light enters it,
  b on which the image is formed,
  c which changes the focal length of the lens.

2 Refraction of light in the eye occurs at

  A the lens only   B the iris   C the cornea only
  D the pupil   E both the cornea and the lens.

3 A short-sighted person has a near point of 15 cm and a far point of 40 cm.
  a Can he see clearly an object at a distance of
    (i) 5 cm,
    (ii) 25 cm,
    (iii) 50 cm?
  b To see clearly an object at infinity what kind of spectacle lens does he need?

4 The near point of a long-sighted person is 50 cm from the eye.
  a Can she see clearly an object at
    (i) a distance of 20 cm,
    (ii) infinity?
  b To read a book held at a distance of 25 cm will she need a convex or a concave spectacle lens?

Checklist

After studying this chapter you should be able to

■ draw and label the main parts of the eye,

■ explain how refraction at the cornea and lens produces an image on the retina,

■ explain how the lens allows accommodation to occur,

■ recall the meaning of **short sight** and describe how it is corrected,

■ recall the meaning of **long sight** and describe how it is corrected,

■ recall the meaning of **astigmatism** and describe how it is corrected,

■ explain the terms **binocular vision** and **persistence of vision**.

9 Colour

◼ *Dispersion*

Colourful clothes, colour television and the flashing coloured lights in a disco all help to make life brighter. It was Newton who, in 1666, set us on the road to understanding how colours may arise. He produced them by allowing sunlight (which is white) to fall on a triangular glass prism, Figure 9.1a. The band of colours obtained, Figure 9.1b, is a **spectrum** and the effect is called **dispersion**. He concluded that

(i) white light is a mixture of many colours of light which the prism separates out, because
(ii) the refractive index of glass is different for each colour, being greatest for violet light.

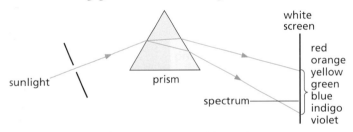

Figure 9.1a Forming a spectrum with a prism

Figure 9.1b White light shining through cut crystal can produce several spectra

Practical work

A pure spectrum

A pure spectrum is one in which the colours do not overlap, as they do when a prism alone is used. It needs a lens to focus each colour as in Figure 9.2a.

Arrange a lens L, Figure 9.2b, so that it forms an image of the vertical filament of a lamp on a screen at S_1, 1 m away. The filament acts as a narrow source of white light. Insert a 60° prism P and move the screen to S_2, keeping it at the same distance from L, to receive the spectrum; rotate P until the spectrum is pure, Figure 9.2c.

Place different colour filters between P and S_2. What happens to the spectrum?

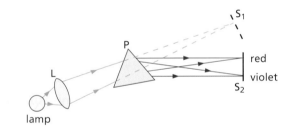

a

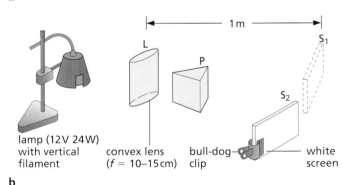

b

Figure 9.2

Figure 9.2c A pure spectrum

27

10 Simple optical instruments

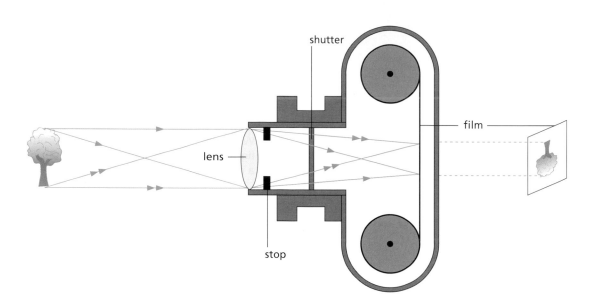

Figure 10.1 Forming an image in a conventional camera

■ *Camera*

A conventional camera is a light-tight box in which a convex lens forms a real image on a film, Figure 10.1. The image is smaller than the object and nearer to the lens. The film contains chemicals that change on exposure to light; it is **developed** to give a negative. From the negative a photograph is made by **printing**. In a digital camera the image is recorded electronically rather than on film (see Figure 60.1a, p. 280).

a) Focusing

In simple cameras the lens is fixed and all distant objects, i.e. beyond about 2 metres, are in reasonable focus. Roughly how far from the film will the lens be if its focal length is 5 cm?

In other cameras exact focusing of an object at a certain distance is done by altering the lens position. For near objects it is moved away from the film, the correct setting being shown by a scale on the focusing ring.

b) Shutter

When a photograph is taken, the shutter is opened for a certain time and exposes the film to light entering the camera. Sometimes exposure times can be varied and are given in fractions of a second, e.g. 1/1000, 1/60, etc. Fast-moving objects require short exposures.

c) Stop

The brightness of the image on the film depends on the amount of light passing through the lens when the shutter is opened and is controlled by the size of the hole (aperture) in the stop. In some cameras this is fixed but in others it can be made larger for a dull scene and smaller for a bright one.

The aperture may be marked in **f-numbers**. The diameter of an aperture with f-number 8 is $\frac{1}{8}$ of the focal length of the lens and so the *larger* the f-number the *smaller* the aperture. The numbers are chosen so that on passing from one to the next higher, e.g. from 8 to 11, the area of the aperture is halved.

Projector

A projector forms a real image on a screen of a slide in a slide projector, or of a film in a cine-projector. The image is larger than the slide or frame of film and is further away from the lens. It is usually so highly magnified that very strong but even illumination of the slide or film is needed if the image is also to be bright. This is achieved by directing light from a small but powerful lamp on to the 'object' by means of a concave mirror and a condenser lens system arranged as in Figure 10.2. The image is produced by the projection lens which can be moved in and out of its mounting to focus the picture.

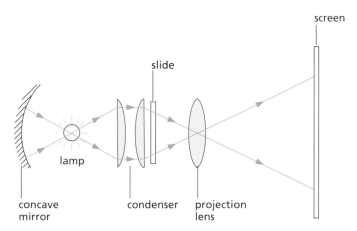

Figure 10.2 Illuminating the slide in a projector

In a projector like that in Figure 10.3 the slide must be inverted to give an upright image, and must be between 2F and F from the lens (see Figure 7.5b).

Figure 10.3 Slide projector

Magnifying glass

The apparent size of an object depends on its actual size and on its distance from the eye. The sleepers on a railway track are all the same length but those nearby seem longer. This is because they enclose a larger angle at your eye than more distant ones: as a result their image on the retina is larger so making them appear bigger.

A convex lens gives an enlarged, upright virtual image of an object placed inside its principal focus F, Figure 10.4a. It acts as a magnifying glass since the angle β made at the eye by the image, formed at the near point, is greater than the angle α made by the object when it is viewed directly at the near point without the magnifying glass, Figure 10.4b.

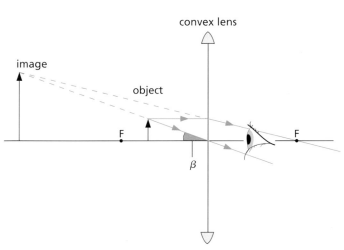

a

b

Figure 10.4 Magnification by a convex lens: $\beta > \alpha$

The fatter (more curved) a convex lens is, the shorter its focal length and the more it magnifies. Too much curvature however distorts the image.

Questions

1 Figure 10.5 shows a camera focused on an object in the middle distance. Should the lens be moved towards or away from the film so that the image of a more distant object is in focus?

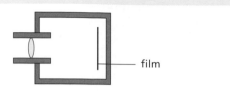
film

Figure 10.5

2 If a projector is moved farther away from the screen on which it was giving a sharp image of a slide, state
a *three* changes that occur in the image,
b how the projection lens must be adjusted to re-focus the image.

3 a Three converging lenses are available having focal lengths of 4 cm, 40 cm and 4 m respectively. Which one would you choose as a magnifying glass?
b An object 2 cm high is viewed through a converging lens of focal length 8 cm. The object is 4 cm from the lens. By means of a ray diagram find the position, nature and magnification of the image.

■ *Checklist*

After studying this chapter you should be able to describe with the aid of diagrams how a single lens is used

■ in a simple camera,
■ in a projector,
■ in a magnifying glass.

11 Telescopes

Practical work
Simple lens telescopes.

Refracting astronomical telescope

Figure 11.1 Using a refracting astronomical telescope

The astronomical telescope consists of two convex lenses, a **long-focus objective** and a **short-focus eyepiece**. Rays from a **point** on a distant object, such as a planet, are nearly parallel on reaching the telescope (see Chapter 7). The objective forms a real, inverted, diminished image I_1 of the object at its principal focus F_o, Figure 11.2. The eyepiece acts as a magnifying glass, treating I_1 as an object and forming a magnified, virtual image.

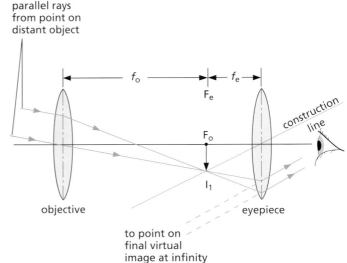

parallel rays from point on distant object

to point on final virtual image at infinity

Figure 11.2 Forming an image in an astronomical telescope

Normally the eyepiece is adjusted to give this final image at infinity, i.e. a long way off, and so I_1 will be at the principal focus F_e of the eyepiece. That is, F_o and F_e coincide. The separation of the lenses is $f_o + f_e$. The object is seen inverted.

A telescope magnifies because the final image it forms subtends a much greater angle at the eye than does the distant object viewed without the telescope. Figure 11.2 shows that the longer the focal length of the objective, the larger is I_1, and so for greatest magnification the objective should have a long focal length and the eyepiece a short one (like any magnifying glass).

Practical work

Simple lens telescopes

Arrange a carbon filament lamp at the far end of the room. Mount a long-focus convex lens (e.g. 50 cm) at eye-level, Figure 11.3a. Point it at the lamp and find its image on a piece of greaseproof paper.

a) Astronomical

Insert a short-focus convex lens (e.g. 5 cm) as the eyepiece and adjust it so that you can see the image on the greaseproof paper clearly, Figure 11.3b. Remove the paper and view the image of the lamp. Look at a book beside the lamp and other distant objects.

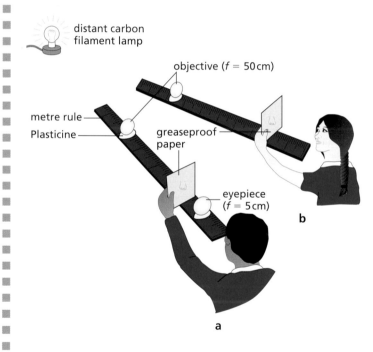

distant carbon filament lamp

objective (f = 50 cm)

metre rule

Plasticine

greaseproof paper

eyepiece (f = 5 cm)

b

a

Figure 11.3

b) Galilean

In an astronomical telescope the final image is inverted. Replace the convex eyepiece with a **short-focus concave** lens and adjust its distance from the objective till you see distant objects clearly. Are their images inverted or upright? This lens combination is used in Galilean telescopes and in opera glasses.

■ *Reflecting astronomical telescope*

The objective of a telescope must have a large diameter (aperture) as well as a large focal length. This (i) gives it good light-gathering power so that faint objects can be seen and (ii) enables it to reveal detail.

The largest lens telescope has an objective of diameter 1 metre; anything more would sag under its own weight. The biggest astronomical telescopes use long-focus **concave mirrors** as objectives; they can be supported at the back, Figure 11.4. Figure 11.5 shows how parallel rays from a point on a distant object are reflected at the objective and then intercepted by a small plane mirror before they form a real image I_1. This image is magnified by the eyepiece.

Figure 11.4 The 3.6 metre diameter reflecting telescope on Mauna Kea, Hawaii

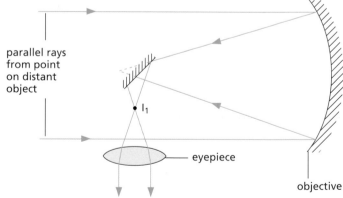

parallel rays from point on distant object

I_1

eyepiece

objective

Figure 11.5 Forming an image in a reflecting telescope

The Hubble Space Telescope, Figure 11.6, launched in 1990, orbits the Earth at a height of 590 km. It has a mirror diameter of 2.4 metres and has sent back many superb pictures of distant galaxies that will enable reliable estimates of the age and size of the Universe (see Chapter 64) to be made. Space telescopes give more detailed images than ground-based telescopes because the distortion introduced when light passes through the Earth's atmosphere is eliminated.

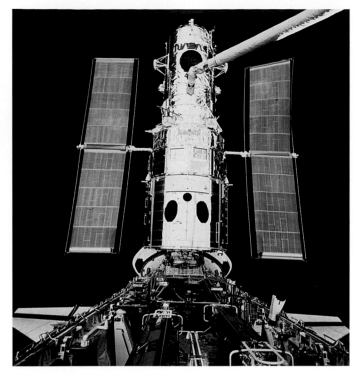

Figure 11.6 The Hubble Space Telescope

Radio telescope

A radio telescope has a similar design to a reflecting astronomical telescope. It uses a large parabolic metal dish (instead of a mirror) to collect radio signals (p. 299) from distant galaxies. The radio waves are focused at the principal focus of the dish and processed electronically to produce an image of the radio source. The larger the diameter of the dish, the greater the detail seen in the image; large arrays of radio telescopes, Figure 11.7, result in even greater image detail.

Radio waves are emitted from events in the Universe such as supernova explosions and by pulsars (p. 332); they can penetrate interstellar dust and gas (which absorb visible light) so are able to give us information about regions we cannot see. Data from radio telescopes is also being used in the SETI project (p. 334) to search for extra-terrestrial intelligent life.

Figure 11.7 Array of radio telescopes in New Mexico

Questions

1 A and B are two convex lenses correctly set up as a telescope to view a distant object, Figure 11.8. One has focal length 5 cm and the other 100 cm.
   a What is A called and what is its focal length?
   b How far from A is the first image of the distant object?
   c What is B called?
   d What acts as the 'object' for B and how far must B be from it if someone looking through the telescope is to see the final image at the same distance as the distant object?
   e What is the distance between A and B with the telescope set up as in **d**?

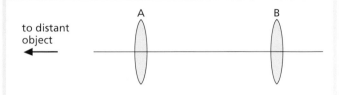

to distant object ←   A   B

Figure 11.8

2 Choose from list **A** to **E** the component most suitable for the objective of **a** a reflecting and **b** a refracting astronomical telescope.

   A A convex lens of focal length 5 cm
   B A convex lens of focal length 100 cm
   C A convex mirror of focal length 200 cm
   D A concave lens of focal length 100 cm
   E A concave mirror of focal length 200 cm

3 **a** Explain why a telescope should have a large diameter objective.
   b Give *one* difference and *one* similarity between a reflecting optical telescope and a radio telescope.
   c What advantage does a space telescope have over a ground-based telescope?

Checklist

After studying this chapter you should be able to describe with the aid of diagrams

- a refracting astronomical telescope,
- a reflecting astronomical telescope.

Light and sight
Additional questions

Light rays; reflection of light; mirrors; refraction; total internal reflection; lenses

1 The image in a plane mirror is

   A upright, real with a magnification of 2
   B upright, virtual with a magnification of 1
   C inverted, real with a magnification of $\frac{1}{2}$
   D inverted, virtual with a magnification of 1
   E inverted, real with a magnification of 2.

2 A parallel beam of light is produced by reflection from a mirror when a point source of light is

   A at the focus of a concave mirror
   B at the focus of a convex mirror
   C inside the focus of a concave mirror
   D inside the focus of a convex mirror
   E in front of a plane mirror.

3 Queen Matilda is looking at King Stephen across the moat. The moat is full of water.

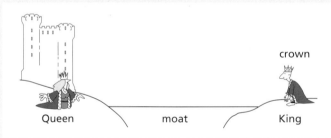

   a The Queen sees the King's crown reflected in the water. Copy the diagram and draw the path of a ray of light to show how this happens. Draw an arrow, on your ray, to show the direction in which the light travels.
   b The King throws his crown into the moat. It sinks to the bottom.

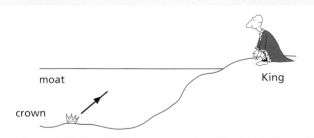

   (i) Copy the diagram and draw the path of the ray of light from the crown to his eyes. It has been started for you.
   (ii) Explain what happens to light when it passes from water into air.
   c The King sees a virtual image of his crown. Write an X on your diagram to show where this image is.
   d The Queen looks in the mirror to see her new belt.

Three of these statements correctly describe the image of the belt that she sees in the mirror. Which *three* are the correct statements?

   The image is as far behind the mirror as the belt is in front.
   The image is larger than the belt.
   The image is real.
   The image is the same size as the belt.
   The image is smaller than the belt.
   The image is upside down.
   The image is virtual.

(OCR Foundation, Summer 99)

4 Light travels up through a pond of water of critical angle 49°. What happens at the surface if the angle of incidence is **a** 30°, **b** 60°?

5 Which diagram shows the ray of light refracted correctly?

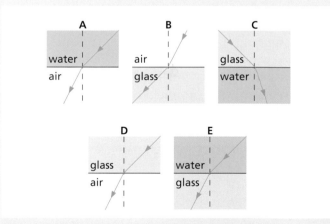

6 Which diagram shows the correct path of the ray through the prism?

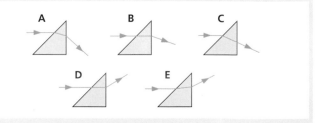

7 a The diagram shows a mirror which can be used by a dentist to see the back of a patient's tooth.

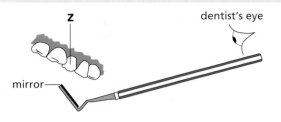

Copy the diagram and draw on it a ray of light to show how the dentist is able to see the tooth labelled Z.

b An air bubble inside a fish tank is hit by two light rays.

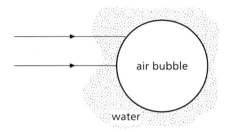

Copy the diagram and complete it to show the path of each light ray through the air bubble and out the other side.

(SEG Foundation, Summer 98)

8 This question is about sunlight and eclipses.
a The diagram shows a solar eclipse.

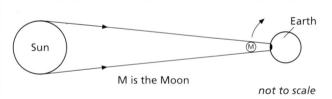

M is the Moon

not to scale

(i) During a solar eclipse part of the Earth is darker than normal. What causes this to happen?
(ii) Copy the diagram and draw *two* more rays on it to show which parts of the Earth are in semi-darkness (penumbra). Label *one* of the areas of semi-darkness with the letter P.
b Mary uses a convex lens to make sunlight burn a hole in a piece of card.

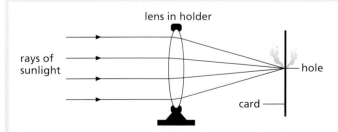

Mary changes the lens in the holder for another lens with *twice* the power. What will she have to do to focus the sunlight on the card?

(OCR Foundation, June 99)

9 An illuminated object is placed at a distance of 20 cm from a convex lens of focal length 15 cm. The image obtained on a screen is

A upright and magnified
B upright and the same size
C inverted and magnified
D inverted and the same size
E inverted and diminished.

10 An object placed 4 cm in front of a convex lens produces a real image 12 cm from the lens.
a What is the magnification of the image?
b Draw a ray diagram to determine the focal length of the lens.

11 By how far does the distance between a girl and her image decrease if she walks from a position where she is 10 m away from a mirror to one where she is 3 m away?

The eye; optical instruments

12 The diagram shows the structure of the eye.
a Choose words from the list below to label the parts A, B, C and D.

ciliary muscles **cornea** **iris** **optic nerve**

pupil **retina**

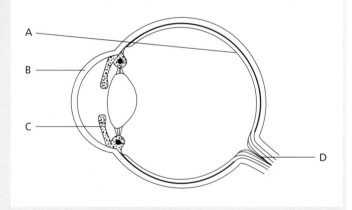

b Complete the following sentences.
(i) The ciliary muscles control the shape of the ......
(ii) To bring light from a distant object to a focus on the back of the eye, the convex lens needs to be made ......
(iii) To bring light from a near object to a focus on the back of the eye, the convex lens needs to be made ......
(iv) The iris controls the amount of ...... entering the eye.
(v) When light is dim, the pupil becomes ......
c Complete the following sentences.
(i) A short-sighted person cannot see ...... objects clearly.
(ii) Short-sightedness can be corrected by using ...... lenses.
(iii) A long-sighted person cannot see ...... objects clearly.
(iv) Long-sightedness can be corrected by using ...... lenses.

d Copy and complete the diagram below to show what happens to the rays of light when they pass through the concave lens.

(AQA (NEAB) Foundation, June 99)

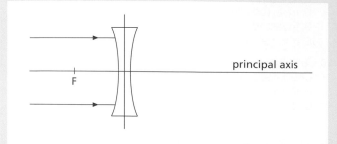

13 A man needs to hold a newspaper at arm's length in order to read it.
  a State a defect of vision which would cause this.
  b What type of spectacle lenses would he require to correct this defect?

14 Which pair of colours **A** to **E** when added together give white light?

  A yellow and blue
  B red and blue
  C green and yellow
  D red and green
  E green and blue

15 To focus a lens camera on a nearer object, is

  A the distance between the lens and film increased
  B the distance between the lens and film decreased
  C the aperture made larger
  D the aperture made smaller
  E the shutter speed changed?

16 An object is placed 10 cm in front of a lens, A; the details of the image are given below. The process is repeated for a different lens, B.

  Lens A Real, inverted, magnified and at a great distance.
  Lens B Real, inverted and same size as the object.

  Estimate the focal length of each lens and state whether it is convex or concave.

17 The ray diagrams below show rays from the object to the lens as used in different optical instruments.

  a Copy and complete both ray diagrams. On each diagram label the image formed.

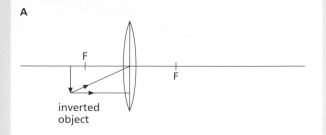

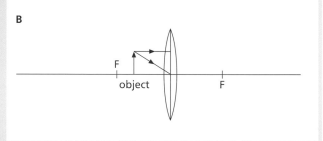

  b (i) Copy and complete the table below to describe the images formed.

	larger or smaller than object	real or virtual
A		
B		

  (ii) Name the optical instruments which use the lenses in each of these ways.

(AQA (NEAB) Higher, June 99)

Waves and sound

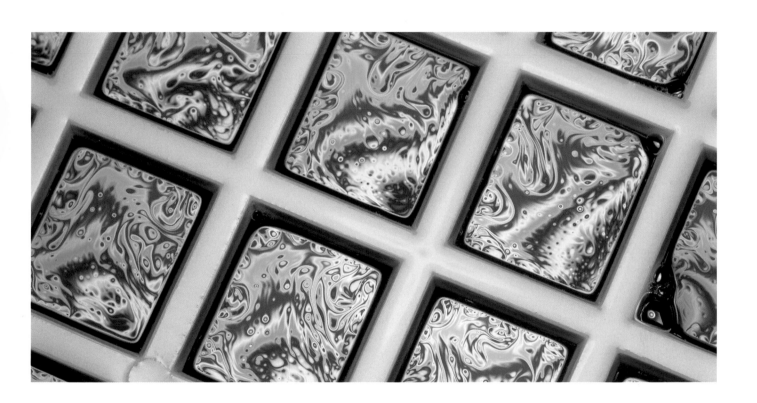

12 Mechanical waves

Types of wave
Describing waves
The wave equation
Reflection
Refraction
Diffraction
Interference
Polarization
Practical work
The ripple tank.

Types of wave

Several kinds of wave occur in physics. **Mechanical waves** are produced by a disturbance, e.g. a vibrating object, in a material medium and are transmitted by the particles of the medium vibrating to and fro. Such waves can be seen or felt and include waves on a rope or spring, water waves and sound waves in air or in other materials.

A **progressive** or travelling wave is a disturbance which carries energy from one place to another without transferring matter. There are two types, **transverse** and **longitudinal** (Chapter 15).

In a transverse wave, the direction of the disturbance is at **right angles** to the direction of travel of the wave. One can be sent along a rope (or a spring) by fixing one end and moving the other rapidly up and down, Figure 12.1. The disturbance generated by the hand is passed on from one part of the rope to the next which performs the same motion but slightly later. The humps and hollows of the wave travel along the rope as each part of the rope vibrates transversely about its undisturbed position.

Water waves are transverse waves.

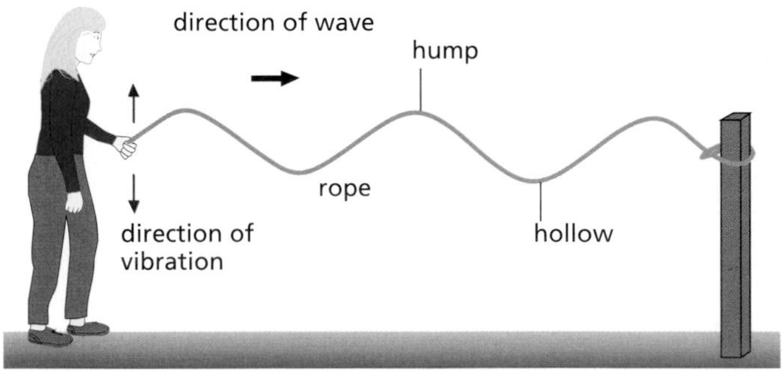

Figure 12.1 A transverse wave

Describing waves

Terms used to describe waves can be explained with the aid of a **displacement–distance** graph, Figure 12.2. It shows the distance moved sideways from their undisturbed positions, of the parts vibrating at different distances from the cause of the wave, at a **certain time**.

a) Wavelength

The wavelength of a wave, represented by the Greek letter λ (lambda), is the distance between successive crests.

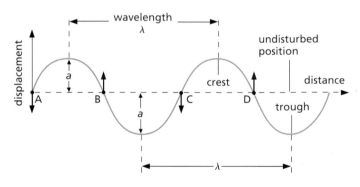

Figure 12.2 Displacement–distance graph for a wave at a particular instant

b) Frequency

The frequency f is the number of complete waves generated per second. If the end of a rope is jerked up and down twice in a second, two waves are produced in this time. The frequency of the wave is 2 vibrations per second or 2 **hertz** (2 Hz; the hertz being the unit of frequency) which is the same as the frequency of jerking of the end of the rope. That is, the frequencies of the wave and its source are equal.

The frequency of a wave is also the number of crests passing a chosen point per second.

c) Speed

The speed v of the wave is the distance moved by a crest or any point on the wave in 1 second.

d) Amplitude

The amplitude a is the height of a crest or the depth of a trough measured from the undisturbed position of what is carrying the wave, e.g. a rope.

e) Phase

The arrows at A, B, C, D on Figure 12.2 show the directions of vibration of the parts of the rope at these points. The parts at A and C have the same speed in the same direction and are **in phase**. At B and D the parts are also in phase but they are **out of phase** with those at A and C because their directions of vibration are opposite.

The wave equation

The faster the end of a rope is waggled, the shorter the wavelength of the wave produced. That is, the higher the frequency of a wave the smaller its wavelength. There is a useful connection between f, λ and v which is true for all types of wave.

Suppose waves of wavelength $\lambda = 20$ cm travel on a long rope and three crests pass a certain point every second. The frequency $f = 3$ Hz. If Figure 12.3 represents this wave motion then if crest A is at P at a particular time, 1 second later it will be at Q, a distance from P of three wavelengths, i.e. $3 \times 20 = 60$ cm. The speed of the wave $v = 60$ cm per second (60 cm/s), obtained by multiplying f by λ. Hence

$$\text{speed of wave} = \text{frequency} \times \text{wavelength}$$

or

$$v = f\lambda$$

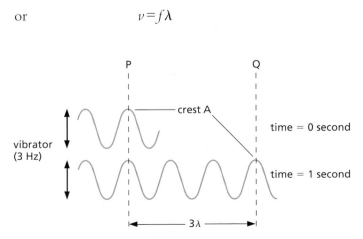

Figure 12.3

Practical work

The ripple tank

The behaviour of water waves can be studied in a ripple tank. It consists of a transparent tray containing water, having a light source above and a white screen below to receive the wave images, Figure 12.4.

Pulses (i.e. short bursts) of ripples are obtained by dipping a finger in the water for circular ones and a ruler for straight ones. **Continuous ripples** are generated by an electric motor and a bar which gives straight ripples if it just touches the water or circular ripples if the bar is raised and a small ball fitted to it.

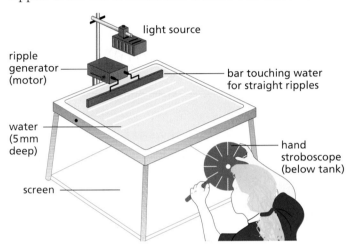

Figure 12.4 A ripple tank

Continuous ripples are studied more easily if they are *apparently* stopped ('frozen') by viewing the screen through a disc with equally spaced slits, that can be spun by hand, i.e. a stroboscope. If the disc speed is such that the waves have advanced one wavelength each time a slit passes your eye, they appear at rest.

Reflection

In Figure 12.5 **straight** water waves are falling on a metal strip placed in a ripple tank at an angle of 60°, i.e. the angle i between the direction of travel of the waves and the normal to the strip is 60°, as is the angle between the wavefront and the strip. The **wavefronts** are represented by straight lines and can be thought of as the crests of the waves. They are at right angles to the direction of travel, i.e. to the **rays**. The angle of reflection r is 60°. Incidence at other angles shows that the angles of reflection and incidence are always equal.

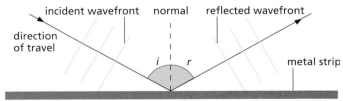

Figure 12.5 Reflection of waves

■ *Refraction*

If a glass plate is placed in a ripple tank so that the water is about 1 mm deep over it but 5 mm elsewhere, continuous straight waves in the shallow region are found to have a shorter wavelength than those in the deeper parts, i.e. the wavefronts are closer together, Figure 12.6. Both sets of waves have the frequency of the vibrating bar and since $v = f\lambda$, if λ has decreased so has v, since f is fixed. Hence **waves travel more slowly in shallow water**.

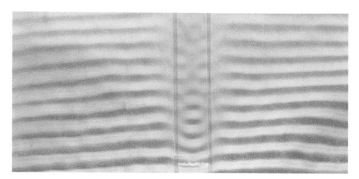

Figure 12.6 Waves in shallower water have a shorter wavelength

When the plate is at an angle to the waves, Figure 12.7a, their direction of travel in the shallow region is bent towards the normal, Figure 12.7b, i.e. refraction occurs. We saw earlier (Chapter 5) that light is refracted because its speed (and wavelength) changes (but not its frequency) when it enters another medium. The refraction of water waves for the same reason seems to suggest that light may be a kind of wave motion.

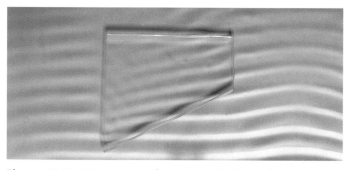

Figure 12.7a Waves are refracted at the boundary between deep and shallow regions

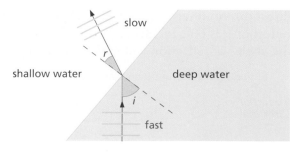

Figure 12.7b The direction of travel is bent towards the normal in the shallow region

■ *Diffraction*

In Figures 12.8a, b, straight water waves in a ripple tank are falling on gaps formed by obstacles. In photograph a the gap width is about the same as the wavelength of the waves (1 cm); those passing through are circular and spread out in all directions. In photograph b the gap is wide (10 cm) compared with the wavelength and the waves continue straight on; some spreading occurs but it is less obvious.

The spreading of waves at the edges of obstacles is called **diffraction**; when designing harbours, engineers use models like that in Figure 12.9 to study it.

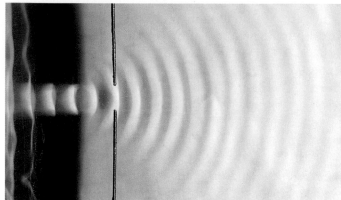

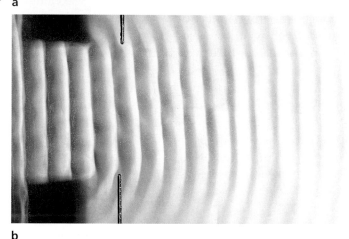

a

b

Figure 12.8 Spreading of waves after passing through a narrow and a wide gap

Figure 12.9 Model of a harbour used to study wave behaviour

Interference

When two sets of continuous **circular** waves cross in a ripple tank, a pattern like that in Figure 12.10 is obtained.

At points where a crest from once source, S_1, arrives at the same time as a crest from the other source, S_2, a bigger crest is formed and the waves are said to be **in phase**. At points where a crest and a trough arrive together, they cancel out (if their amplitudes are equal); the waves are exactly **out of phase** (due to travelling different distances from S_1 and S_2) and the water is undisturbed; in Figure 12.10 the blurred lines radiating from between S_1 and S_2 join such points.

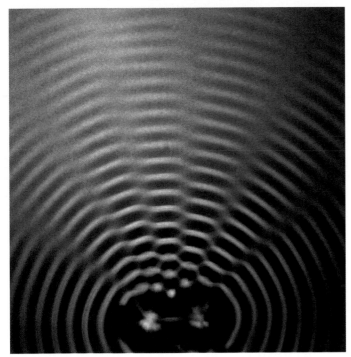

Figure 12.10 Interference of circular waves

Interference or **superposition** is the combination of waves to give a larger or a smaller wave, Figure 12.11a. Figure 12.11b shows how the pattern in Figure 12.10 is formed. All points on AB are equidistant from S_1 and S_2 and since these vibrate in phase, crests (or troughs) from S_1 arrive at the same time as crests (or troughs) from S_2. Hence along AB reinforcement occurs by superposition and a wave of double amplitude is obtained. Points on CD are half a wavelength nearer to S_1 than to S_2, i.e. there is a path difference of half a wavelength. Therefore crests (or troughs) from S_1 arrive simultaneously with troughs (or crests) from S_2 and the waves cancel. Along EF the difference of distances from S_1 and S_2 to any point is one wavelength, making EF a line of reinforcement.

Study this effect with two ball 'dippers' about 3 cm apart on the bar of a ripple tank. Also observe the effect of changing (i) the frequency and (ii) the separation of the 'dippers'; use a stroboscope when necessary. You will find that if the frequency is increased, i.e. the

wavelength decreased, the blurred lines are closer together. Increasing the separation has the same effect.

Similar patterns are obtained if straight waves fall on two small gaps: interference occurs between the sets of emerging (circular) diffracted waves.

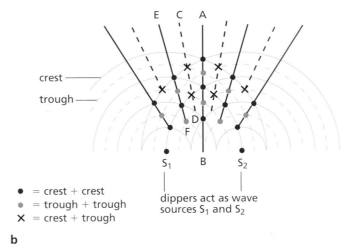

a

b

Figure 12.11

Polarization

This effect occurs only with transverse waves. It can be shown by fixing a rope at one end, D in Figure 12.12, and passing it through two slits B and C. If end **A** is moved to and fro in all directions (as shown by the short arrowed lines), vibrations of the rope occur in every plane and transverse waves travel towards B.

At B only waves due to vibrations in a vertical plane can emerge from the vertical slit. The wave between B and C is said to be **plane polarized** (in the vertical plane containing the slit at B). By contrast the waves between A and B are unpolarized. If the slit at C is vertical, the wave travels on, but if it is horizontal as shown, the wave is stopped and the slits are said to be 'crossed'.

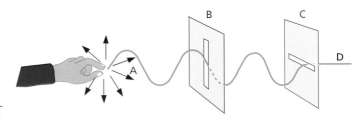

Figure 12.12 Polarizing waves on a rope

43

Questions

1 The lines in Figure 12.13 are crests of straight ripples.
   a What is the wavelength of the ripples?
   b If ripple A occupied 5 seconds ago the position now occupied by ripple F, what is the frequency of the ripples?
   c What is the speed of the ripples?

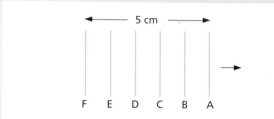

Figure 12.13

2 During the refraction of a wave which *two* of the following properties change?

   A the speed
   B the frequency
   C the wavelength

3 One side of a ripple tank ABCD is raised slightly, Figure 12.14, and a ripple started at P by a finger. After a second the shape of the ripple is as shown.
   a Why is it not circular?
   b Which side of the tank has been raised?

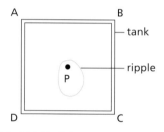

Figure 12.14

4 Figure 12.15 gives a full-scale representation of the water in a ripple tank 1 second after the vibrator was started. The coloured lines represent crests.
   a What is represented at A at this instant?
   b Estimate
      (i) the wavelength,
      (ii) the speed of the waves, and
      (iii) the frequency of the vibrator.
   c Sketch a suitable attachment which could have been vibrated up and down to produce this wave pattern.
   d Explain how the waves combine
      (i) at B, and
      (ii) at C.

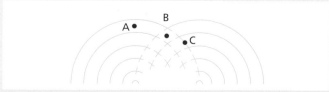

Figure 12.15

5 Copy Figure 12.16 and show on it what happens to the waves as they pass through the gap, if the water is much shallower on the right-hand side than on the left.

Figure 12.16

Checklist

After studying this chapter you should be able to

■ describe the production of pulses and progressive transverse waves on ropes, springs and ripple tanks,

■ recall the meaning of **wavelength, frequency, speed, amplitude** and **phase**,

■ represent a transverse wave on a displacement–distance graph and extract information from it,

■ recall the wave equation $v = f\lambda$ and use it to solve problems,

■ describe an experiment to show reflection of waves,

■ recall that the angle of reflection equals the angle of incidence and draw a diagram for the reflection of straight wavefronts at a plane surface,

■ describe experiments to show refraction of waves,

■ recall that refraction at a straight boundary is due to change of wave speed but *not* of frequency,

■ draw a diagram for the refraction of straight wavefronts at a straight boundary,

■ explain the term **diffraction**,

■ describe experiments to show diffraction of waves,

■ draw diagrams for the diffraction of straight wavefronts at single slits of different widths,

■ predict the effect of changing the wavelength or the size of the gap on diffraction of waves at a single slit,

■ describe an experiment to show interference of water waves using two point sources,

■ draw interference patterns for waves from two point sources,

■ explain interference as the superposition of crests or troughs at points where waves arrive **in phase** or **out of phase**,

■ predict the effect on the interference pattern of changing the separation of the two sources or the wavelength of the waves,

■ recall the meaning of a **plane polarized** wave.

13 Light waves

Diffraction
Interference
Everyday interference effects
Colour, wavelength and frequency

Red sunsets; blue sky
Polarization
Lasers and holograms

Although we cannot see how light travels, it displays the properties of waves.

Diffraction

Diffraction occurs when light passes the edge of an object, but it is not easy to detect. This suggests that light has a very small wavelength, since we saw that diffraction of water waves is most obvious at a gap when its width is comparable with the wavelength of the waves. Figure 13.1 is the diffraction pattern of a very narrow vertical slit (1/100 mm or less) and shows how light has spread into regions that would be in shadow if it went exactly in straight lines.

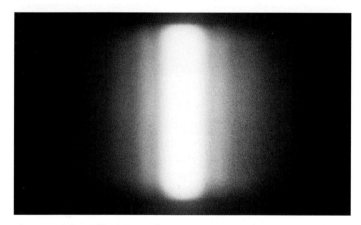

Figure 13.1 Diffraction of white light at a narrow slit

Interference

A steady interference pattern is obtained with water waves from two sources in a ripple tank because both sets of waves have the same frequency and wavelength and are exactly in phase when they leave the sources. They are said to be **coherent**. It is impossible to obtain a steady interference pattern with light using two separate lamps since most light sources emit light waves of many wavelengths in short erratic bursts, each out of phase with the next. The two sets of waves are not coherent.

These difficulties were overcome by Young in 1801 by allowing light from *one* lamp in a darkened room to fall on two narrow parallel slits very close together (about 0.5 mm separation), as shown in the simple arrangement of Figure 13.2. A pattern of equally spaced bright and dark bands, called **fringes**, is obtained on a screen. The waves leaving the slits are coherent since any phase changes in the bursts of light from the lamp affect both sets of waves at the same time.

The bright bands are coloured, except for the central one which is white. If a red filter is placed in front of the lamp, the bright bands are red and are spaced farther apart than those given by a blue filter; see Figures 13.3a, b, overleaf.

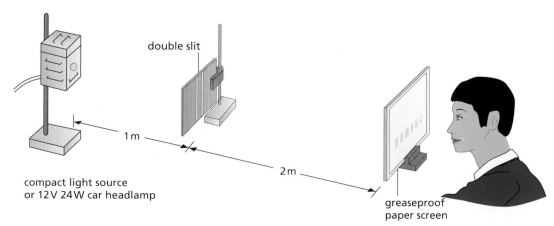

double slit

1 m

2 m

compact light source
or 12 V 24 W car headlamp

greaseproof
paper screen

Figure 13.2 Viewing Young's double-slit interference

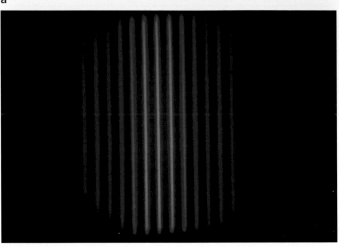

Figure 13.3 The fringes with red light are spaced farther apart than with blue light

The fringes can be explained by assuming that diffraction occurs at each slit, and in the region where the two diffracted beams cross there is interference, Figure 13.4. At points on the screen where a 'crest' from one slit arrives at the same time as a 'crest' from the other, the waves are in phase and there are bright bands. Dark bands occur where 'crests' and 'troughs' arrive simultaneously and the waves cancel. We then have: light + light = darkness. This makes sense only if we regard light as having a wave nature. Figure 12.11b, which we used to explain the water wave interference pattern, is also helpful when thinking about how light produces interference effects.

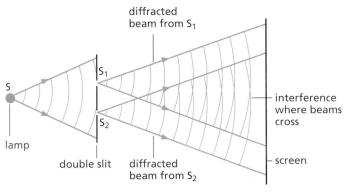

Figure 13.4 The two slits S_1 and S_2 act as coherent sources

Everyday interference effects

The colour effects produced by thin oil films on a wet road, Figure 13.5a, and in soap bubbles, Figure 13.5b, are due to interference between light reflected from the two surfaces of the film or bubble. Figure 13.6 shows this for a film of oil on water.

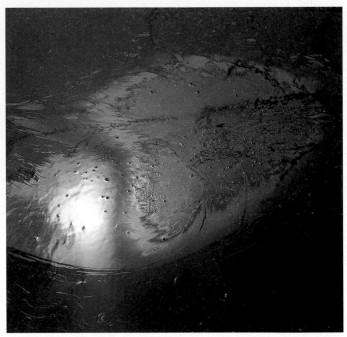

a

b

Figure 13.5 Interference of light reflected from thin oil films and soap bubbles gives rise to beautiful coloured patterns

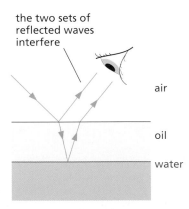

Figure 13.6 Reflection of light from an oil film on water

Colour, wavelength and frequency

Red light has the longest wavelength of about 0.0007 mm (7×10^{-7} m $= 0.7$ μm), while violet light has the shortest wavelength of about 0.0004 mm (4×10^{-7} m $= 0.4$ μm). Colours between these in the spectrum of white light have intermediate values. Light of one colour and so of one wavelength is called **monochromatic** light.

Remembering that $v = f\lambda$ for all waves including light, it follows that red light has a lower frequency, f, than violet light since (i) the wavelength, λ, of red light is greater and (ii) all colours travel with the same speed, v, of 3×10^8 m/s in air (strictly a vacuum). It is the **frequency** of light which decides its colour, rather than its wavelength which is different in different media, as is its speed (Chapter 5).

Different frequencies of light travel at different speeds through a transparent medium and so are refracted by different amounts. This explains dispersion (Chapter 9), i.e. why the refractive index of a material depends on the wavelength of the light.

The **amplitude** of a light (or any other) wave is greater the greater the **intensity** of the source, i.e. the brighter it is in the case of light.

Red sunsets; blue sky

The wavelength of red light is about twice that of blue light. The longer the wavelength the more penetrating the light, while the shorter the wavelength the more easily it is scattered.

When the Sun is setting the light from it has to travel through a greater thickness of the Earth's atmosphere and only the longer wavelength red light is able to get through. Sunsets are therefore red, Figure 13.7.

The shorter wavelengths, like blue, are scattered in all directions by the atmosphere, which is why the sky looks blue.

Figure 13.7 Red light penetrates further through the atmosphere than blue light

Polarization

Interference and diffraction need a wave model to explain the effects they produce but they do not show whether the waves are transverse or longitudinal. Polarization suggests that light waves are transverse.

a) Using polarizing material

If a lamp is viewed through a piece of polarizing material such as Polaroid (used in sunglasses), apart from it seeming slightly less bright, there is no effect when the Polaroid is rotated. However, using two pieces, one of which is kept at rest and the other rotated slowly, Figure 13.8, the light is cut off more or less completely in one position. The Polaroids are then 'crossed'. Rotation through a further 90° allows maximum light through.

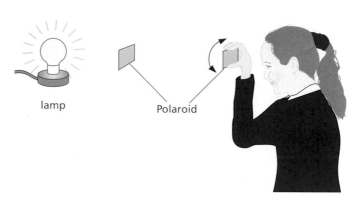

lamp   Polaroid

Figure 13.8 Polaroid polarizes the light waves

This experiment is similar to that shown in Figure 12.12 where transverse waves are sent along a rope to two slits B and C. Slit B produces a wave that is **polarized** in a vertical plane. This wave cannot pass through horizontal slit C. In this position slits B and C are 'crossed'. A longitudinal wave, e.g. vibrations along a coiled spring, would emerge from both slits whatever their positions, i.e. it cannot be polarized.

The experiment with the Polaroids can be explained if we regard light as a transverse wave motion. Light from the Sun and other sources is unpolarized and consists of 'vibrations' in every plane at right angles to the direction of travel of the wave. The Polaroid nearest the lamp polarizes the light by transmitting only the 'vibrations' in one particular plane. The other Polaroid transmits or absorbs the plane-polarized light falling on it depending on its orientation with respect to the first Polaroid.

b) By reflection

When unpolarized light falls on glass, water or a polished surface, the reflected ray is, in general, partly plane polarized. At a certain angle of incidence the polarization is complete.

Glare caused by light reflected from a smooth surface can be reduced by using polarizing material such as Polaroid. Polarizing discs, suitably orientated, are used in sunglasses and also in photography as 'filters' in front of the camera lens, enabling detail to be seen that would otherwise be hidden by glare, Figure 13.9.

Figure 13.9 Glare can be reduced with a polarizing filter in front of the camera

Lasers and holograms

A laser produces a very intense beam of light which is monochromatic, coherent and parallel. A hologram is a three-dimensional image of an object recorded on a special photographic plate. The image is formed by the interference of two beams of light from a laser, one of which has been reflected from the object. When developed and illuminated, the image looks 'real', appears to float in space and moves when the viewer does.

Phonecards contain a hologram which keeps a check on the number of units used. Credit cards contain holograms to prevent forgery, Figure 13.10a. Holograms are also used to make safety checks on car tyres and engines, Figure 13.10b, to design artificial limbs, and to display and market goods.

Figure 13.10a Hologram on a credit card

Figure 13.10b Hologram of an engine

Questions

1 In Figure 13.11, L is a source of light, wavelength λ, S_1S_2 a double slit and O, O_1 and O_2 are points on a screen.

  a If the central bright band is formed at O, how do the distances S_1O and S_2O compare?

  b If the first dark band next to the central bright band is formed at O_1, how do S_1O_1 and S_2O_1 compare?

  c If the first bright band occurs at O_2 next to the first dark band, how do S_1O_2 and S_2O_2 compare?

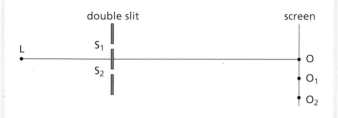

Figure 13.11

2 **a** Why is the diffraction of light not easy to detect?

  b Why is it not possible to produce a steady interference pattern using two light sources?

3 Give the approximate wavelength in micrometres (μm) of

  a red light,

  b violet light.

4 **a** State *two* ways of producing polarized light.

  b State *one* use of polarized light.

  c Is it possible to polarize longitudinal waves?

Checklist

After studying this chapter you should be able to

- explain why diffraction of light is not normally observed,
- describe a simple Young's double-slit experiment to show light has a wave-like nature,
- recall that the colour of light depends on its frequency, that red light has a lower frequency (but longer wavelength) than blue light and that all colours travel at the same speed in air,
- describe some evidence that light is a transverse wave,
- explain briefly how a hologram is made.

14 Electromagnetic radiation

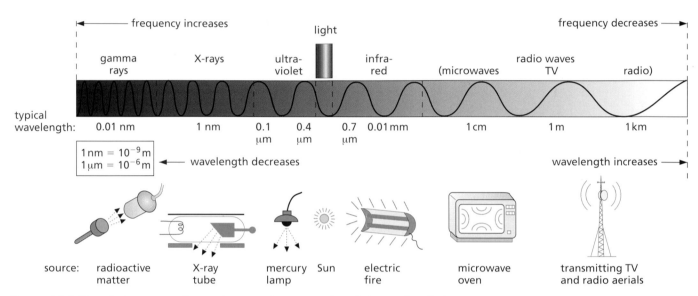

frequency increases ← → frequency decreases

| gamma rays | X-rays | ultra-violet | light | infra-red | (microwaves | radio waves TV | radio) |

typical wavelength: 0.01 nm   1 nm   0.1 μm   0.4 μm   0.7 μm   0.01 mm   1 cm   1 m   1 km

$1\,nm = 10^{-9}\,m$
$1\,\mu m = 10^{-6}\,m$

← wavelength decreases   wavelength increases →

source: radioactive matter | X-ray tube | mercury lamp | Sun | electric fire | microwave oven | transmitting TV and radio aerials

Figure 14.1 The electromagnetic spectrum and sources of each type of radiation

Light is one member of a family of electromagnetic radiation, which forms a continuous spectrum beyond both ends of the visible (light) spectrum, Figure 14.1. While each type of radiation has a different source, all result from electrons in atoms undergoing an energy change and all have certain properties in common.

Properties

1 All types of electromagnetic radiation **travel through space at 300 000 km/s** ($3 \times 10^8\,$m/s), i.e. with the speed of light.
2 They **exhibit interference, diffraction and polarization**, which suggests they have a transverse wave nature.
3 They **obey the wave equation** $v = f\lambda$ where v is the speed of light, f is the frequency of the waves and λ is the wavelength. Since v is constant for a particular medium, it follows that large f means small λ.
4 They **carry energy from one place to another and can be absorbed by matter to cause heating and other effects**. The higher the frequency and the smaller the wavelength of the radiation, the greater is the energy carried, i.e. gamma rays are more 'energetic' than radio waves.

This is shown by the **photoelectric effect** in which electrons are ejected from metal surfaces when electromagnetic waves fall on them. As the frequency of the waves increases so too does the speed (and energy) with which electrons are emitted.

Because of its electrical origin, its ability to travel in a vacuum (e.g. from the Sun to the Earth) and its wave-like properties (i.e. **2** above), electromagnetic radiation is regarded as a **progressive transverse wave**. The wave is a combination of travelling electric and magnetic fields. The fields vary in value and are directed at right angles to each other and to the direction of travel of the wave, as shown by the representation in Figure 14.2.

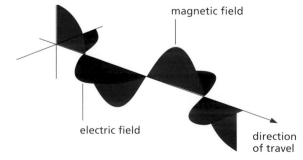

magnetic field

electric field

direction of travel

Figure 14.2 An electromagnetic wave

■ Infrared radiation

Our bodies detect infrared radiation (IR) by its heating effect on the skin. It is sometimes called 'radiant heat' or 'heat radiation'.

Anything which is hot but not glowing, i.e. below 500°C, emits IR alone. At about 500°C a body becomes red-hot and emits red light as well as IR – the heating element of an electric fire, a toaster or a grill are examples. At about 1500°C things such as lamp filaments are white-hot and radiate IR and white light, i.e. all the colours of the visible spectrum.

Infrared is also detected by special temperature-sensitive photographic films which allow pictures to be taken in the dark. Infrared sensors are used on satellites and aircraft for weather forecasting, monitoring of land use, Figure 14.3, assessing heat loss from buildings, and locating victims of earthquakes.

Figure 14.3 Infrared aerial photograph of Washington DC

Infrared lamps are used to dry the paint on cars during manufacture and in the treatment of muscular complaints. Remote control keypads for televisions contain a small infrared transmitter for changing programmes.

■ Ultraviolet radiation

Ultraviolet (UV) rays have shorter wavelengths than light. They cause sun-tan and produce vitamins in the skin but can penetrate deeper causing skin cancer. Dark skin is able to absorb more UV so reducing the amount reaching deeper tissues. Exposure to the harmful UV rays present in sunlight can be reduced by wearing protective clothing such as a hat or using sunscreen lotion.

Ultraviolet causes fluorescent paints and clothes washed in some detergents to fluoresce, Figure 14.4. They glow by re-radiating as light the energy they absorb as UV. This may be used to verify 'invisible' signatures on bank documents.

Figure 14.4 White clothes fluorescing violet at a disco

A UV lamp used for scientific or medical purposes contains mercury vapour and this emits UV when an electric current passes through it. Fluorescent tubes also contain mercury vapour and their inner surfaces are coated with special powders called phosphors which radiate light.

Radio waves

Radio waves have the longest wavelengths in the electromagnetic spectrum. They are radiated from aerials and used to 'carry' sound, pictures and other information over long distances.

a) Long, medium and short waves (wavelengths of 2 km to 10 m)

These diffract round obstacles so can be received when hills etc. are in their way, Figure 14.5a. They are also reflected by layers of electrically charged particles in the upper atmosphere (the **ionosphere**), which makes long-distance radio reception possible, Figure 14.5b.

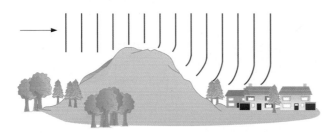

a Diffraction of radio waves

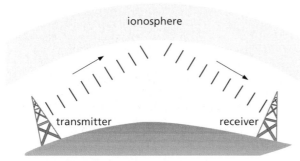

b Reflection of radio waves

Figure 14.5

b) VHF (very high frequency) and UHF (ultra high frequency) waves (wavelengths of 10 m to 10 cm)

These shorter wavelength radio waves need a clear, straight-line path to the receiver. They are not reflected by the ionosphere. They are used for local radio and for television.

c) Microwaves (wavelengths of a few cm)

These are used for international telecommunications and television relay via geostationary satellites (see Figure 2c, p. x) and for mobile phone networks via microwave aerial towers and low-orbit satellites (Chapter 34). The microwave signals are transmitted through the ionosphere by dish aerials, amplified by the satellite and sent back to a dish aerial in another part of the world.

Microwaves are also used for **radar** detection of ships and aircraft, and in police speed traps.

Microwaves can be used for cooking since they cause water molecules in the moisture of the food to vibrate vigorously at the frequency of the microwaves. As a result, heating occurs inside the food which cooks itself.

Living cells can be damaged or killed by the heat produced when microwaves are absorbed by water in the cells. There is some debate at present as to whether their use in mobile phones is harmful; 'hands-free' mode, where separate earphones are used, may be safer.

Practical work

Wave nature of microwaves

The 3 cm microwave transmitter and receiver shown in Figure 14.6 can be used. The three metal plates can be set up for double-slit interference with 'slits' about 3 cm wide; the reading on the meter connected to the horn receiver rises and falls as it is moved across behind the plates.

Diffraction can be shown similarly using the two wide metal plates to form a single slit.

If the grid of vertical metal wires is placed in front of the transmitter, the signal is absorbed but transmission occurs when the wires are horizontal, showing that the microwaves are vertically polarized. This can also be shown by rotating the receiver through 90° in a vertical plane from the maximum signal position, when the signal decreases to a minimum.

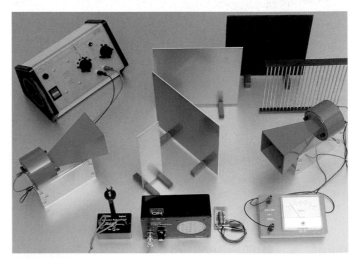

Figure 14.6

X-rays

These are produced when high-speed electrons are stopped by a metal target in an X-ray tube. X-rays have smaller wavelengths than UV.

They are absorbed to some extent by living cells but can penetrate some solid objects and affect a photographic film. With materials like bones, teeth and metals which they do not pass through easily, shadow pictures can be taken, like that in Figure 14.7 of someone shaving. In industry X-ray photography is used to inspect welded joints.

X-ray machines need to be shielded with lead since normal body cells can be killed by high doses and made cancerous by lower doses.

Gamma rays (Chapter 58) are more penetrating and dangerous than X-rays. They are used to kill cancer cells, and also harmful bacteria in food and on surgical instruments.

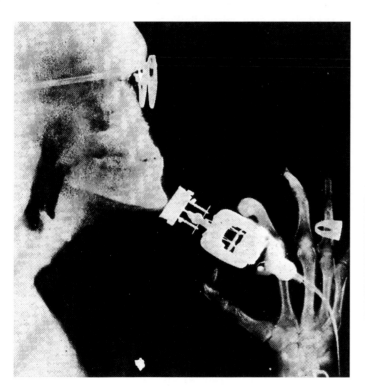

Figure 14.7 X-rays cannot penetrate bone and metal

Questions

1 The electromagnetic spectrum includes radio waves, microwaves, infra-red, visible light, ultraviolet, X-rays and gamma rays.
   a Which statement about electromagnetic waves is correct?

   A They all have the same wavelength.
   B They all have the same frequency.
   C They are all transverse waves.

   b Name *two* types of electromagnetic wave used to cook food.
   c Name *one* type of electromagnetic wave used to detect a break in a bone.
   d Ultraviolet radiation reaching us from the Sun may be increasing.
   (i) Describe the danger of ultraviolet radiation to human health.
   (ii) Suggest *two* ways in which people can reduce the health risk from ultraviolet radiation.

   (*London Foundation, June 99*)

2 Which of the following types of radiation has
   a the longest wavelength,
   b the highest frequency?

   A UV
   B radio waves
   C light
   D X-rays
   E IR

3 Name one type of electromagnetic radiation which
   a causes sun-tan,
   b is used for satellite communication,
   c is used to sterilize surgical instruments,
   d is used in TV remote control.

4 A VHF radio station transmits on a frequency of 100 MHz (1 MHz = 10^6 Hz). If the speed of radio waves is 3×10^8 m/s,
   a what is the wavelength of the waves,
   b how long does the transmission take to travel 60 km?

Checklist

After studying this chapter you should be able to

- recall the types of electromagnetic radiation,
- recall that all electromagnetic waves have the same speed in space and are progressive transverse waves,
- distinguish between **infrared radiation, ultraviolet radiation, radio waves** and **X-rays** in terms of their wavelengths, properties and uses,
- be aware of the harmful effects of different types of electromagnetic radiation and of how exposure to them can be reduced.

15 Sound waves

Origin and transmission of sound
Longitudinal waves
The ear

Reflection and echoes
Speed of sound

Refraction, diffraction and
interference
Ultrasonics

■ Origin and transmission of sound

Sources of sound all have some part which **vibrates**. A guitar has strings, Figure 15.1, a drum has a stretched skin and the human voice has vocal cords. The sound travels through the air to our ears and we hear it. That the air is necessary may be shown by pumping out a glass jar containing a ringing electric bell, Figure 15.2; the sound disappears though the striker can still be seen hitting the gong. Evidently sound cannot travel in a vacuum as light can. Other materials, including solids and liquids, transmit sound.

Sound also gives interference and diffraction effects. Because of this and its other properties, we believe it is a form of energy (as the damage from supersonic booms shows) which travels as a progressive wave, but of a type called **longitudinal**.

Figure 15.1 A guitar string vibrating. The sound waves produced are amplified when they pass through the circular hole into the guitar's sound box

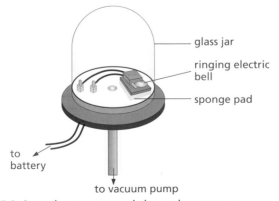

glass jar

ringing electric bell

sponge pad

to battery

to vacuum pump

Figure 15.2 Sound cannot travel through a vacuum

■ Longitudinal waves

a) Waves on a spring

In a progressive longitudinal wave the particles of the transmitting medium vibrate to and fro along the same line as that in which the wave is travelling and not at right angles to it as in a transverse wave. A longitudinal wave can be sent along a spring, stretched out on the bench and fixed at one end, if the free end is repeatedly pushed and pulled sharply. Compressions C (where the coils are closer together) and rarefactions R (where the coils are farther apart), Figure 15.3, travel along the spring.

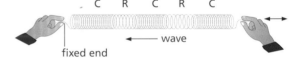

fixed end

wave

Figure 15.3 A longitudinal wave

b) Sound waves

A sound wave, produced for example by a loudspeaker, consists of a train of compressions ('squashes') and rarefactions ('stretches') in the air, Figure 15.4.

The speaker has a cone which is made to vibrate in and out by an electric current. When the cone moves out the air in front is compressed; when it moves in the air is rarefied (goes 'thinner'). The wave progresses through the air but the air as a whole does not move. The air particles (molecules) vibrate backwards and forwards a little as the wave passes. When the wave enters your ear the compressions and rarefactions cause small, rapid pressure changes on the eardrum and you experience the sensation of sound.

The number of compressions produced per second is the frequency f of the sound wave (and equals the frequency of the vibrating loudspeaker cone); the distance between successive compressions is the wavelength λ. As for transverse waves the speed $v = f\lambda$.

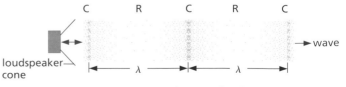

loudspeaker cone

wave

Figure 15.4 Sound travels as a longitudinal wave

54

The ear

The ear has three main parts, Figure 15.5.

a) Outer ear

The **ear canal** collects and directs the sound waves on to a thin membrane called the **eardrum**. This is made to vibrate at the same frequency by the air vibration of the sound waves.

b) Middle ear

This contains three tiny bones, the **ossicles**, the outermost being joined to the eardrum and the innermost fitting into a small hole in the skull called the **oval window**. When the eardrum vibrates so too do the ossicles.

c) Inner ear

The vibrations of the ossicles are passed on to the sensitive part of the inner ear. This is a fluid in a coiled tube known as the **cochlea** containing the **auditory nerve endings** which send off impulses to the brain when the fluid in the cochlea vibrates. Nerve endings in the first part of the cochlea respond to high frequency vibrations and those in the last part to low frequencies, enabling the brain to distinguish between high- and low-pitched sounds.

Humans hear only sounds with frequencies from about 20 Hz to 20 000 Hz. These are the **limits of audibility**; the upper limit decreases with age.

The **semicircular canals** are part of the inner ear but are not concerned with hearing. The information received by the brain from their sense organs enables us to stand upright when still and to keep our balance when moving around.

A hearing aid can be used to send an amplified sound to the eardrum when someone can hear only quite loud sounds. If the eardrum or middle ear has worse damage, an aid can pass on vibrations to the cochlea directly through the skull bones. Deafness occurs if the cochlea or auditory nerve is damaged.

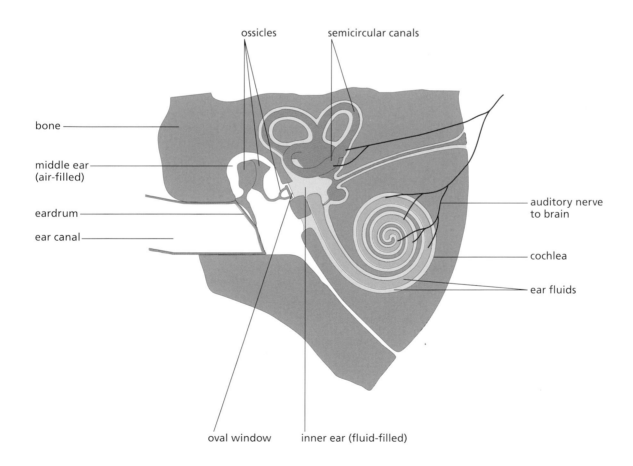

Figure 15.5 The structure of the ear

◼ *Reflection and echoes*

Sound waves are reflected well from hard, flat surfaces such as walls or cliffs and obey the same laws of reflection as light. The reflected sound forms an **echo**.

If the reflecting surface is nearer than 15 m, the echo joins up with the original sound which then seems to be prolonged. This is called **reverberation**. Some is desirable in a concert hall to stop it sounding 'dead', but too much causes 'confusion'. Modern concert halls are designed for the optimum amount of reverberation. Seats and some wall surfaces are covered with sound-absorbing material.

◼ *Speed of sound*

The speed of sound depends on the material through which it is passing. It is greater in solids than in liquids or gases because the molecules in a solid are closer together than in a liquid or a gas. Some values are given in Table 15.1.

Table 15.1 Speed of sound in different materials

Material	air (0 °C)	water	concrete	steel
Speed/ metres/second	330	1400	5000	6000

In air the speed **increases with temperature** and at high altitudes, where the temperature is lower, it is less than at sea-level. Changes of atmospheric pressure do not affect it.

An estimate of the speed of sound can be made directly if you stand about 100 metres from a high wall or building and clap your hands. Echoes are produced. When the clapping rate is such that each clap coincides with the echo of the previous one, the sound has travelled to the wall and back in the time between two claps, i.e. one interval. By timing 30 intervals with a stopwatch, the time t for one interval can be found. Also, knowing the distance d to the wall, a rough value is obtained from

$$\text{speed of sound in air} = \frac{2d}{t}$$

◼ *Refraction, diffraction and interference*

a) Refraction

When light crosses the boundary between two transparent materials its speed changes and refraction occurs for angles of incidence greater than zero (Figure 5.1, p. 14). Similarly when a sound wave enters a medium of different density its speed changes and it undergoes refraction unless it strikes the boundary normally. Seismic waves (Chapter 62) are refracted when they pass through layers of different density inside the Earth.

b) Diffraction

Audible sounds have wavelengths from about 1.5 centimetres (frequency 20 kHz) up to 15 metres (20 Hz) and suffer diffraction by objects of similar size, e.g. a doorway 1 metre wide. This explains why we hear sound round corners. Low frequency (longer wavelength) notes are diffracted more than higher frequencies (shorter wavelengths) such as those caused by traffic noise.

c) Interference

In Figure 15.6 sound waves of the same frequency from two loudspeakers (supplied by one signal generator) produce a steady interference pattern. The resulting variations in loudness of the sound, due to the waves reinforcing and cancelling one another, can be heard when you walk past the loudspeakers.

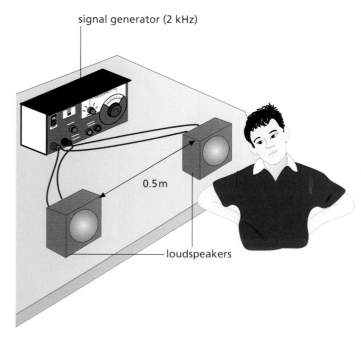

Figure 15.6 Sound gives interference effects

Ultrasonics

Sound waves with frequencies above 20 kHz are called **ultrasonic** waves; their frequency is too high to be detected by the human ear but they can be detected electronically and displayed on a cathode ray oscilloscope (CRO, see Chapter 57).

a) Quartz crystal oscillators

Ultrasonic waves are produced by a quartz crystal which is made to vibrate electrically at the required frequency; they are emitted in a narrow beam in the direction in which the crystal oscillates. An ultrasonic receiver also consists of a quartz crystal but it works in reverse, i.e. when it is set into vibration by ultrasonic waves it generates an electrical signal which is then amplified. The same quartz crystal can act as both a transmitter and a receiver.

b) Ultrasonic echo techniques

Ultrasonic waves are partially or totally reflected from surfaces at which the density of the medium changes; this property is exploited in techniques such as the non-destructive testing of materials, sonar and medical ultrasound imaging. A bat emitting ultrasonic waves can judge the distance away of an object from the time taken by the reflected wave or 'echo' to return. We can use a CRO to display and measure the time interval, t, between a transmitted pulse and the return of the echo, as shown in Figure 15.7 where reflection of an ultrasonic pulse occurs from a flaw in a rod. Rearranging the equation $v = 2d/t$ gives

$$d = \frac{vt}{2}$$

so the distance d of the flaw along the rod can be calculated if t is measured and v (the speed of sound in the material) is known.

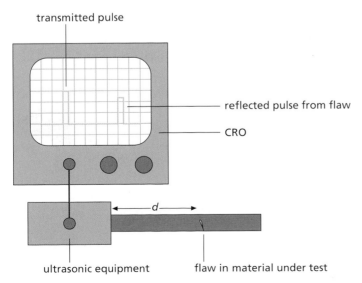

transmitted pulse

reflected pulse from flaw

CRO

d

ultrasonic equipment

flaw in material under test

Figure 15.7 Ultrasonic testing of a material

Ships with **sonar** can determine the depth of a shoal of fish or the sea-bed, Figure 15.8, in the same way; motion sensors (Chapter 28) also work on this principle.

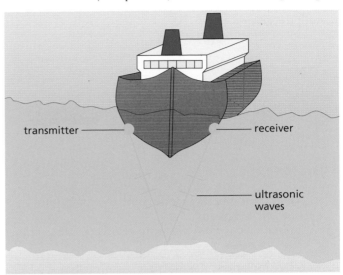

transmitter

receiver

ultrasonic waves

Figure 15.8 A ship using sonar

In **medical ultrasound imaging**, used in pre-natal clinics to monitor the health and sometimes to determine the sex of an unborn baby, an ultrasonic transmitter/receiver is scanned over the mother's abdomen and a detailed image of the fetus is built up, Figure 15.9. Reflection of the ultrasonic pulses occurs from boundaries of soft tissue, in addition to bone, so images of internal organs not seen by using X-rays can be obtained. Less detail of bone structure is seen than with X-rays, as the wavelength of ultrasonic waves is larger, typically about 1 mm, but ultrasound has no harmful effects on human tissue.

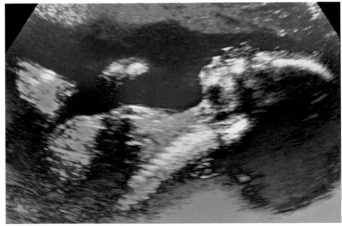

Figure 15.9 Checking the development of a fetus using ultrasound imaging

c) Other uses

Ultrasound can also be used in ultrasonic drills to cut holes of any shape or size in hard materials such as glass and steel. Jewellery, or more mundane objects such as street lamp covers, can be cleaned by immersion in a tank of solvent which has an ultrasonic vibrator in the base.

Questions

1 If 5 seconds elapse between a lightning flash and the clap of thunder how far away is the storm? Speed of sound = 330 m/s.

2

Figure 15.10

A ship in front of a cliff sounds one blast of its horn. Later an echo is heard.

a Choose a word from the list to complete the sentence below.

    radiation   reflection   refraction

The echo is caused by ......

b The ship is 1020 metres from the cliff. The echo is heard 6 seconds later.

(i) Choose a distance from the list to complete the sentence below.

    510 m   1020 m   2040 m

The distance the sound wave travels from the ship to the cliff and back to the ship is ......

(ii) Calculate the speed of the sound wave.
(AQA (NEAB) Foundation, June 99)

3 When the frequency of sound is changed the wavelength also changes. The table shows the results of an experiment to measure the wavelength of sound at different frequencies.

Wavelength (m)	0.7	1.0	1.5	2.5	4.0
Frequency (Hz)	460	320	210	130	80

a Draw the graph of wavelength against frequency, using axes like those in Figure 15.11.

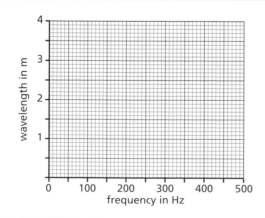

Figure 15.11

b Complete the sentence below.

When the frequency of the sound is increased, the wavelength ......

c (i) Use your graph to find the wavelength of sound with a frequency of 200 Hz.
(ii) Calculate the speed of the sound.
(London Foundation, June 99)

4 a A girl stands 160 m away from a high wall and claps her hands at a steady rate so that each clap coincides with the echo of the one before. If she makes 60 claps in 1 minute, what value does this give for the speed of sound?

b If she moves 40 m closer to the wall she finds the clapping rate has to be 80 per minute. What value do these measurements give for the speed of sound?

c If she moves again and finds the clapping rate becomes 30 per minute, how far is she from the wall if the speed of sound is the value you found in a?

5 a What properties of sound suggest it is a wave motion?

b How does a progressive transverse wave differ from a longitudinal one? Which type is sound?

6 What are ultrasonic waves and how are they produced? Outline their use in medical ultrasound imaging. Why are they preferred to X-rays for monitoring unborn babies?

▧ *Checklist*

After studying this chapter you should be able to

▧ recall that sound is produced by vibrations,

▧ describe an experiment to show that sound is not transmitted through a vacuum,

▧ describe how sound travels in a medium as progressive longitudinal waves,

▧ draw and label the structure of the human ear and describe the functions of its parts,

▧ recall the limits of audibility (i.e. the range of frequencies) for the normal human ear,

▧ explain echoes and reverberation,

▧ describe a simple method of estimating the speed of sound and recall its approximate value,

▧ solve problems using the speed of sound, e.g. thundercloud proximity,

▧ describe experiments to show diffraction and interference of sound waves,

▧ recall some uses of ultrasonics.

16 Musical notes

Pitch
Loudness

Quality
Vibrating strings; stationary waves

Resonance
Noise pollution

Irregular vibrations such as those of motor engines cause **noise**; regular vibrations such as occur in the instruments of a brass band, Figure 16.1, produce **musical notes** which have three properties – pitch, loudness and quality.

Figure 16.1 Musical instruments produce regular sound vibrations

Pitch

The pitch of a note depends on the frequency of the sound wave reaching the ear, i.e. on the frequency of the source of sound. A high-pitched note has a high frequency and a short wavelength. The frequency of middle C (in 'scientific pitch') is 256 vibrations per second or 256 Hz and that of upper C is 512 Hz. Notes are an **octave** apart if the frequency of one is twice that of the other. Pitch is like colour in light; both depend on the frequency.

Notes of known frequency can be produced in the laboratory by a signal generator supplying alternating electric current (a.c.) to a loudspeaker. The cone of the speaker vibrates at the frequency of the a.c. which can be varied and read off a scale on the generator. A set of tuning forks with frequencies marked on them can also be used. A tuning fork, Figure 16.2, has two steel prongs which vibrate when struck; the prongs move in and out *together*, generating compressions and rarefactions.

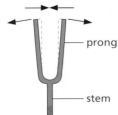

Figure 16.2 A tuning fork

Loudness

A note becomes louder when more sound energy enters our ears per second than before and is caused by the source vibrating with a larger amplitude. If a violin string is bowed more strongly, its amplitude of vibration increases as does that of the resulting sound wave and the note heard is louder because more energy has been used to produce it.

Quality

The same note on different instruments sounds different; we say the notes differ in **quality** or **timbre**. The difference arises because no instruments (except a tuning fork and a signal generator) emit a 'pure' note, i.e. of one frequency. Notes consist of a main or **fundamental** frequency mixed with others, called **overtones**, which are usually weaker and have frequencies that are exact multiples of the fundamental. The number and strength of the overtones decides the quality of a note. A violin has more and stronger higher overtones than a piano. Overtones of 256 Hz (middle C) are 512 Hz, 768 Hz and so on.

The **waveform** of a note played near a microphone connected to a CRO can be displayed on the CRO screen. Those for the *same* note on three instruments are given in Figure 16.3. Their different shapes show that while they have the same fundamental frequency, their quality differs. The 'pure' note of a tuning fork has a **sine** waveform and is the simplest kind of sound wave.

Note Although the waveform on the CRO screen is transverse it represents a longitudinal sound wave.

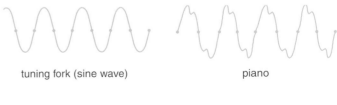

tuning fork (sine wave)   piano

violin

Figure 16.3 Notes of the same frequency (pitch) but different quality

59

Vibrating strings; stationary waves

In a string instrument such as a guitar the 'string' is a tightly stretched wire or length of gut. When it is plucked transverse waves travel to both ends, which are fixed, and are reflected. Interference occurs between the incident and reflected waves and a **stationary** or **standing** wave pattern is formed. In this, certain points on the string, called **nodes**, are at rest while points midway between each pair of nodes are in continuous vibration with maximum amplitude; they are called **antinodes**.

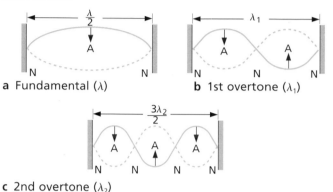

a Fundamental (λ)   **b** 1st overtone (λ_1)

c 2nd overtone (λ_2)

Figure 16.4 Vibrations on a string

A string can vibrate in various ways. If the standing wave has one loop, Figure 16.4a, the fundamental note is emitted. If there is more than one loop, Figures 16.4b, c, overtones are produced which have more nodes and higher frequency than the fundamental note. In all cases the **separation of the nodes N and the antinodes A is one-quarter of the wavelength** of the wave on the string causing the note. In an instrument, a string vibrates in several ways at the same time depending on where it is plucked; this decides the quality of the note.

The standing wave patterns of a vibrating string (or rubber cord) may be viewed as in Figure 16.5.

The frequency of the fundamental note emitted by a vibrating string depends on its

(i) **length**: short strings emit high notes and halving the length doubles the frequency,
(ii) **tension**: tight wires produce high notes, and
(iii) **mass per unit length**: thin strings give high notes.

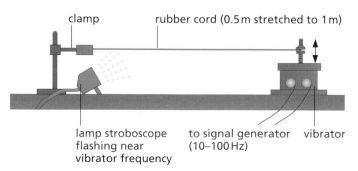

Figure 16.5 Viewing standing wave patterns on a string

Resonance

All objects have a natural frequency of vibration. The vibration can be started and increased by another object vibrating at the same frequency. The effect is called **resonance**. For example, when the heavy pendulum X in Figure 16.6 is set swinging, it forces all the light ones to swing at the same frequency, but D, which has the same length as X, does so with a much larger amplitude, i.e. D **resonates** with X.

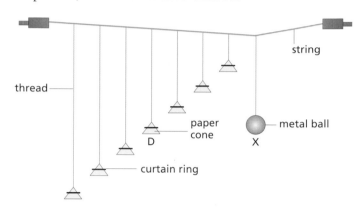

Figure 16.6 Demonstrating resonance

a) Advantages

A playground swing can be made to swing high by someone pushing in time with the free swinging, Figure 16.7.

Figure 16.7 Pushing at the right time makes the swing go higher

Resonance occurs in sound when, for example, the column of air in a wind instrument or the air in the hollow body of a string instrument is made to vibrate. The air resonates and this creates the volume of sound.

Electrical resonance is used in radio and television tuners so that only transmissions of a particular frequency are picked up.

'Stones' of calcium compounds can grow in the kidneys and cause severe pain. One treatment focuses beams of ultrasonic waves on them at their resonant frequency in order to break them up.

b) Disadvantages

If a large structure such as a tall tower or a suspension bridge starts vibrating at its natural frequency, the result can be disastrous. The collapse of the Tacoma Narrows Bridge, USA, in 1940 may have been due to a cross-wind causing resonant vibrations, Figure 16.8. Models of new bridges are now tested in wind tunnels to check that their natural frequencies are outside the range of vibrations caused by wind. The new Millennium Bridge in London, Figure 16.9, has had to be adapted to prevent swaying when pedestrians walk across it in step.

Some singers who can produce very high frequency notes are said to be able to break wine glasses when the notes have the same frequency as the natural frequency of the glass.

Figure 16.8 The Tacoma Narrows Bridge after its collapse

Figure 16.9 The Millenium Bridge on its day of opening in 2000

▮ *Noise pollution*

Unpleasant sounds are called noises. High-pitched noises are usually more annoying than low-pitched ones. Noise can damage the ears, cause tiredness and loss of concentration and, if it is very loud, result in sickness and temporary deafness. Sudden increases in loudness cause most damage. Some of the main 'noise polluters' are aircraft, road traffic, greatly amplified music and many types of machinery including domestic appliances.

Ways of reducing unwanted noise include designing quieter engines. For example, rotating shafts in machinery can be balanced better so that they do not cause vibration. Car engines are often mounted on metal brackets via rubber blocks which absorb vibrations and do not pass them on to the car body. Vehicle exhaust systems are fitted with 'silencers'.

The use in the home of sound-insulating materials, such as carpets and curtains, and of double-glazed windows also helps. The farther away the noise originates the weaker it is, so distance is a natural barrier, as are trees between houses and a noisy road. Tractor drivers, factory workers, pneumatic drill operators and others exposed regularly to noise often have to wear ear protectors.

Noise levels are measured in **decibels** (dB) by a noise meter. The sound level the average human ear can just detect, called the **threshold of hearing**, is taken as 0 dB. Normal conversation is about 60 dB, a jet plane overhead is 100 dB and the **threshold of pain** (which explains itself) is 120 dB.

Questions

1 a Draw the waveform of
   (i) a loud, low-pitched note, and
   (ii) a soft, high-pitched note.
 b If the speed of sound is 340 m/s what is the wavelength of a note of frequency
   (i) 340 Hz,
   (ii) 170 Hz?

2 Most young people can hear sounds in the frequency range 20 Hz to 20 000 Hz.
 a Which statement best describes frequency?

   A the maximum disturbance caused by a wave
   B the number of complete vibrations per second
   C the distance between one crest of a wave and the next one
   D the distance travelled by a wave in 1 second

 b Diagram **X** shows a trace on an oscilloscope screen.

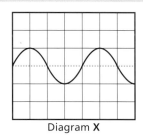
Diagram **X**

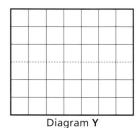
Diagram **Y**

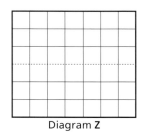
Diagram **Z**

Figure 16.10

   (i) Draw a trace, on a grid like that in diagram **Y**, which has a higher frequency than that shown in diagram **X**.
   (ii) Draw a trace, on a grid like that in diagram **Z**, which has a larger amplitude than that shown in diagram **X**.
 c Choose words from the list below to complete the following sentences.

   higher   louder   lower   quieter

   (i) A musical note with a high frequency sounds ...... than one with a low frequency.
   (ii) A noise of small amplitude sounds ...... than one with large amplitude.
 d (i) Write down the name given to sound waves of frequency higher than 20 000 Hz.
   (ii) Write down *two* uses of these very high frequency sound waves.

(*NEAB Foundation, June 98*)

Checklist

After studying this chapter you should be able to

- use the terms **pitch**, **loudness** and **quality** (timbre) and connect them to wave properties,
- describe stationary (standing) waves and explain how they are produced in string instruments,
- recall the factors affecting the frequency of the note emitted by a vibrating string,
- describe resonance and its effects,
- discuss noise pollution constructively.

Waves and sound
Additional questions

Mechanical waves

1 The lines in the diagram are the crests of straight ripples produced in a ripple tank by a wave generator.
  a What is the wavelength of the ripples if there are 6 complete waves in a distance of 30 cm?
  b If the wave generator makes 4 vibrations per second what is the frequency of the ripples?
  c What is the equation connecting the wavelength (λ), the frequency (f) and the speed (v) of the ripples?
  d Calculate the speed of the ripples.

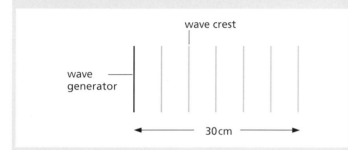

2 The diagrams **a** and **b** show ripples approaching metal plates with gaps in them. Gap **a** is narrow compared with the wavelength and gap **b** is wide compared with the wavelength.

Copy the diagrams and draw the shapes of the ripples after passing through the gaps.

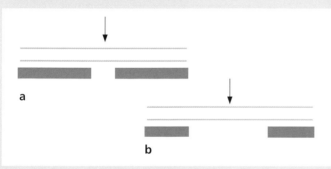

3 The blue curve shows the position of a water wave travelling to the right. X, Y and Z are corks floating on the surface and the black dotted line is the undisturbed water surface. As the wave moves forward does (i) X, (ii) Y, (iii) Z, move up or down or stay where it is?

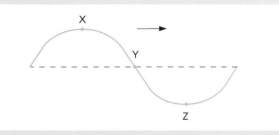

Light waves

4 In the diagram light waves are incident on an air–glass boundary. Some are reflected and some are refracted in the glass. One of the following is the same for the incident wave and the refracted wave. Which?

  A speed   **B** wavelength   **C** direction
  D brightness   **E** frequency

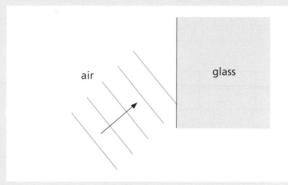

5 A double slit is formed on a blackened glass plate by drawing two lines close together. When a blue light is viewed through the double slit, an interference pattern like that in diagram **a** is seen.

  a What can be concluded about the nature of light from this pattern?
  b If the blue light is replaced by one giving a different colour of light, the pattern in diagram **b** is seen. Account for its being more spread out.
  c What could be the colour of the light in the second case?

Electromagnetic spectrum

6 a The diagram shows a wave pattern.

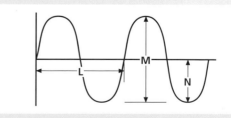

  Which letter, **L**, **M** or **N** shows:
  (i) the wavelength?
  (ii) the amplitude?

b Copy the chart below and draw a line to join each electromagnetic wave to its use or effect. One line has been drawn for you.

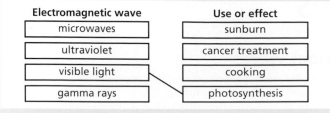

Electromagnetic wave	Use or effect
microwaves	sunburn
ultraviolet	cancer treatment
visible light	cooking
gamma rays	photosynthesis

c Describe how you could show that visible light travels in straight lines. You may wish to draw a diagram to help explain your answer.

(SEG Foundation, Summer 98)

7 This question is about the electromagnetic spectrum.
   a The diagram shows parts of the electromagnetic spectrum in order. Some parts have been named.

radio waves	K	visible light	L	X-rays

(i) Write down the name of part **K**.
(ii) Write down the name of part **L**.
(iii) Look at the diagram. Which part of the spectrum has the *shortest* wavelength?

b Finish the sentences by choosing the *best* words from this list. The first one has been done for you.

infra-red   microwaves   radio waves   ultra-violet visible light

Food can be cooked using *microwaves*.

Thermometers can detect ......

Holiday photographs taken with an ordinary camera use ......

Skin cancer can be caused by ......

c Describe *one* medical use of **gamma rays**.
d X-rays are used to take a photograph of a broken bone in a leg.

A sheet of film is placed under the leg. The X-ray machine is turned on. An image of the broken bone is produced on the film.

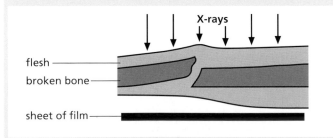

(i) Explain how X-rays produce an image of the bone on the film.
(ii) X-rays are dangerous. How does the X-ray machine operator protect himself from them?

(OCR Foundation, Summer 99)

Sound waves

8 An echo-sounder in a trawler receives an echo from a shoal of fish 0.4s after it was sent. If the speed of sound in water is 1500 m/s, how deep is the shoal?

A 150 m   **B** 300 m   **C** 600 m   **D** 7500 m
E 10 000 m

9 At Heathrow Airport in London, tests are being done to reduce the noise made by aircraft waiting on the runway. This is achieved by the process of destructive interference.

Anti-noise generators are placed on the ground under the jet engines. The sound produced by the engines is analysed. The frequency of the loudest sound is selected.

jet engine
anti-noise generator

a The diagram below represents the waveform of the selected frequency produced by the jet engine. The sound output of the anti-noise generator is adjusted so that it *cancels out* the noise of this frequency. Copy the diagram and draw the wave produced by the anti-noise generator.

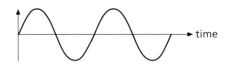

time

b The frequency of the anti-noise generator changes slightly. The diagram below represents the waveform produced by the jet engine and the generator.

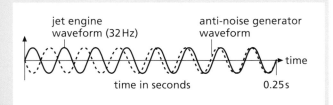

jet engine waveform (32 Hz)   anti-noise generator waveform

time in seconds   0.25 s

(i) The frequency of the jet engine waveform is 32 Hz. The speed of sound is 320 metres per second (m/s). Calculate the wavelength of the sound produced by the jet engine. You *must* show how you work out your answer.
(ii) Use the information on the diagram to calculate the frequency of the anti-noise generator waveform. You must show how you work out your answer.
(iii) The two waveforms combine. Describe what you would hear over the next second.

(OCR Higher, June 99)

Matter and molecules

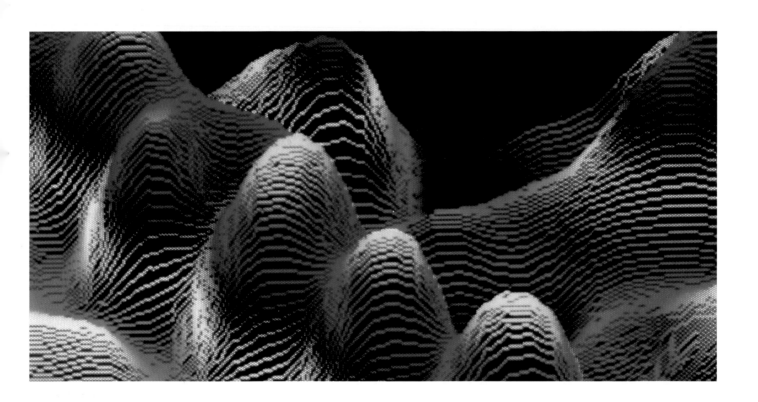

17 Measurements

Units and basic quantities
Powers of ten shorthand
Length
Significant figures
Area

Volume
Mass
Time
Practical work
Oscillations of a mass–spring system.

■ Units and basic quantities

Before a measurement can be made, a standard or **unit** must be chosen. The size of the quantity to be measured is then found with an instrument having a scale marked in the unit.

Three basic quantities we have to measure in physics are **length**, **mass** and **time**. Units for other quantities are based on them. The SI (Système International d'Unités) system is a set of metric units now used in many countries. It is a decimal system in which units are divided or multiplied by 10 to give smaller or larger units.

Figure 17.1 Measuring instruments on the flight deck of the space shuttle Discovery provide the crew with information about the performance of the shuttle, here during re-entry

■ Powers of ten shorthand

This is a neat way of writing numbers especially if they are large or small. It works like this:

$$
\begin{aligned}
4000 &= 4 \times 10 \times 10 \times 10 &&= 4 \times 10^3 \\
400 &= 4 \times 10 \times 10 &&= 4 \times 10^2 \\
40 &= 4 \times 10 &&= 4 \times 10^1 \\
4 &= 4 \times 1 &&= 4 \times 10^0 \\
0.4 &= 4/10 &&= 4/10^1 &&= 4 \times 10^{-1} \\
0.04 &= 4/100 &&= 4/10^2 &&= 4 \times 10^{-2} \\
0.004 &= 4/1000 &&= 4/10^3 &&= 4 \times 10^{-3}
\end{aligned}
$$

The small figures 1, 2, 3, etc., are called **powers of ten.** The power gives the number of times the number has to be multiplied by 10 if it is greater than 0 or divided by 10 if it is less than 0. Note that 1 is written as 10^0.

This way of writing numbers is called **standard form**.

■ Length

The unit of length is the **metre** (m) and is the distance travelled by light in a vacuum during a specific time interval. At one time it was the distance between two marks on a certain metal bar. Submultiples are:

$$
\begin{aligned}
1 \text{ centimetre (cm)} &= 10^{-2}\,\text{m} \\
1 \text{ millimetre (mm)} &= 10^{-3}\,\text{m} \\
1 \text{ micrometre } (\mu\text{m}) &= 10^{-6}\,\text{m} \\
1 \text{ nanometre (nm)} &= 10^{-9}\,\text{m}
\end{aligned}
$$

A multiple for large distances is

$$1 \text{ kilometre (km)} = 10^3\,\text{m} \;(\tfrac{5}{8} \text{ mile approx.})$$

Many length measurements are made with rulers; the correct way to read one is shown in Figure 17.2. The reading is 76 mm or 7.6 cm. Your eye must be right over the mark on the scale or the thickness of the ruler causes parallax errors.

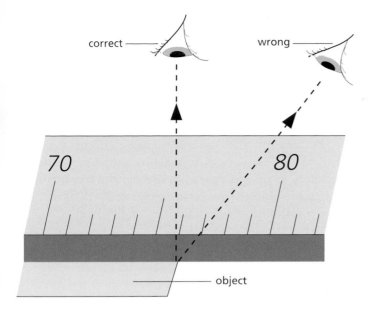

Figure 17.2 The correct way to measure with a ruler

Significant figures

Every measurement of a quantity is an attempt to find its true value and is subject to errors arising from limitations of the apparatus and the experimenter. The number of figures, called **significant** figures, given for a measurement indicates how accurate we think it is and more figures should not be given than is justified.

For example, a value of 4.5 for a measurement has two significant figures; 0.0385 has three significant figures, 3 being the most significant and 5 the least, i.e. it is the one we are least sure about since it might be 4 or it might be 6. Perhaps it had to be estimated by the experimenter because the reading was between two marks on a scale.

When doing a calculation your answer should have the same number of significant figures as the measurements used in the calculation. For example, if your calculator gave an answer of 3.4185062, this would be written as 3.4 if the measurements had two significant figures. It would be written as 3.42 for three significant figures. Note that in deciding the least significant figure you look at the next figure. If it is less than 5 you leave the least significant figure as it is (hence 3.41 becomes 3.4) but if it equals or is greater than 5 you increase the least significant figure by 1 (hence 3.418 becomes 3.42).

If a number is expressed in standard form, the number of significant figures is the number of digits before the power of ten. For example, 2.73×10^3 has three significant figures.

Area

The area of the square in Figure 17.3a with sides 1 cm long is 1 square centimetre (1 cm²). In Figure 17.3b the rectangle measures 4 cm by 3 cm and has an area of $4 \times 3 = 12 \text{ cm}^2$ since it has the same area as twelve squares each of area 1 cm². The **area of a square** or **rectangle** is given by

$$\text{area} = \text{length} \times \text{breadth}$$

The SI unit of area is the square metre (m²) which is the area of a square with sides 1 m long. Note that

$$1 \text{ cm}^2 = \frac{1}{100} \text{ m} \times \frac{1}{100} \text{ m} = \frac{1}{10\,000} \text{ m}^2 = 10^{-4} \text{ m}^2$$

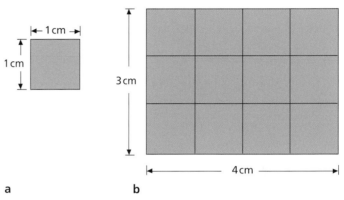

a   b

Figure 17.3

Sometimes we need to know the **area of a triangle** (Chapter 29). It is given by

$$\text{area of triangle} = \tfrac{1}{2} \times \text{base} \times \text{height}$$

For example in Figure 17.4

$$\begin{aligned}\text{area } \triangle ABC &= \tfrac{1}{2} \times AB \times AC \\ &= \tfrac{1}{2} \times 4 \text{ cm} \times 6 \text{ cm} = 12 \text{ cm}^2\end{aligned}$$

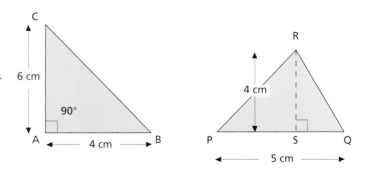

Figure 17.4

and

$$\begin{aligned}\text{area } \triangle PQR &= \tfrac{1}{2} \times PQ \times SR \\ &= \tfrac{1}{2} \times 5 \text{ cm} \times 4 \text{ cm} = 10 \text{ cm}^2\end{aligned}$$

The **area of a circle** of radius r is πr^2 where $\pi = 22/7$ or 3.14; its **circumference** is $2\pi r$.

Volume

Volume is the amount of space occupied. The unit of volume is the **cubic metre** (m³) but as this is rather large, for most purposes the **cubic centimetre** (cm³) is used. The volume of a cube with 1 cm edges is 1 cm³. Note that

$$1\,cm^3 = \frac{1}{100}\,m \times \frac{1}{100}\,m \times \frac{1}{100}\,m$$

$$= \frac{1}{1\,000\,000}\,m^3 = 10^{-6}\,m^3$$

For a regularly shaped object such as a rectangular block, Figure 17.5 shows that

$$volume = length \times breadth \times height$$

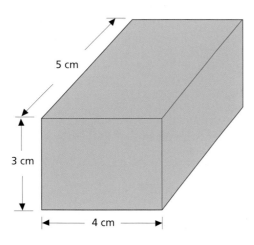

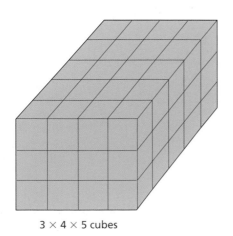

3 × 4 × 5 cubes

Figure 17.5

The **volume of a sphere** of radius r is $\frac{4}{3}\pi r^3$ and that of a **cylinder** of radius r and height h is $\pi r^2 h$.

The **volume of a liquid** may be obtained by pouring it into a measuring cylinder, Figure 17.6a. A known volume can be run off accurately from a burette, Figure 17.6b. When making a reading both vessels must be upright and the eye must be level with the bottom of the curved liquid surface, i.e. the meniscus. The meniscus formed by mercury is curved oppositely to that of other liquids and the top is read.

Liquid volumes are also expressed in litres (l); 1 litre = 1000 cm³. One millilitre (1 ml) = 1 cm³.

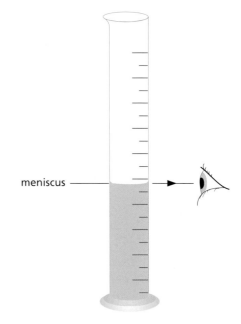

meniscus

a A measuring cylinder

b A burette

Figure 17.6

Mass

The mass of an object is the measure of the amount of matter in it. The unit of mass is the **kilogram** (kg) and is the mass of a piece of platinum–iridium alloy at the Office of Weights and Measures in Paris. The gram (g) is one-thousandth of a kilogram.

$$1\,\mathrm{g} = 1/1000\,\mathrm{kg} = 10^{-3}\,\mathrm{kg} = 0.001\,\mathrm{kg}$$

The term **weight** is often used when mass is really meant. In science the two ideas are distinct and have different units as we shall see later. The confusion is not helped by the fact that mass is found on a balance by a process we unfortunately call 'weighing'!

There are several kinds of balance. In the **beam balance** the unknown mass in one pan is balanced against known masses in the other pan. In the **lever balance** a system of levers acts against the mass when it is placed in the pan. A direct reading is obtained from the position on a scale of a pointer joined to the lever system. A digital **top-pan balance** is shown in Figure 17.7.

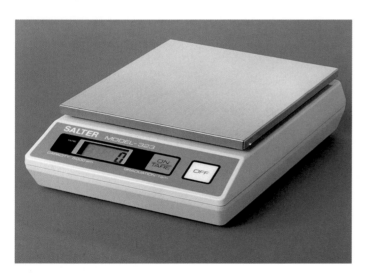

Figure 17.7 A digital top-pan balance

Time

The unit of time is the **second**(s) which used to be based on the length of a day, this being the time for the Earth to revolve once on its axis. However, days are not all of exactly the same duration and the second is now defined as the time interval for a certain number of energy changes to occur in the caesium atom.

Time-measuring devices rely on some kind of constantly repeating oscillations. In traditional clocks and watches a small wheel (the balance wheel) oscillates to and fro; in digital clocks and watches the oscillations are produced by a tiny quartz crystal. A swinging pendulum controls a pendulum clock.

Practical work

Oscillations of a mass–spring system

In this investigation you have to make time measurements using a stopwatch or clock.

Support a spiral steel spring as in Figure 17.8a. Hang a mass (e.g. 50 g) from its lower end so that it is stretched several centimetres. Pull the mass *vertically* downwards a few centimetres, Figure 17.8b, and release it so that it oscillates up and down above and below its rest position.

Find the time for the mass to make several complete oscillations; one oscillation is from A to O to B to O to A, Figure 17.8c. Repeat the timing a few times for the same number of oscillations and work out the average. The time for one oscillation is the **period** T. What is it for your system? The **frequency** f of the oscillations is the number of complete oscillations per second and equals 1/T. Calculate f.

How does the amplitude of the oscillations, Figure 17.8b, change with time?

Investigate the effect on T of (i) a greater mass, (ii) a smaller mass, (iii) a different spring. A motion sensor connected to a datalogger and computer (Chapter 28) could be used instead of a stopwatch for these investigations.

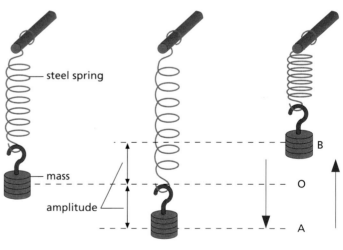

a   b   c

Figure 17.8

Mechanical oscillations (vibrations) have their uses (e.g. to produce sound) but if they build up in a system (e.g. in a bridge, Chapter 16) they can cause damage.

Questions

1 How many millimetres are there in
   a 1 cm, b 4 cm, c 0.5 cm, d 6.7 cm, e 1 m?

2 What are these lengths in metres:
   a 300 cm, b 550 cm, c 870 cm, d 43 cm, e 100 mm?

3 a Write the following as powers of ten with one
     figure before the decimal point:

   100 000   3500   428 000 000   504   27 056

   b Write out the following in full:

   10^3 2×10^6 6.9×10^4 1.34×10^2 10^9

4 a Write these fractions as powers of ten:

   1/1000   7/100 000   1/10 000 000   3/60 000

   b Express the following decimals as powers of ten
     with one figure before the decimal point:

   0.5   0.084   0.000 36   0.001 04

5 The pages of a book are numbered 1 to 200 and each
   leaf is 0.10 mm thick. If each cover is 0.20 mm thick,
   what is the thickness of the book?

6 How many significant figures are there in a length
   measurement of
   a 2.5 cm, b 5.32 cm, c 7.180 cm, d 0.042 cm?

7 A rectangular block measures 4.1 cm by 2.8 cm by
   2.1 cm. Calculate its volume giving your answer to an
   appropriate number of significant figures.

8 A metal block measures 10 cm × 2 cm × 2 cm. What is
   its volume? How many blocks each 2 cm × 2 cm × 2 cm
   have the same total volume?

9 How many blocks of ice cream each
   10 cm × 10 cm × 4 cm can be stored in the
   compartment of a freezer measuring
   40 cm × 40 cm × 20 cm?

10 A Perspex container has a 6 cm square base and
   contains water to a height of 7 cm, Figure 17.9.
   a What is the volume of the water?
   b A stone is lowered into the water so as to be
     completely covered and the water rises to a height
     of 9 cm. What is the volume of the stone?

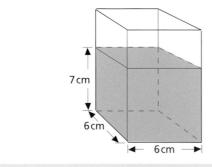

7 cm

6 cm

6 cm

Figure 17.9

Checklist

After studying this chapter you should be able to

- recall three basic quantities in physics,
- write a number in powers of ten (standard) form,
- recall the unit of length and the meaning of the prefixes **kilo, centi, milli, micro, nano**,
- use a ruler to measure length so as to minimize errors,
- give a result to an appropriate number of significant figures,
- measure areas of squares, rectangles, triangles and circles,
- measure the volume of regular solids and of liquids,
- recall the unit of mass and how mass is measured,
- recall the unit of time and how time is measured,
- describe an experiment to find the period of a mass–spring system.

18 Density

Calculations
Simple density measurements

Floating and sinking

In everyday language lead is said to be 'heavier' than wood. By this it is meant that a certain volume of lead is heavier than the same volume of wood. In science such comparisons are made by using the term **density**. This is the **mass per unit volume** of a substance and is calculated from

$$\text{density} = \frac{\text{mass}}{\text{volume}}$$

The density of lead is 11 grams per cubic centimetre ($11\,\text{g/cm}^3$) and this means that a piece of lead of volume $1\,\text{cm}^3$ has mass $11\,\text{g}$. A volume of $5\,\text{cm}^3$ of lead would have mass $55\,\text{g}$. Knowing the density of a substance the mass of *any* volume can be calculated. This enables engineers to work out the weight of a structure if they know from the plans the volumes of the materials to be used and their densities. Strong enough foundations can then be made.

The SI unit of density is the **kilogram per cubic metre**. To convert a density from g/cm^3, normally the most suitable unit for the size of sample we use, to kg/m^3, we multiply by 10^3. For example the density of water is $1.0\,\text{g/cm}^3$ or $1.0 \times 10^3\,\text{kg/m}^3$.

The approximate densities of some common substances are given in Table 18.1.

Table 18.1 Densities of some common substances

Solids	Density/ g/cm³	Liquids	Density/ g/cm³
aluminium	2.7	paraffin	0.80
copper	8.9	petrol	0.80
iron	7.9	pure water	1.0
gold	19.3	mercury	13.6
glass	2.5	**Gases**	**kg/m³**
wood (teak)	0.80	air	1.3
ice	0.92	hydrogen	0.09
polythene	0.90	carbon dioxide	2.0

Calculations

Using the symbols d for density, m for mass and V for volume, the expression for density is

$$d = \frac{m}{V}$$

Rearranging the expression gives

$$m = V \times d \qquad \text{and} \qquad V = \frac{m}{d}$$

These are useful if d is known and m or V have to be calculated. If you do not see how they are obtained refer to the *Mathematics for physics* section on p. 358. The triangle in Figure 18.1 is an aid to remembering them. If you cover the quantity you want to know with a finger, e.g. m, it equals what you can still see, i.e. $d \times V$. To find V, cover V and you get $V = m/d$.

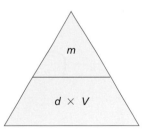

Figure 18.1

Worked example

Taking the density of copper as $9\,\text{g/cm}^3$, find **a** the mass of $5\,\text{cm}^3$ and **b** the volume of $63\,\text{g}$.

a $d = 9\,\text{g/cm}^3$, $V = 5\,\text{cm}^3$ and m is to be found.

$$\therefore \qquad m = V \times d = 5\,\text{cm}^3 \times 9\,\text{g/cm}^3 = 45\,\text{g}$$

b $d = 9\,\text{g/cm}^3$, $m = 63\,\text{g}$ and V is to be found.

$$\therefore \qquad V = \frac{m}{d} = \frac{63\,\text{g}}{9\,\text{g/cm}^3} = 7\,\text{cm}^3$$

71

Simple density measurements

If the mass m and volume V of a substance are known its density can be found from $d = m/V$.

a) Regularly shaped solid

The mass is found on a balance and the volume by measuring its dimensions with a ruler.

b) Irregularly shaped solid, e.g. pebble or glass stopper

The solid is weighed and its volume measured by one of the methods shown in Figures 18.2a, b. In diagram a the volume is the difference between the first and second readings. In diagram b it is the volume of water collected in the measuring cylinder.

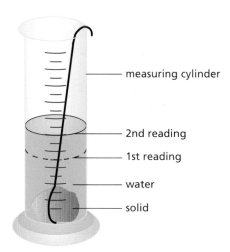

- measuring cylinder
- 2nd reading
- 1st reading
- water
- solid

a

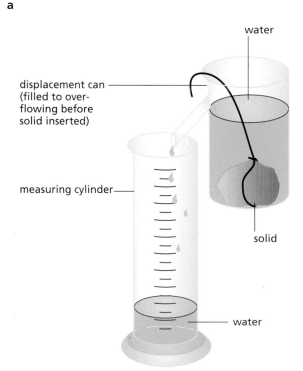

water

displacement can (filled to over-flowing before solid inserted)

measuring cylinder

solid

water

b

Figure 18.2 Measuring the volume of an irregular solid

c) Liquid

A known volume is transferred from a burette or a measuring cylinder into a weighed beaker which is re-weighed to give the mass of liquid.

d) Air

A $500\,cm^3$ round-bottomed flask is weighed full of air and then after removing the air with a vacuum pump; the difference gives the mass of air in the flask. The volume of air is found by filling the flask with water and pouring it into a measuring cylinder.

Floating and sinking

An object sinks in a liquid of smaller density than its own; otherwise it floats, partly or wholly submerged. For example, a piece of glass of density $2.5\,g/cm^3$ sinks in water (density $1.0\,g/cm^3$) but floats in mercury (density $13.6\,g/cm^3$). An iron nail sinks in water but an iron ship floats because its average density is less than that of water. You will find more about floating and sinking in Chapter 26.

Figure 18.3 Why is it easy to float in the Dead Sea?

Questions

1 a If the density of wood is 0.5 g/cm³ what is the mass of
    (i) 1 cm³,
    (ii) 2 cm³,
    (iii) 10 cm³?
  b What is the density of a substance of
    (i) mass 100 g and volume 10 cm³,
    (ii) volume 3 m³ and mass 9 kg?
  c The density of gold is 19 g/cm³. Find the volume of
    (i) 38 g,
    (ii) 95 g of gold.

2 A piece of steel has a volume of 12 cm³ and a mass of 96 g. What is its density in
  a g/cm³,
  b kg/m³?

3 What is the mass of 5 m³ of cement of density 3000 kg/m³?

4 What is the mass of air in a room measuring 10 m × 5.0 m × 2.0 m if the density of air is 1.3 kg/m³?

5 When a golf ball is lowered into a measuring cylinder of water, the water level rises by 30 cm³ when the ball is completely submerged. If the ball weighs 33 g in air, find its density.

Checklist

After studying this chapter you should be able to

■ define **density** and perform calculations using $d = m/V$,

■ describe experiments to measure the density of solids, liquids and air,

■ relate floating and sinking to density.

19 Weight and stretching

Force
Weight
The newton

Hooke's law
Practical work
Stretching a spring.

Force

A **force** is a push or a pull. It can cause a body at rest to move, or if the body is already moving it can change its speed or direction of motion. It can also change its shape or size.

Figure 19.1 A weightlifter in action exerts first a pull and then a push

Weight

We all constantly experience the force of gravity, i.e. the pull of the Earth. It causes an unsupported body to fall from rest to the ground.

> The weight of a body is the force of gravity on it.

The nearer a body is to the centre of the Earth the more the Earth attracts it. Since the Earth is not a perfect sphere but is flatter at the poles, the weight of a body varies over the Earth's surface. It is greater at the poles than at the equator.

Gravity is a force which can act through space, i.e. there does not need to be contact between the Earth and the object on which it acts as there does when we push or pull something. Other action-at-a-distance forces which, like gravity, decrease with distance are:

(i) **magnetic** forces between magnets, and
(ii) **electric** forces between electric charges.

The newton

The unit of force is the **newton** (N). It will be defined later (Chapter 31); the definition is based on the change of speed a force can produce in a body. Weight is a force and therefore should be measured in newtons.

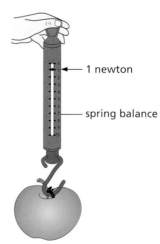

Figure 19.2 The weight of an average-sized apple is about 1 newton

The weight of a body can be measured by hanging it on a **spring balance** marked in newtons, Figure 19.2, and letting the pull of gravity stretch the spring in the balance. The greater the pull the more the spring stretches. On most of the Earth's surface:

The weight of a body of mass 1 kg is 9.8 N.

Often this is taken as 10 N. A mass of 2 kg has a weight of 20 N and so on. The mass of a body is the same wherever it is and does not depend on the presence of the Earth as does weight.

Practical work

Stretching a spring

Arrange a steel spring as in Figure 19.3. Read the scale opposite the bottom of the hanger. Add 100 g loads one at a time (thereby increasing the stretching force by steps of 1 N) and take the readings after each one. Enter the readings in a table for loads up to 500 g.

Stretching force/N	Scale reading/ mm	Total extension/ mm

Do the results suggest any rule about how the spring behaves when it is stretched?

Sometimes it is easier to discover laws by displaying the results on a graph. Do this on graph paper by plotting **stretching force** readings along the y-axis (vertical axis) and **total extension** readings along the x-axis (horizontal axis). Every pair of readings will give a point; mark them by small crosses and draw a smooth line through them. What is its shape?

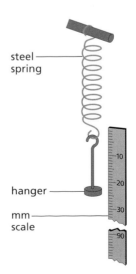

steel spring

hanger

mm scale

Figure 19.3

Hooke's law

Springs were investigated by Robert Hooke nearly 350 years ago. He found that the extension was proportional to the stretching force so long as the spring was not permanently stretched. This means that doubling the force doubles the extension, trebling the force trebles the extension and so on. Using the sign for proportionality we can write **Hooke's law** as

extension ∝ stretching force

It is true only if the **elastic limit** of the spring is not exceeded. The spring returns to its original length when the force is removed.

The graph of Figure 19.4 is for a spring stretched beyond its elastic limit E. OE is a straight line passing through the origin O and is graphical proof that Hooke's law holds over this range. If the force for point A on the graph is applied to the spring, the elastic limit is passed and on removing the force some of the extension (OS) remains. Over which part of the graph does a spring balance work?

The **force constant** k of a spring is the force needed to cause unit extension, i.e. 1 m. If a force F produces extension e then

$$k = \frac{F}{e}$$

Hooke's law also holds when a force is applied to a straight metal wire or an elastic band, as long as they are not permanently stretched. Force–extension graphs similar to Figure 19.4 are obtained. For a rubber band, a small force causes a large extension.

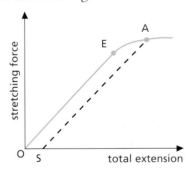

Figure 19.4 Force–extension graph for a spring

Worked example

A spring is stretched 10 mm (0.01 m) by a weight of 2.0 N. Calculate **a** the force constant k and **b** the weight W of an object which causes an extension of 80 mm (0.08 m).

a $k = \dfrac{F}{e} = \dfrac{2.0\,\text{N}}{0.01\,\text{m}} = 200\,\text{N/m}$

b $W = $ stretching force $F = k \times e$
     $= 200\,\text{N/m} \times 0.08\,\text{m} = 16\,\text{N}$

Questions

1 A body of mass 1 kg has weight 10 N at a certain place. What is the weight of
a 100 g,
b 5 kg,
c 50 g?

2 The force of gravity on the Moon is said to be one-sixth of that on the Earth. What would a mass of 12 kg weigh
a on the Earth, and
b on the Moon?

3 What is the force constant of a spring which is stretched
a 2 mm by a force of 4 N,
b 4 cm by a mass of 200 g?

4 The spring in Figure 19.5 stretches from 10 cm to 22 cm when a force of 4 N is applied. If it obeys Hooke's law, its total length in cm when a force of 6 N is applied is

A 28   **B** 42   **C** 50   **D** 56   **E** 100

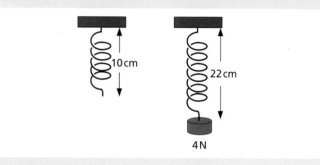

Figure 19.5

Checklist

After studying this chapter you should be able to

- recall that a force can cause a change in the motion, size or shape of a body,
- recall that the weight of a body is the force of gravity on it,
- recall the unit of force and how force is measured,
- describe an experiment to study the relation between force and extension for springs,
- draw conclusions from force–extension graphs,
- recall Hooke's law and solve problems using it,
- explain the term **force constant**.

20 *Molecules*

Matter is made up of tiny particles or **molecules** which are too small for us to see directly. But they can be 'seen' by scientific 'eyes'. One of these is the electron microscope: Figure 20.1 is a photograph taken with such an instrument showing molecules of a protein. Molecules consist of even smaller particles called **atoms** and are in continuous motion.

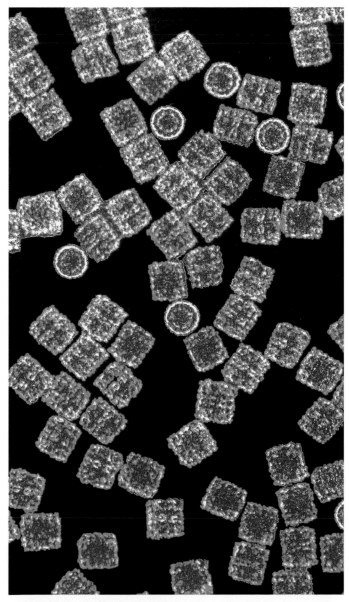

Figure 20.1 Protein molecules

Practical work

Brownian motion

The apparatus is shown in Figure 20.2a. First fill the glass cell with smoke using a match, Figure 20.2b. Replace the lid on the apparatus and set it on the microscope platform. Connect the lamp to a 12 V supply; the glass rod acts as a lens and focuses light on the smoke.

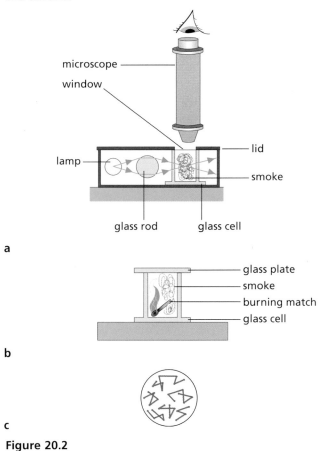

Figure 20.2

Carefully adjust the microscope until you see bright specks dancing around haphazardly, Figure 20.2c. The specks are smoke particles seen by reflected light; their random motion is due to collisions with fast-moving air molecules in the cell. The effect is called **Brownian motion**.

Kinetic theory of matter

As well as being in continuous motion, molecules also exert strong electric forces on one another when they are close together. The forces are both attractive and repulsive. The former hold molecules together and the latter cause matter to resist compression. The **kinetic theory** can explain the existence of the solid, liquid and gaseous states.

a) Solids

The theory states that in solids the molecules are close together and the attractive and repulsive forces between neighbouring molecules balance. Also each molecule vibrates to and fro about a fixed position.

It is just as if springs, representing the electric forces between molecules, hold the molecules together, Figure 20.3. This enables the solid to keep a definite shape and volume, while still allowing the individual molecules to vibrate backwards and forwards. The theory shows that the molecules in a solid could be arranged in a regular, repeating pattern like those formed by crystalline substances.

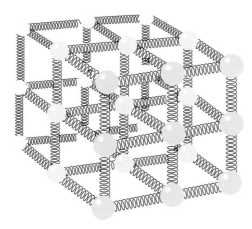

Figure 20.3 The electric forces between molecules in a solid can be represented by springs

b) Liquids

The theory considers that in liquids the molecules are slightly farther apart than in solids but still close enough together to have a definite volume. As well as vibrating, they can at the same time move rapidly over short distances, slipping past each other in all directions. They are never near another molecule long enough to get trapped in a regular pattern which would stop them flowing and taking the shape of the vessel containing them.

A model to represent the liquid state can be made by covering about a third of a tilted tray with marbles ('molecules'), Figure 20.4. It is then shaken to and fro and the motion of the marbles observed. They are able to move around but most stay in the lower half of the tray, so the liquid has a fairly definite volume. A few energetic ones 'escape' from the 'liquid' into the space above. They represent molecules that have 'evaporated' from the 'liquid' surface and become 'gas' or 'vapour' molecules. The thinning out of the marbles near the 'liquid' surface can also be seen.

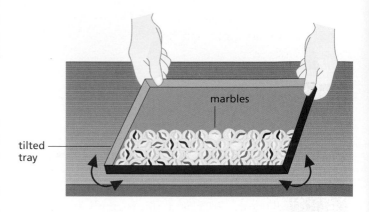

Figure 20.4 A model of molecular behaviour in a liquid

c) Gases

The molecules in gases are much farther apart than in solids or liquids (about ten times) and so gases are much less dense and can be squeezed (compressed) into a smaller space. The molecules dash around at very high speed (about 500 m/s for air molecules at 0 °C) in all the space available. It is only during the brief spells when they collide with other molecules or with the walls of the container that the molecular forces act.

A model of a gas is shown in Figure 20.5. The faster the vibrator works the more often the ball-bearings have collisions with the lid, the tube and with each other, representing a gas at a higher temperature. Adding more ball-bearings is like pumping more air into a tyre; it increases the pressure. If a polystyrene ball (1 cm diameter) is dropped into the tube its irregular motion represents Brownian motion.

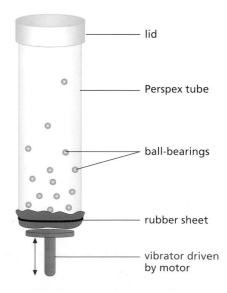

Figure 20.5 A model of molecular behaviour in a gas

Crystals

Crystals have hard, flat sides and straight edges. Whatever their size, crystals of the same substance have the same shape. This can be seen by observing through a microscope very small cubic salt crystals growing as water evaporates from salt solution on a glass slide, Figure 20.6.

Figure 20.6 Salt crystals viewed under a microscope with polarized light

Figure 20.7 Splitting a calcite crystal

A calcite crystal will split cleanly if a trimming knife, held exactly parallel to one side of the crystal, is struck by a hammer, Figure 20.7.

These facts suggest crystals are made of small particles (e.g. atoms) arranged in an orderly way in planes. Metals have crystalline structures, but many other common solids such as glass, plastics and wood do not.

Diffusion

Smells, pleasant or otherwise, travel quickly and are caused by rapidly moving molecules. The spreading of a substance of its own accord is called **diffusion** and is due to molecular motion.

Diffusion of gases can be shown if some brown nitrogen dioxide gas is made by pouring a mixture of equal volumes of concentrated nitric acid and water on copper turnings in a gas jar. When the action has stopped, a gas jar of air is inverted over the bottom jar, Figure 20.8. The brown colour spreads into the upper jar showing that nitrogen dioxide molecules diffuse upwards against gravity. Air molecules also diffuse into the lower jar.

The speed of diffusion of a gas depends on the speed of its molecules and is greater for light molecules. The apparatus of Figure 20.9 shows this. When hydrogen surrounds the porous pot, the liquid in the U-tube moves in the direction of the arrows. This is due to the lighter, faster molecules of hydrogen diffusing into the pot faster than the heavier, slower molecules of air diffuse out. The opposite happens when carbon dioxide surrounds the pot. Why?

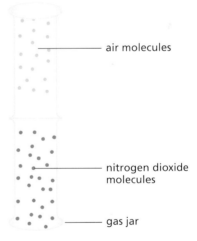

air molecules

nitrogen dioxide molecules

gas jar

Figure 20.8 Demonstrating diffusion of a gas

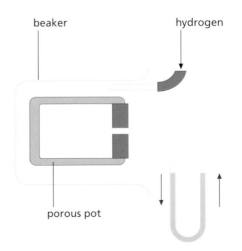

beaker   hydrogen

porous pot

Figure 20.9 The speed of diffusion is greater for lighter molecules

Diffusion in liquids can be seen using the arrangement in Figure 20.10. After 24 hours the blue copper sulphate solution has diffused upwards into the water.

Figure 20.10 Demonstrating diffusion of a liquid

Questions

1 Which one of the following statements is *not* true?

   A The molecules in a solid vibrate about a fixed position.
   B The molecules in a liquid are arranged in a regular pattern.
   C The molecules in a gas exert negligibly small forces on each other, except during collisions.
   D The densities of most liquids are about 1000 times greater than those of gases because liquid molecules are much closer together than gas molecules.
   E The molecules of a gas occupy all the space available.

2 Using what you know about the compressibility (squeezability) of the different states of matter, explain why
   a air is used to inflate tyres,
   b steel is used to make railway lines.

3 Explain in terms of molecular behaviour why bromine molecules which travel at about 200 m/s in a vacuum take several minutes to travel 40 cm inside a tube of air, Figure 20.11.

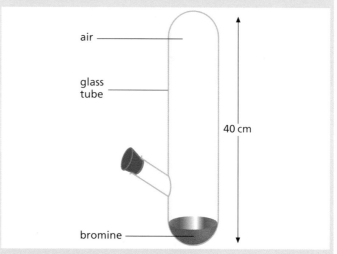

Figure 20.11

4 Graphite consists of **layers** of carbon atoms that are $2\frac{1}{2}$ times farther apart than the atoms in the layers. Why is it soft and flaky and used as a lubricant and in pencils?

5 Explain why
   a diffusion occurs more quickly in a gas than in a liquid,
   b diffusion is still quite slow even in a gas,
   c an inflated balloon gradually goes down even when tied.

Checklist

After studying this chapter you should be able to

   - describe and explain an experiment to show Brownian motion,
   - use the kinetic theory to explain the physical properties of solids, liquids and gases,
   - recall how crystals support the view that matter is made of small particles,
   - describe and explain simple experiments on diffusion.

Matter and molecules
Additional questions

Measurements; density

1 a Name the basic units of: length, mass, time.
   b What is the difference between two measurements with values of 3.4 and 3.42?
   c Write expressions for
      (i) the area of a circle,
      (ii) the volume of a sphere,
      (iii) the volume of a cylinder.

2 If 200 cm³ of water (density 1.0 g/cm³) is mixed with 300 cm³ of methylated spirit (density 0.80 g/cm³), what is the density of the mixture?

Weight and stretching

3 The springs in the diagram are identical. If the extension produced in **a** is 4 cm, what are the extensions in **b** and **c**?

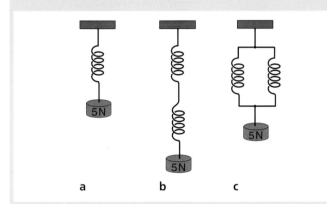

a   b   c

4 A student set up the following apparatus.

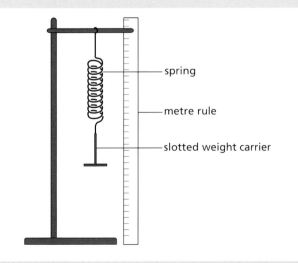

spring

metre rule

slotted weight carrier

The student measured the length of the spring when different weights were added. The student's results are given in the table.

Weight in newtons	Length in millimetres
0	20
1	23
2	26
3	29
4	32
5	35
6	49
7	66

a Draw a graph of these results. The first 3 points have been plotted below.

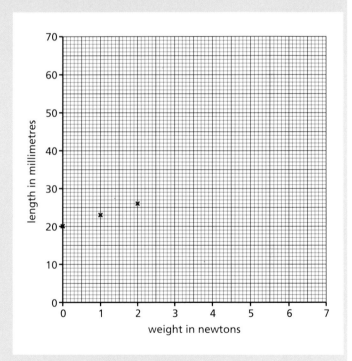

b By how many millimetres did the 2 newton weight cause the spring to stretch?
c Write the letter **Q** at any point on your graph which is past the spring's elastic limit.
d In the following sentence, *two* lines in the box are wrong. Indicate the correct line.

When the 7 newton weight is taken off the spring,

the length of the spring will be
| longer than |
| equal to |
| shorter than |

20 millimetres.

(SEG Foundation, June 98)

5 The mattress of a bed contains springs. The diagrams show the change that takes place when a person lies on the bed.

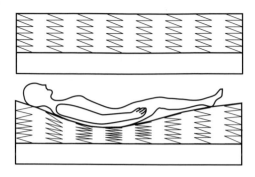

a (i) How do the springs change when a person lies on the bed?
(ii) Sketch the second diagram and circle the spring that has the greatest force on it.
(iii) How can you tell that this spring has the greatest force acting on it?

b A manufacturer makes a mattress that sags less in the middle when a person lies on it. Suggest *two* ways of doing this.

c One force acting on the person is the upward push of the springs.
(i) Another force acts on the person. Draw an arrow on your diagram to show the direction of this force.
(ii) Use words from the box to complete the sentence.

| downward | Earth | mattress | upward |

The other force on the person is the ......
pull of the ......

(London Foundation, June 99)

Molecules

6 This question is about particles.
a The diagrams show how particles are arranged in solids, liquids and gases.

(i) Which diagram shows how particles are arranged in solids?
(ii) Which diagram shows how particles are arranged in liquids?

b Harry uses a microscope to look at smoke particles. He sees the smoke particles moving randomly. What makes the smoke particles move randomly?

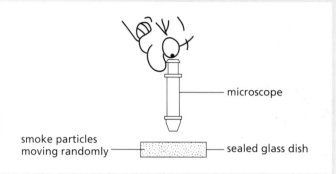

c The temperature inside the dish rises.
(i) What effect does this have on how the smoke particles move?
(ii) What effect does the rise in temperature have on the gas pressure inside the dish?
(iii) Explain how a gas exerts pressure. Use your ideas of particles.

(OCR Foundation, June 99)

Forces and pressure

21 Moments and levers

Moment of a force

The handle on a door is at the outside edge so that it opens and closes easily. A much larger force would be needed if the handle were near the hinge. Similarly it is easier to loosen a nut with a long spanner than with a short one.

The **turning effect** or **moment of a force** depends on both the size of the force and how far it is applied from the pivot or **fulcrum**. It is measured by multiplying the force by the **perpendicular** distance of the line of action of the force from the fulcrum. The unit is the **newton metre** (N m).

> moment of a force = force × perpendicular distance of the line of action of the force from fulcrum

In Figure 21.1a a force F acts on a gate at its edge, and in b at the centre.

In a:

$$\text{moment of } F \text{ about } O = 5\,\text{N} \times 3\,\text{m} = 15\,\text{N m}$$

In b:

$$\text{moment of } F \text{ about } O = 5\,\text{N} \times 1.5\,\text{m} = 7.5\,\text{N m}$$

The turning effect of F is greater in the first case; this agrees with the fact that a gate opens most easily when pushed or pulled at the edge furthest from the hinge.

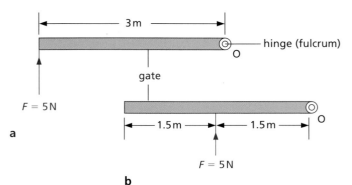

a

b

84 ■ Figure 21.1

Practical work

Law of moments

Balance a half-metre ruler at its centre, adding Plasticine to one side or the other until it is horizontal.

Hang unequal loads m_1 and m_2 from either side of the ruler and alter their distances d_1 and d_2 from the centre until the ruler is again balanced, Figure 21.2. Forces F_1 and F_2 are exerted by gravity on m_1 and m_2 and so on the ruler; on 100 g the force is 1 N. Record the results in a table and repeat for other loads and distances.

m_1 /g	F_1 /N	d_1 /cm	$F_1 \times d_1$ /N cm	m_2 /g	F_2 /N	d_2 /cm	$F_2 \times d_2$ /N cm

F_1 is trying to turn the ruler anticlockwise and $F_1 \times d_1$ is its moment. F_2 is trying to cause clockwise turning and its moment is $F_2 \times d_2$. When the ruler is balanced or, as we say, **in equilibrium**, the results should show that the anticlockwise moment $F_1 \times d_1$ equals the clockwise moment $F_2 \times d_2$.

The **law of moments** (also called the **law of the lever**) is stated as follows.

> When a body is in equilibrium the sum of the clockwise moments about any point equals the sum of the anticlockwise moments about the same point.

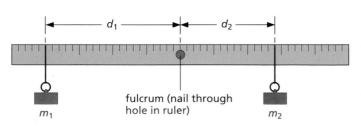

Figure 21.2

Worked example

The see-saw in Figure 21.3 balances when Shani of weight 320 N is at A, Tom of weight 540 N is at B and Harry of weight W is at C. Find W.

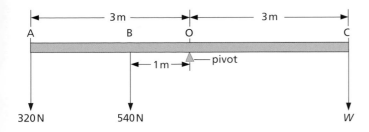

Figure 21.3

Taking moments about the fulcrum O:

anticlockwise moment
$$= 320\,\text{N} \times 3\,\text{m} + 540\,\text{N} \times 1\,\text{m} = 1500\,\text{N\,m}$$

clockwise moment $= W \times 3\,\text{m}$

By the law of moments,

clockwise moments = anticlockwise moments

∴ $$W \times 3\,\text{m} = 1500\,\text{N\,m}$$

∴ $$W = \frac{1500\,\text{N\,m}}{3\,\text{m}} = 500\,\text{N}$$

Levers

A lever is any device which can turn about a pivot. In a working lever a force called the **effort** is used to overcome a resisting force called the **load**. The pivotal point is called the **fulcrum**.

When we use a crowbar to move a heavy boulder, Figure 21.4a, our hands apply the effort at one end of the bar and the load is the force exerted by the boulder on the other end. If distances from the fulcrum O are as shown and the load is 1000 N (i.e. the part of the weight of the boulder supported by the crowbar), the effort can be calculated from the law of moments. As the boulder *just begins* to move we can say, taking moments about O, that

clockwise moment = anticlockwise moment

effort $\times 200\,\text{cm} = 1000\,\text{N} \times 10\,\text{cm}$

$$\text{effort} = \frac{10\,000\,\text{N\,cm}}{200\,\text{cm}} = 50\,\text{N}$$

Force and distance multipliers

Figure 21.4b shows that the crowbar has enabled an effort (E) to raise a load (L) 20 times as large, i.e. it is a **force multiplier**, but E has had to move 20 times as far as L. The lever has a **mechanical advantage** (MA) of 20 and a **velocity ratio** (VR) of 20 where

$$\text{MA} = \frac{L}{E} \quad \text{and} \quad \text{VR} = \frac{\text{distanced moved by } E}{\text{distance moved by } L}$$

Other examples of levers are shown in Figures 21.5a to d, overleaf. In diagram a the load is between the effort and the fulcrum; in this case as in Figure 21.4 the effort is less than the load. In b the effort (applied by the biceps muscle) is between the load and the fulcrum and is greater than the load. The forearm is a type of lever called a **distance multiplier** because the load moves farther than the effort, i.e. its VR (and its MA) is less than 1.

How does the effort compare with the load for scissors and a spanner in Figures 21.5c and d?

OA = 10 cm
OB = 200 cm

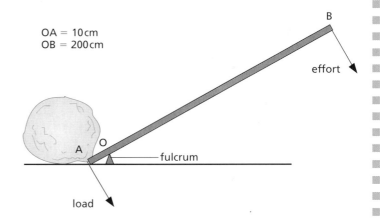

a Crowbar

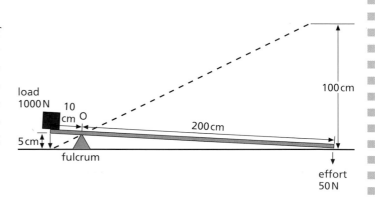

b

Figure 21.4

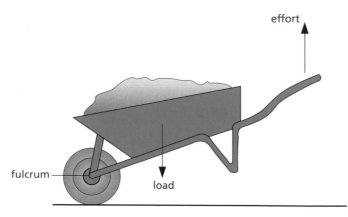

a Wheelbarrow

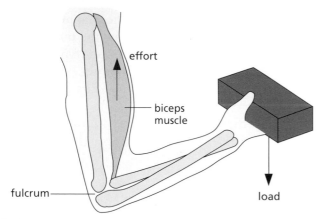

b Forearm

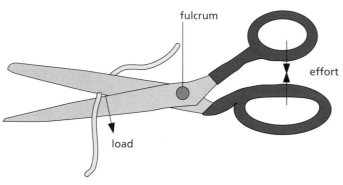

c Scissors

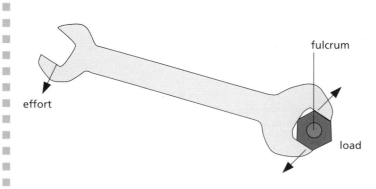

d Spanner

Figure 21.5 Other levers

■ *Conditions for equilibrium*

Sometimes a number of parallel forces act on a body so that it is in equilibrium. We can then say:

(i) The sum of the forces in one direction equals the sum of the forces in the opposite direction.
(ii) The law of moments must apply.

As an example consider two decorators of weights 500 N and 700 N standing at A and B on a plank resting on two trestles, Figure 21.6. In the next chapter we will see that the whole weight of the plank (400 N) may be taken to act vertically downwards at its centre, O. If P and Q are the upward forces exerted by the trestles on the plank (called **reactions**) then we have from (i)

$$P + Q = 500\,\text{N} + 400\,\text{N} + 700\,\text{N} = 1600\,\text{N} \qquad (1)$$

Moments can be taken about any point but if we take them about C the moment due to Q is zero.

$$\text{clockwise moment} = P \times 4\,\text{m}$$

anticlockwise moments

$$= 700\,\text{N} \times 1\,\text{m} + 400\,\text{N} \times 2\,\text{m} + 500\,\text{N} \times 5\,\text{m}$$

$$= 700\,\text{N m} + 800\,\text{N m} + 2500\,\text{N m} = 4000\,\text{N m}$$

Since the plank is in equilibrium we have from (ii)

$$P \times 4\,\text{m} = 4000\,\text{N m}$$

$$\therefore \qquad P = \frac{4000\,\text{N m}}{4\,\text{m}} = 1000\,\text{N}$$

From (1), $\qquad Q = 600\,\text{N}$

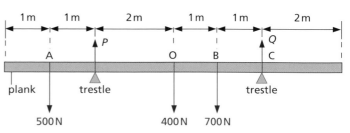

Figure 21.6

Beams and structures

a) Beams

Stretching a rod or bar puts it in **tension**, Figure 21.7a; squeezing it puts it in **compression**, b; twisting it subjects it to **shearing**, c. The greater the cross-section area of a rod, the greater are the balancing internal resisting forces it sets up to tensile, compressive and shearing forces. Resistance to such forces also depends on shape, so different beams are made for different purposes.

When a beam bends, either under its own weight or because it is loaded, one side is compressed, the other is stretched and the centre is unstressed (a neutral plane), Figure 21.8a. I-shaped steel girders, Figure 21.8b, are used in large structures. They are, in effect, beams that have had material removed from the neutral plane and so weigh less. The top and bottom flanges withstand the compression and tension forces due to loading. In a hollow tube the removal of unstressed material gives similar advantages.

Concrete, though strong in compression, is weak in tension but can be made stronger if it is reinforced by steel rods. Steel is strong in tension as well as in compression.

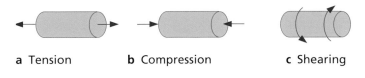

a Tension   **b** Compression   **c** Shearing

Figure 21.7

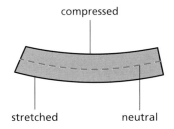

a A bent beam

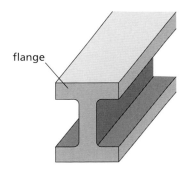

b An I-shaped girder

Figure 21.8

b) Bridges

In an arched stone bridge, the stone is in compression, since it is weak in tension, Figure 21.9a. A steel girder bridge has little material in the neutral plane; it is strengthened by diagonal bars, Figure 21.9b. One is shown in Figure 21.10.

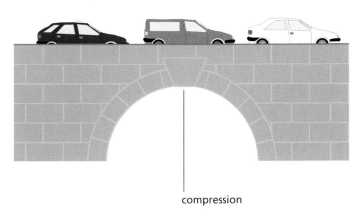

a A stone bridge

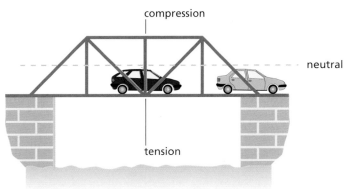

b A steel girder bridge

Figure 21.9

Figure 21.10 The Forth Bridge is a steel girder bridge

Questions

1 The metre rule in Figure 21.11 is pivoted at its centre. If it balances, the mass of *M* is given by the equation

A $M + 50 = 40 + 100$
B $M \times 40 = 100 \times 50$
C $M/50 = 100/40$
D $M/50 = 40/100$
E $M \times 50 = 100 \times 40$

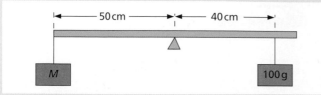

Figure 21.11

2 The uniform beam of weight 30 N and length 4 m is hinged at end A, as in Figure 21.12. The force *F*, in *N*, which must be applied vertically upwards at a distance of 1 m from end B so that the beam is horizontal is

A 60   **B** 50   **C** 40   **D** 30   **E** 20

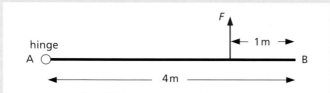

Figure 21.12

3 Figure 21.13 shows three positions of the pedal on a bicycle which has a crank 0.20 m long. If the cyclist exerts the same vertically downward push of 25 N with his foot, in which case is the turning effect
(i) $25 \times 0.2 = 5 \, \text{N m}$,
(ii) 0,
(iii) between 0 and 5 N m?
Explain your answers.

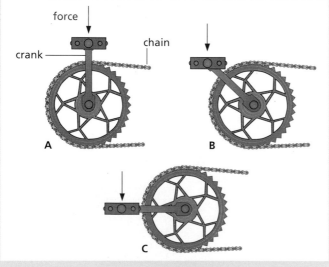

Figure 21.13

4 Is a bicycle a 'distance multiplier' or a 'force multiplier'?

Checklist

After studying this chapter you should be able to

■ define the **moment** of a force about a point,
■ describe an experiment to study turning effects on a body in equilibrium,
■ state the law of moments and use it to solve problems,
■ explain the action of common tools and devices as levers,
■ recognize a machine as a force multiplier or a distance multiplier,
■ define the terms **mechanical advantage** (MA) and **velocity ratio** (VR) of a machine and solve problems using them,
■ state the conditions for equilibrium when parallel forces act on a body,
■ describe the behaviour of beams and bridges when loaded.

22 *Centres of gravity*

A body behaves as if its whole weight were concentrated at one point, called its **centre of gravity** (c.g.) or **centre of mass**, even though the Earth attracts every part of it. The c.g. of a uniform ruler is at its centre and when supported there it can be balanced, Figure 22.1a. If it is supported at any other point it topples because the moment of its weight W about the point of support is not zero, Figure 22.1b.

Your c.g. is near the centre of your body and the vertical line from it to the floor must be within the area enclosed by your feet or you will fall over. You can test this by standing with one arm and the side of one foot pressed against a wall, Figure 22.2. Now try to raise the other leg sideways.

A tight-rope walker has to keep his c.g. exactly above the rope. Some carry a long pole to help them to balance, Figure 22.3. The combined weight of the walker and pole is then spread out more and if the walker begins to topple to one side he moves the pole to the other side.

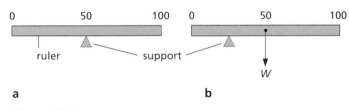

ruler   support W

a   **b**

Figure 22.1

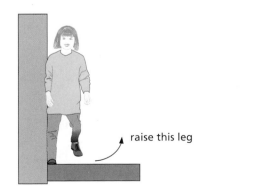

raise this leg

Figure 22.2 Can you do this without falling over?

The c.g. of a regularly shaped body of the same density throughout is at its centre. In other cases it can be found by experiment.

Figure 22.3 A tight-rope walker using a long pole

Practical work

Centre of gravity using a plumb line

Suppose we have to find the c.g. of an irregularly shaped lamina (a thin sheet) of cardboard.

Make a hole A in the lamina and hang it so that it can *swing freely* on a nail clamped in a stand. It will come to rest with its c.g. vertically below A. To locate the vertical line through A tie a plumb line (a thread and a weight) to the nail, Figure 22.4, and mark its position AB on the lamina. The c.g. lies on AB.

Hang the lamina from another position C and mark the plumb line position CD. The c.g. lies on CD and must be at the point of intersection of AB and CD. Check this by hanging the lamina from a third hole. Also try balancing it at its c.g. on the tip of your forefinger.

Devise a method using a plumb line for finding the c.g. of a tripod.

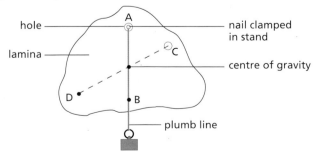

hole   nail clamped in stand

lamina

centre of gravity

plumb line

Figure 22.4

■ *Toppling*

The position of the c.g. of a body affects whether or not it topples over easily. This is important in the design of such things as tall vehicles (which tend to overturn when rounding a corner), racing cars, reading lamps and even drinking glasses.

A body topples when the vertical line through its c.g. falls outside its base, Figure 22.5a. Otherwise it remains stable, Figure 22.5b.

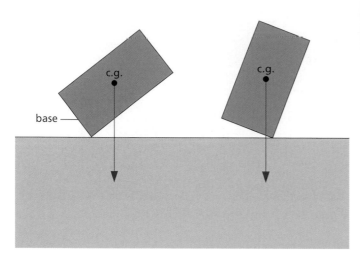

a Topples   **b** Stable

Figure 22.5

Toppling can be investigated by placing an empty can on a plank (with a rough surface to prevent slipping) which is slowly tilted. The angle of tilt is noted when the can falls over. This is repeated with a mass of 1 kg in the can. How does this affect the position of the c.g.? The same procedure is followed with a second can of the same height as the first but of greater width. It will be found that the second can with the weight in it can be tilted through the greater angle.

The stability of a body is therefore increased by

(i) lowering its c.g., and
(ii) increasing the area of its base.

In Figure 22.6a the c.g. of a tractor is being found. It is necessary to do this when testing a new design since tractors are often driven over sloping surfaces and any tendency to overturn must be discovered.

The stability of double-decker buses is being tested in Figure 22.6b. When the top deck only is fully laden with passengers (represented by sand bags in the test), it must not topple if tilted through an angle of 28°.

Racing cars have a low c.g. and a wide wheel base for maximum stability.

Figure 22.6a A tractor under test to find its c.g.

Figure 22.6b A double-decker bus being tilted to test its stability

Stability

Three terms are used in connection with stability.

a) Stable equilibrium

A body is in 'stable equilibrium' if when slightly displaced and then released it returns to its previous position. The ball at the bottom of the dish in Figure 22.7a is an example. Its c.g. rises when it is displaced. It rolls back because its weight has a moment about the point of contact.

b) Unstable equilibrium

A body is in 'unstable equilibrium' if it moves farther away from its previous position when slightly displaced. The ball in Figure 22.7b behaves in this way. Its c.g. falls when it is displaced slightly because there is a moment which increases the displacement. Similarly in Figure 22.1a (p. 89) the balanced ruler is in unstable equilibrium.

c) Neutral equilibrium

A body is in 'neutral equilibrium' if it stays in its new position when displaced, Figure 22.7c. Its c.g. does not rise or fall because there is no moment to increase or decrease the displacement.

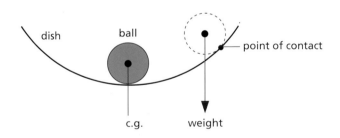

a Stable

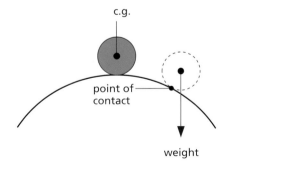

b Unstable

c Neutral

Figure 22.7 States of equilibrium

Balancing tricks and toys

Some tricks that you can try or toys you can make are shown in Figure 22.8. In each case the c.g. is vertically below the point of support and equilibrium is stable.

a Balancing a needle on its point

b The perched parrot

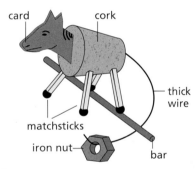

c A rocking horse

Figure 22.8 Balancing tricks

A self-righting toy, Figure 22.9, has a heavy base and when tilted, the weight acting through the c.g. has a moment about the point of contact. This restores it to the upright position.

Figure 22.9 A self-righting toy

91

Questions

1 a A thin sheet of cardboard is cut to the shape in Figure 22.10. Describe, with a diagram, an experiment to find its centre of mass.

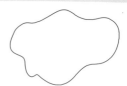

Figure 22.10

b Copy Figure 22.11 below and label with an **X** the centre of mass of each of the three objects.

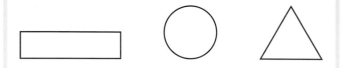

Figure 22.11

c Explain why a mechanic would choose a long spanner to undo a tight nut.

(NEAB Foundation, June 98)

2 Figure 22.12 shows a Bunsen burner in three different positions. State in which position it is in

A stable equilibrium,
B unstable equilibrium,
C neutral equilibrium.

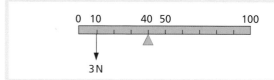

Figure 22.12

3 The weight of the uniform bar in Figure 22.13 is 10 N. Does it balance, tip to the right or tip to the left?

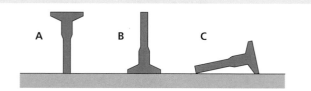

Figure 22.13

4 A uniform metre rule is balanced at the 30 cm mark when a load of 0.80 N is hung at the zero mark, Figure 22.14.
(i) At what point on the rule is the centre of gravity of the rule?
(ii) Show with an arrow drawn on a copy of the diagram the weight of the rule acting through the centre of gravity.
(iii) Calculate the weight of the rule.

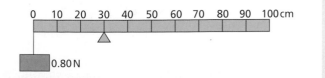

Figure 22.14

5 The following paragraphs appeared in a newspaper.

JEEP FAILS GOVERNMENT TEST

A car manufacturer confirmed yesterday that one of its four-wheel drive mini-jeeps rolled over at 38 mph during stability tests conducted by the Government.

Testing has been halted until safety cages can be fitted to seven makes of mini jeeps which the Department of Transport has agreed to test after repeated complaints from the Consumers' Association. The Association claims its own tests show that the narrow track, short-wheelbase vehicles are prone to rolling over. The Government's tests highlight how passengers raise the centre of mass. All seven vehicles passed the test unladen, although one raised two wheels.

a Write down *two* factors mentioned in the newspaper article which affect the stability of vehicles.
b Figure 22.15 shows a tilted vehicle.

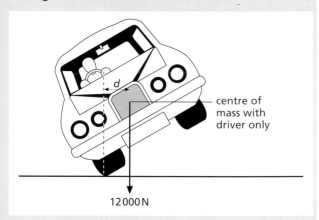

Figure 22.15

The distance *d* shown in the diagram is 50 cm. Calculate the moment of the force about the point of contact with the road.
c Explain how passengers make the vehicles more likely to roll over (less stable). You may use diagrams if you wish.

(NEAB Higher, June 98)

Checklist

After studying this chapter you should be able to

- recall that an object behaves as if its whole weight were concentrated at its centre of gravity,
- describe an experiment to find the centre of gravity of an object,
- connect the stability of an object to the position of its centre of gravity.

23 *Adding forces*

Forces and resultants

Force has both magnitude (size) and direction. It is represented in diagrams by a straight line with an arrow to show its direction of action.

Usually more than one force acts on an object. As a simple example, an object resting on a table is pulled downwards by its weight W and pushed upwards by a force R due to the table supporting it, Figure 23.1. Since the object is at rest, the forces must balance, i.e. $R = W$.

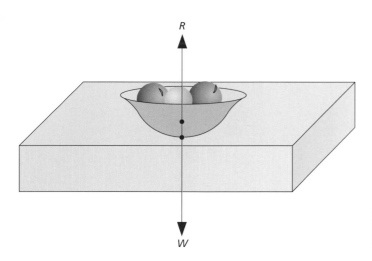

Figure 23.1

Figure 23.2 The design of an off-shore oil platform requires an understanding of the combination of many forces

In structures like a giant oil platform, Figure 23.2, two or more forces may act at the same point. It is then often useful for the design engineer to know the value of the single force, i.e. the **resultant**, which has exactly the same effect as these forces. If the forces act in the same straight line the resultant is found by simple addition or subtraction as shown in Figure 23.3, but if they do not they are added by the 'parallelogram law'.

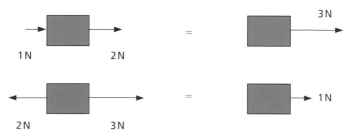

Figure 23.3 The resultant of forces acting in the same straight line is found by addition or subtraction

Practical work

Parallelogram law

Arrange the apparatus as in Figure 23.4a with a sheet of paper behind it on a vertical board. We have to find the resultant of forces P and Q.

Read the values of P and Q from the spring balances. Mark on the paper the directions of P, Q and W as shown by the strings. Remove the paper and using a scale of 1 cm to represent 1 N, draw OA, OB and OD to represent the three forces P, Q and W which act at O, Figure 23.4b. (W = weight of the 1 kg mass = 9.8 N, therefore OD = 9.8 cm.)

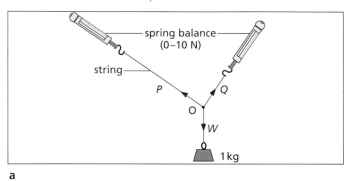

a

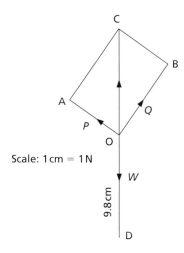

b

Figure 23.4 Finding a resultant by the parallelogram law

P and Q together are balanced by W and so their resultant must be a force equal and opposite to W.

Complete the parallelogram OACB. Measure the diagonal OC; if it is equal in size (i.e. 9.8 cm) and opposite in direction to W then it represents the resultant of P and Q.

The **parallelogram law** for adding two forces is:

> If two forces acting at a point are represented in size and direction by the sides of a parallelogram drawn from the point, their resultant is represented in size and direction by the diagonal of the parallelogram drawn from the point.

Worked example

Find the resultant of two forces of 4.0 N and 5.0 N acting at an angle of 45° to each other.

Using a scale of 1.0 cm = 1.0 N, draw parallelogram ABCD with AB = 5.0 cm, AC = 4.0 N and angle CAB = 45°, Figure 23.5. By the parallelogram law, the diagonal AD represents the resultant in magnitude and direction; it measures 8.3 cm and angle BAD = 21°.

∴ Resultant is a force of 8.3 N acting at an angle of 21° to the force of 5.0 N.

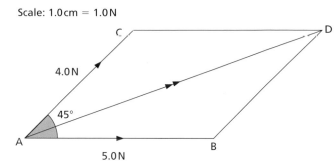

Figure 23.5

Examples of addition of forces

1. Two people carrying a heavy bucket. The weight of the bucket is balanced by the resultant F of F_1 and F_2, Figure 23.6a.

2. Two tugs pulling a ship. The resultant of T_1 and T_2 is forwards, Figure 23.6b, and so the ship moves forwards (as long as the resultant is greater than the resistance to motion of the sea and the wind).

a

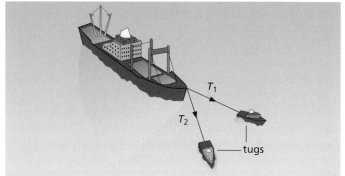

b

Figure 23.6

Vectors and scalars

A **vector** quantity is one such as force which is described completely only if both its size (magnitude) and direction are stated. It is not enough to say, for example, a force of 10 N, but rather a force of 10 N acting vertically downwards.

A vector can be represented by a straight line whose length represents the magnitude of the quantity and whose direction gives its line of action. An arrow on the line shows which way along the line it acts.

A **scalar** quantity has magnitude only. Mass is a scalar and is completely described when its value is known. Scalars are added by ordinary arithmetic; vectors are added geometrically, taking account of their directions as well as their magnitudes.

Friction

Friction is the force that opposes one surface moving, or trying to move, over another. It can be a help or a hindrance. We could not walk if it did not exist between the soles of our shoes and the ground. Our feet would slip backwards, as they tend to if we walk on ice. On the other hand, engineers try to reduce friction to a minimum in the moving parts of machinery by using lubricating oils and ball-bearings.

When a gradually increasing force P is applied through a spring balance to a block on a table, Figure 23.7, it does not move at first. This is because an equally increasing but opposing frictional force F acts where the block and table touch. At any instant P and F are equal and opposite.

If P is increased further the block eventually moves; as it does so F has its maximum value, called **starting** or **static** friction. When it is moving at a steady speed the balance reading is slightly less than that for starting friction. **Sliding** or **dynamic** friction is therefore less than starting or static friction.

A weight on the block increases the force pressing the surfaces together and increases friction.

When work is done against friction, the temperatures of the bodies in contact rise (as you can test by rubbing your hands together); mechanical energy is being changed into heat energy (see Chapter 24).

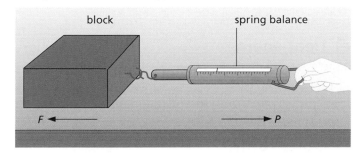

Figure 23.7 Friction opposes motion between surfaces in contact

Questions

1 Jo, Daniel and Helen are pulling a metal ring. Jo pulls with a force of 100 N in one direction and Daniel with a force of 140 N in the opposite direction. If the ring does not move, what force does Helen exert if she pulls in the same direction as Jo?

2 A boy drags a suitcase along the ground with a force of 100 N. If the frictional force opposing the motion of the suitcase is 50 N, what is the resultant forward force on the suitcase?

3 A picture is supported by two vertical strings; if the weight of the picture is 50 N what is the force exerted by each string?

4 Using a scale of 1 cm to represent 10 N find the size and direction of the resultant of forces of 30 N and 40 N acting at right angles to each other.

Checklist

After studying this chapter you should be able to

- combine forces to find their resultant,
- distinguish between vectors and scalars and give examples of each,
- recall that friction opposes motion between surfaces in contact,
- recall that heat energy is produced when work is done against friction.

24 Energy transfer

Energy is a theme that pervades all branches of science. It links a wide range of phenomena and enables us to explain them. It exists in different forms and when something happens, it is likely to be due to energy being transferred from one form to another. **Energy transfer** is needed to enable people, computers, machines and other devices to work and to enable processes and changes to occur. For example, in Figure 24.1, the water skier can only be pulled along by the boat if there is energy transfer in its engine from the burning petrol to its rotating propeller.

Figure 24.1 Energy transfer in action

Forms of energy

a) Chemical energy

Food and fuels, like oil, gas, coal and wood, are concentrated stores of chemical energy (see Chapter 42). The energy of food is released by chemical reactions in our bodies, and during the transfer to other forms we are able to do useful jobs. Fuels cause energy transfers when they are burnt in an engine or a boiler.

Batteries are compact sources of chemical energy, which in use is transferred to electrical energy.

b) Potential energy (p.e.)

This is the energy a body has because of its position or condition. A body above the Earth's surface, like the water in a mountain reservoir, has p.e. stored in the form of **gravitational potential energy**.

Wound-up springs and stretched rubber bands have **elastic potential energy** because of their 'strained' condition.

c) Kinetic energy (k.e.)

Any moving body has k.e. and the faster it moves the more k.e. it has. As a hammer strikes a nail in a piece of wood, there is a transfer of energy from the k.e. of the moving hammer to other forms of energy as the nail is driven in.

d) Electrical energy

This is produced by energy transfers at power stations and in batteries. It is the commonest form of energy used in homes and industry because of the ease of transmission and transfer to other forms.

e) Heat energy

This is also called **thermal** or **internal energy** and is the ultimate fate of other forms of energy. It is frequently transferred by conduction, convection or radiation, as we will see later.

f) Other forms

These include light and other forms of electromagnetic radiation, sound and nuclear energy.

Energy transfers

a) Demonstration

The apparatus in Figure 24.2 can be used to show a battery changing **chemical energy** to **electrical energy** which becomes **k.e.** in the electric motor. The motor raises a weight, giving it **p.e.** If the changeover switch is joined to the lamp and the weight allowed to fall, the motor acts as a generator in which there is an energy transfer from **k.e.** to **electrical energy**. When this is supplied to the lamp, it produces a transfer to **heat** and **light energy**.

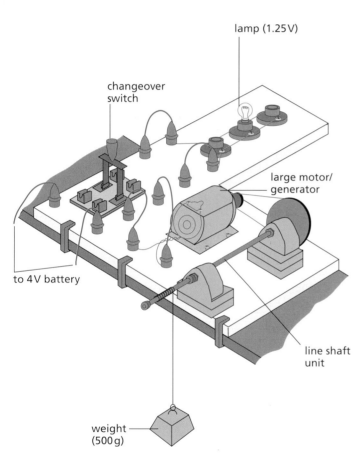

Figure 24.2 Demonstrating energy transfers

b) Other examples

Study the energy transfers shown in Figures 24.3a to d. Some devices have been invented to cause particular energy transfers. For example, a **microphone** changes sound energy into electrical energy; a **loudspeaker** does the reverse. Belts, chains or gears are used to transfer energy between moving parts, e.g. in a bicycle.

a Potential energy to kinetic energy

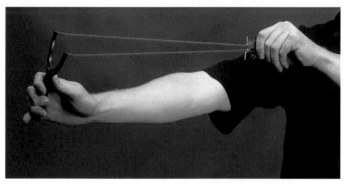

b Electrical energy to heat and light energy

c Chemical energy (from muscles of hand) to p.e. (strain energy of catapult)

d Potential energy of water to kinetic energy of turbine to electrical energy from generator

Figure 24.3 Some energy transfers

Energy measurements

a) Work

In science the word **work** has a different meaning from its everyday one. **Work is done when a force moves.** No work is done in the scientific sense by someone standing still holding a heavy pile of books: an upward force is exerted, but no motion results.

If a building worker carries ten bricks up to the first floor of a building he does more work than if he carries only one brick because he has to exert a larger force. Even more work is required if he carries the ten bricks to the second floor. The amount of work done depends on the size of the force applied and the distance it moves. We therefore measure work by

> work = force × distance moved in direction of
> force   (1)

The unit of work is the **joule** (J) and is **the work done when a force of 1 newton (N) moves through 1 metre** (m). For example, if you have to pull with a force of 50 N to move a crate steadily 3 m in the direction of the force, Figure 24.4a, the work done is 50 N × 3 m = 150 N m = 150 J. That is

> joules = newtons × metres

If you lift a mass of 3 kg vertically through 2 m, Figure 24.4b, you have to exert a vertically upward force equal to the weight of the body, i.e. 30 N (approximately) and the work done is 30 N × 2 m = 60 N m = 60 J.

Note that we must always take the distance in the direction in which the force acts.

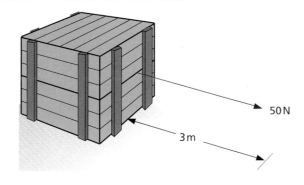

50 N

3 m

a

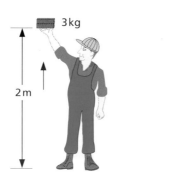

3 kg

2 m

b

Figure 24.4

b) Measuring energy transfers

In an energy transfer work is done. **The work done is a measure of the amount of energy transferred.** For example, if you have to exert an upward force of 10 N to raise a stone steadily through a vertical distance of 1.5 m, the work done is 15 J. This is also the amount of chemical energy transferred from your muscles to p.e. of the stone. All forms of energy, as well as work, are measured in joules.

c) Power

The more powerful a car is the quicker it can accelerate or climb a hill, i.e. the more rapidly it does work. The power of a device is the work it does per second, i.e. the rate at which it does work. This is the same as **the rate at which it transfers energy from one form to another**.

$$\text{power} = \frac{\text{work done}}{\text{time taken}} = \frac{\text{energy transfer}}{\text{time taken}} \qquad (2)$$

The unit of power is the **watt** (W) and is **a rate of working of 1 joule per second**, i.e. 1 W = 1 J/s. Larger units are the kilowatt (kW) and the megawatt (MW):

$$1 \text{ kW} = 1000 \text{ W} = 10^3 \text{ W}$$
$$1 \text{ MW} = 1\,000\,000 \text{ W} = 10^6 \text{ W}$$

If a machine does 500 J of work in 10 s its power is 500 J/10 s = 50 J/s = 50 W. A small car develops a maximum power of about 25 kW.

Practical work

Measuring power

a) Your own power

Get someone with a stopwatch to time you running up a flight of stairs, the more steps the better. Find your weight (in newtons). Calculate the total vertical height (in metres) you have climbed by measuring the height of one step and counting the number of steps.

The work you do (in joules) in lifting your weight to the top of the stairs is (your weight) × (vertical height of stairs). Calculate your power (in watts) from equation (2). About 0.5 kW is good. (1 horsepower = 0.75 kW)

b) Electric motor

This experiment is described in Chapter 49.

Energy conservation

a) Principle of conservation of energy

This is one of the basic laws of physics and is stated as follows.

> Energy cannot be created or destroyed; it is always conserved.

However, energy is continually being transferred from one form to another. Some forms, such as electrical and chemical energy, are more easily transferred than others, such as heat for which it is hard to arrange a useful transfer. Ultimately all energy transfers result in the surroundings being heated (e.g. as a result of doing work against friction) and the energy is wasted, i.e. spread out and increasingly more difficult to use. For example, when a brick falls its p.e. becomes k.e.; as it hits the ground, its temperature rises and heat and sound are produced. If it seems in a transfer that some energy has disappeared, the 'lost' energy is often converted into non-useful heat. This appears to be the fate of all energy in the Universe and is one reason why new sources of useful energy have to be developed (Chapter 42).

b) Efficiency of energy transfers

The efficiency of a device is the percentage of the energy supplied to it which is usefully transferred. It is calculated from the expression:

$$\text{efficiency} = \frac{\text{useful energy transferred by device}}{\text{total energy supplied to device}} \times 100\%$$

For example, for the lever shown in Figure 21.4 (p. 85)

$$\text{efficiency} = \frac{\text{work done on load}}{\text{work done by effort}} \times 100\%$$

It can be shown that

$$\text{efficiency} = \frac{\text{MA}}{\text{VR}} \times 100\%$$

This will be less than 100% if there is friction in the fulcrum.

Table 24.1 lists the efficiencies of some devices and the energy transfers involved.

Table 24.1

Device	% Efficiency	Energy transfer
large electric motor	90	electrical to k.e.
large electric generator	90	k.e. to electrical
domestic gas boiler	75	chemical to heat
compact fluorescent lamp	50	electrical to light
steam turbine	45	heat to k.e.
car engine	25	chemical to k.e.
filament lamp	10	electrical to light

Energy of food

When food is eaten it reacts with the oxygen we breathe into our lungs and is slowly 'burnt'. As a result chemical energy stored in food becomes thermal energy to warm the body and mechanical energy for muscular movement.

> The **energy value** of a food substance is the amount of energy released when 1 kg is completely oxidized.

Energy value is measured in J/kg. The energy values of some foods are given in Figure 24.5 in megajoules per kilogram.

margarine 26   eggs 7   cheese 21   butter 31   potatoes 4   beef 10

sugar 16   apples 2.6   milk 2.9   flour 15   ice cream 9   carrots 1.7

Figure 24.5 Energy values of some foods in MJ/kg

Food with high values are 'fattening' and if more food is eaten than the body really needs, the extra is stored as fat. The average adult requires about 10 MJ per day.

Our muscles change chemical energy into mechanical energy when we exert a force – to lift a weight, for example. Unfortunately, they are not too good at doing this; of every 100 J of chemical energy they use, they can convert only 25 J into mechanical energy – that is, they are only 25% efficient at changing chemical energy into mechanical energy. The other 75 J becomes thermal energy, much of which the body gets rid of by sweating.

Combustion of fuels

Fuels can be solids like wood and coal, liquids like fuel oil and paraffin, or gases like methane and butane.

Some fuels are better than others for certain jobs. For example, fuels for cooking or keeping us warm should, as well as being cheap, have a high **heating value**. This means that every gram of fuel should produce a large amount of heat energy when burnt. A fuel for a space rocket (e.g. liquid hydrogen) must also burn very quickly so that the gases created expand rapidly and leave the rocket at high speed.

The heating values of some fuels are given in Table 24.2 in kilojoules per gram.

Table 24.2 Heating values of fuels/kJ/g

Solids	Value	Liquids	Value	Gases	Value
Wood	17	Fuel oil	45	Methane	55
Coal	25–33	Paraffin	48	Butane	50

The thick dark liquid called **petroleum** or **crude oil** is the source of most liquid and gaseous fuels. It is obtained from underground deposits at oil wells in many parts of the world. Natural gas (methane) is often found with it. In an oil refinery different fuels are obtained from petroleum, including fuel oil for industry, diesel oil for lorries, paraffin (kerosene) for jet engines, petrol for cars, as well as butane (bottled gas).

Questions

1 Name the energy transfers which occur when
   a an electric bell rings,
   b someone speaks into a microphone,
   c a ball is thrown upwards,
   d there is a picture on a television screen,
   e a torch is on.

2 Name the forms of energy represented by the letters A, B, C and D in the following statement.

   In a coal-fired power station, the (A) energy of coal becomes (B) energy which changes water into steam. The steam drives a turbine which drives a generator. A generator transfers (C) energy into (D) energy.

3 How much work is done when a mass of 3 kg (weighing 30 N) is lifted vertically through 6 m?

4 A hiker climbs a hill 300 m high. If she weighs 50 kg calculate the work she does in lifting her body to the top of the hill.

5 In loading a lorry a man lifts boxes each of weight 100 N through a height of 1.5 m.
   a How much work does he do in lifting one box?
   b How much energy is transferred when one box is lifted?
   c If he lifts 4 boxes per minute at what power is he working?

6 A boy whose weight is 600 N runs up a flight of stairs 10 m high in 12 s. What is his average power?

7 a When the energy input to a gas-fired power station is 1000 MJ, the electrical energy output is 300 MJ. What is the efficiency of the power station in changing the energy in gas into electrical energy?
   b What form does the 700 MJ of 'lost' energy take?
   c What is the fate of the 'lost' energy?

8 A load of 500 N is raised 0.20 m by a machine in which an effort of 150 N moves 1.0 m. What is
   a the work done on the load,
   b the work done by the effort,
   c the efficiency?

Checklist

After studying this chapter you should be able to

- recall the different forms of energy,
- describe energy transfers in given examples,
- use the relation **work done = force × distance moved** to calculate energy transfer,
- define the unit of work,
- recall that power is the rate of energy transfer, give its unit and solve problems,
- describe an experiment to measure your own power,
- state the principle of conservation of energy,
- recall the meaning of **efficiency** of a device,
- recall the meaning of **energy value** of food,
- explain what is meant by a fuel having a **high heating value**.

25 Pressure in liquids

■ Pressure

To make sense of some effects in which a force acts on a body we have to consider not only the force but also the area on which it acts. For example, wearing skis prevents you sinking into soft snow because your weight is spread over a greater area. We say the **pressure** is less.

Pressure is the force (or thrust) **acting on unit area** (i.e. $1\,m^2$) and is calculated from

$$\text{pressure} = \frac{\text{force}}{\text{area}}$$

The unit of pressure is the **pascal** (Pa); it equals 1 newton per square metre (N/m^2) and is quite a small pressure. An apple in your hand exerts about $1000\,Pa$.

The greater the area over which a force acts, the less the pressure. The pressure exerted on the floor by the same box, (a) standing on end, (b) lying flat, is shown in Figure 25.1. This is why a tractor with wide wheels can move over soft ground. The pressure is large when the area is small and accounts for a nail being given a sharp point. Walnuts can be broken in the hand by squeezing two together but not one. Why?

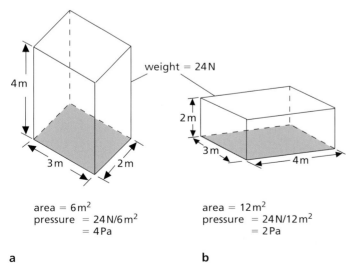

weight = 24 N

area = $6\,m^2$
pressure = $24\,N/6\,m^2$
 = $4\,Pa$

area = $12\,m^2$
pressure = $24\,N/12\,m^2$
 = $2\,Pa$

a   b

Figure 25.1

■ Liquid pressure

1. Pressure in a liquid increases with depth because the farther down you go the greater the weight of liquid above. In Figure 25.2a water spurts out fastest and furthest from the lowest hole.

2. Pressure at one depth acts equally in all directions. The can of water in Figure 25.2b has similar holes all round it at the same level. Water comes out equally fast and equally far from each hole. Hence the pressure exerted by the water at this depth is the same in all directions.

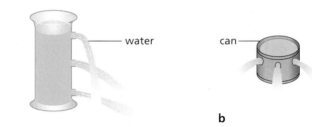

water   can

a   b
Figure 25.2

3. A liquid finds its own level. In the U-tube of Figure 25.3a the liquid pressure at the foot of P is greater than at the foot of Q because the left-hand column is higher than the right-hand one. When the clip is opened the liquid flows from P to Q until the pressure and the levels are the same, i.e. the liquid 'finds its own level'. Although the weight of liquid in Q is now greater than in P, it acts over a greater area since tube Q is wider.

In Figure 25.3b the liquid is at the same level in each tube and confirms that the pressure at the foot of a liquid column depends only on the *vertical* depth of the liquid and not on the tube width or shape.

4. Pressure depends on the density of the liquid. The denser the liquid, the greater the pressure at any given depth.

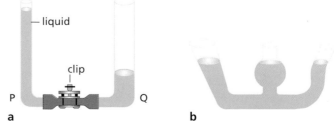

liquid

clip

P   Q

a   b
Figure 25.3

Water supply system

A town's water supply often comes from a reservoir on high ground. Water flows from it through pipes to any tap or storage tank that is below the level of water in the reservoir, Figure 25.4. The lower the place supplied, the greater the water pressure. In very tall buildings it may be necessary first to pump the water to a large tank in the roof.

Reservoirs for water supply or for hydroelectric power stations are often made in mountainous regions by building a dam at one end of a valley. The dam must be thicker at the bottom than at the top due to the large water pressure at the bottom.

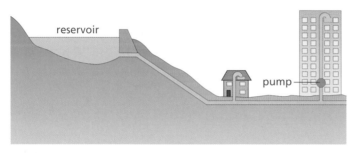

Figure 25.4 Water supply system. Why is the pump needed in the high-rise building?

Hydraulic machines

Liquids are almost incompressible (i.e. their volume cannot be reduced by squeezing) and they 'pass on' any pressure applied to them. Use is made of these facts in hydraulic machines. Figure 25.5 shows the principle on which they work.

Suppose a downward force f acts on a piston of area a. The pressure transmitted through the liquid is

$$\text{pressure} = \frac{\text{force}}{\text{area}} = \frac{f}{a}$$

This pressure acts on a second piston of larger area A producing an upward force $F = \text{pressure} \times \text{area} = f \times A/a$. So

$$F = f \times \frac{A}{a}$$

Since A is larger than a, F must be larger than f and the hydraulic system is a force multiplier; the **multiplying factor** is A/a.

For example if $f = 1\,\text{N}$, $a = \frac{1}{100}\,\text{m}^2$ and $A = \frac{1}{2}\,\text{m}^2$ then

$$F = 1\,\text{N} \times \frac{(1/2)\,\text{m}^2}{(1/100)\,\text{m}^2}$$

$$= 50\,\text{N}$$

A force of 1 N could lift a load of 50 N; the hydraulic system multiplies the force 50 times.

A **hydraulic jack**, Figure 25.6, has a platform on top of piston B and is used in garages to lift cars. Both valves open only to the right and they allow B to be raised a long way when A moves up and down repeatedly. Hydraulic fork-lift trucks and similar machines such as loaders, Figure 25.7, work in the same way. In a hydraulic press there is a fixed plate above piston B, and sheets of steel placed between B and the plate can be forged.

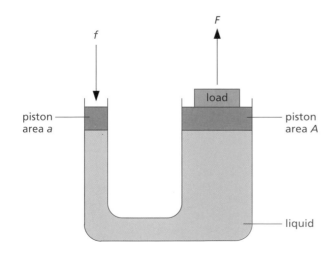

Figure 25.5 The hydraulic principle

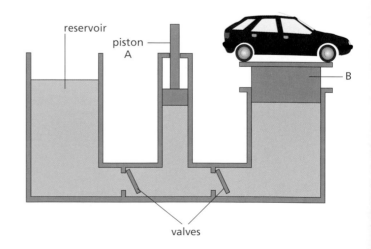

Figure 25.6 A hydraulic jack

Figure 25.7 A hydraulic machine in action

Hydraulic car brakes are shown in Figure 25.8. When the brake pedal is pushed the piston in the master cylinder exerts a force on the brake fluid and the resulting pressure is transmitted equally to eight other pistons (four are shown). These force the brake shoes or pads against the wheels and stop the car.

Expression for liquid pressure

In designing a dam an engineer has to calculate the pressure at various depths below the water surface.

An expression for the pressure at a depth h in a liquid of density d can be found by considering a horizontal area A, Figure 25.9. The force acting vertically downwards on A equals the weight of a liquid column of height h and cross-sectional area A above it. Then

$$\text{volume of liquid column} = hA$$

Since mass = volume × density we can say

$$\text{mass of liquid column} = hAd$$

Taking a mass of 1 kg to have weight 10 N,

$$\text{weight of liquid column} = 10\,hAd$$
$$\therefore \qquad \text{force on area } A = 10\,hAd$$
$$\therefore \qquad \text{pressure} = \text{force/area} = 10\,hAd/A$$
$$= 10\,hd$$

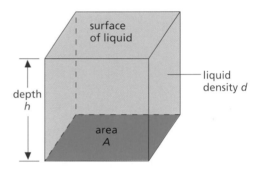

Figure 25.9

This pressure acts equally in all directions at depth h and depends only on h and d. Its value will be in Pa if h is in m and d in kg/m^3.

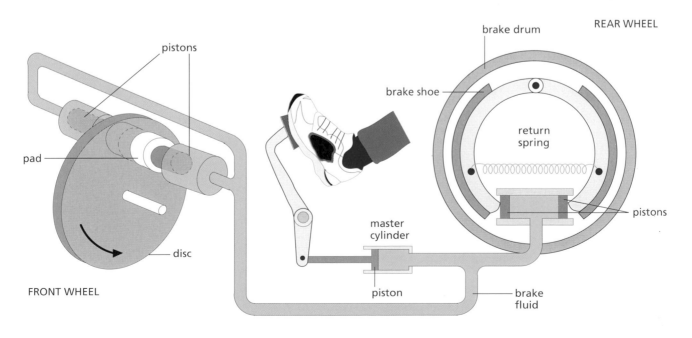

Figure 25.8 Hydraulic brakes

Pressure gauges

These measure the pressure exerted by a fluid, i.e. by a liquid or a gas.

a) Bourdon gauge

This works like the toy in Figure 25.10. The harder you blow into the paper tube, the more it uncurls. In a Bourdon gauge, Figure 25.11, when a fluid pressure is applied, the curved metal tube tries to straighten out and rotates a pointer over a scale. Car oil–pressure gauges and the gauges on gas cylinders are of this type.

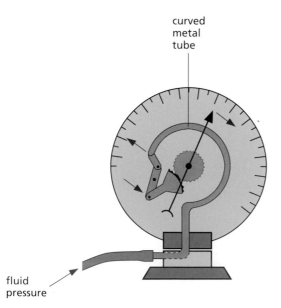

Figure 25.10 The harder you blow, the greater the pressure and the more it uncurls

fluid
pressure

Figure 25.11 A Bourdon gauge

b) U-tube manometer

In Figure 25.12a each surface of the liquid is acted on equally by atmospheric pressure and the levels are the same. If one side is connected to, for example, the gas supply, Figure 25.12b, the gas exerts a pressure on surface A and level B rises until

pressure of gas = atmospheric pressure
+ pressure due to liquid column BC

The pressure of the liquid column BC therefore equals the amount by which the gas pressure *exceeds* atmospheric pressure. It equals $10\,hd$ (in Pa) where h is the vertical height of BC (in m) and d is the density of the liquid (in kg/m³). The height h is called the **head of liquid** and sometimes, instead of stating a pressure in Pa, we say that it is so many cm of water (or mercury for higher pressures).

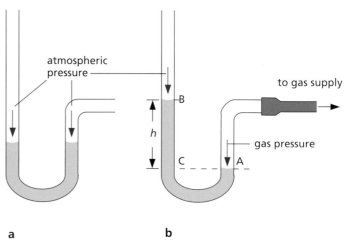

a   b

Figure 25.12 A U-tube manometer

Questions

1 A student investigated the grip of three types of sports shoe shown in Figure 25.13.

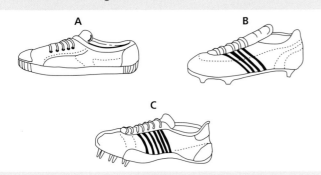

Figure 25.13

a Describe how the student can measure the area of contact between shoe **A** and the ground.

b A girl is wearing a pair of shoes like **A**. She weighs 550 N. The area of contact for shoe **A** is 160 cm². Calculate the pressure exerted on the ground by the girl.

c Explain why shoe **C** should **not** be worn in a gym.

(London Foundation, June 98)

2 a What is the pressure on a surface when a force of 50 N acts on an area of
(i) 2.0 m²,
(ii) 100 m²,
(iii) 0.50 m²?

b A pressure of 10 Pa acts on an area of 3.0 m². What is the force acting on the area?

3 To support the weight of a house the walls are built on concrete foundations. Figure 25.14 shows two different types of foundations used by builders.

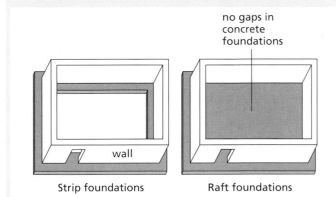

Figure 25.14

a (i) Write down the equation which links area, force and pressure.
(ii) The materials needed to build a house weigh 216 000 N. Calculate the pressure the house will exert on the ground if the total area of the foundations is 18 m². Show clearly how you work out your answer.

b Which type of foundations should be used for a house built on very soft ground? Explain the reason for your choice.

(AQA (SEG) Foundation, Summer 99)

4 In a hydraulic press a force of 20 N is applied to a piston of area 0.20 m². The area of the other piston is 2.0 m². What is
a the pressure transmitted through the liquid,
b the force on the other piston?

5 a Why must a liquid and not a gas be used as the 'fluid' in a hydraulic machine?

b On what other important property of a liquid do hydraulic machines depend?

6 What is the pressure 100 m below the surface of sea water of density 1150 kg/m³?

7 Figure 25.15 shows a hydraulic braking system on a bike.

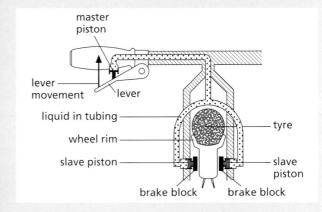

Figure 25.15

a When the brake lever is pulled towards the handlebars the brake blocks push against the wheel rim.
(i) Why is a liquid used in the tubing?
(ii) There are two brake blocks, one on each side of the rim. Explain how the system ensures that the same force is applied by each brake block.

b The force between one brake block and the wheel rim is 500 N. The area of the brake block is 2 cm². Calculate the pressure exerted by the brake block on the wheel rim.

c Explain how this hydraulic braking system is a force multiplier.

(AQA (NEAB) Higher, June 99)

Checklist

After studying this chapter you should be able to

▪ define **pressure** and recall its unit,

▪ connect the pressure in a fluid with its depth and density,

▪ recall that pressure is transmitted through a liquid and use it to explain the hydraulic jack and hydraulic car brakes,

▪ calculate the multiplying factor for a hydraulic system,

▪ use pressure = 10 *hd* to solve problems,

▪ describe how a Bourdon gauge or a U-tube manometer may be used to measure fluid pressure.

26 Floating, sinking and flying

A ship gets support from the water and floats because its average density is less than that of the water (see Chapter 18). Any object in a liquid, whether floating or submerged, is acted on by an upward force or **upthrust**. This makes it seem to weigh less than normal.

Archimedes' principle

In Figure 26.1a the block hanging from the spring balance weighs 10 N in air. When it is completely immersed in water the reading becomes 6 N, Figure 26.1b. The loss in weight of the block is $10 - 6 = 4$ N: the upthrust of the water on it is therefore 4 N.

If a can like that in Figure 26.1c is used, full of water to the level of the spout, the water displaced (the overflow) can be collected and weighed. Its weight here is 4 N, the same as the upthrust. (Its volume is exactly equal to the volume of the block.) Experiments with other liquids and also with gases lead to the general conclusion called **Archimedes' principle**.

> When a body is wholly or partly submerged in a fluid the upthrust equals the weight of fluid displaced (i.e. pushed aside).

A fluid means either a liquid or a gas. The case of gases will be dealt with later.

Floating

A stone held below the surface of water sinks when released, a cork rises. The weight of the stone is greater than the upthrust on it (i.e. the weight of water displaced) and there is a net or resultant downward force on it, Figure 26.2a. If the cork has the same volume as the stone, it will displace the same weight (and volume) of water. The upthrust on it is therefore the same as for the stone but it is greater than the weight of the cork. The resultant upward force on the cork makes it rise, Figure 26.2b.

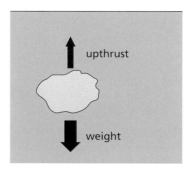

a Stone sinks

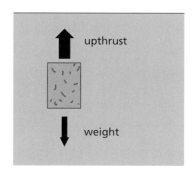

b Cork rises

Figure 26.2

When a body floats in water the upthrust equals the weight of the body. The net force on the body is zero. This is a case of the **principle of flotation**.

> A floating body displaces its own weight of fluid.

For example, a floating block of wood of weight 10 N displaces an amount of water (or any other liquid in which it floats) having weight 10 N.

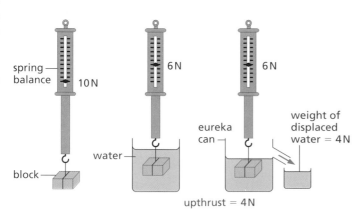

a   **b**   **c**

Ships, submarines, balloons

a) Ships

A floating ship displaces a weight of water equal to its own weight including that of the cargo. The load lines (called the Plimsoll mark) on the side of a ship show how low in the water it can lie and still be safely and legally loaded under different conditions, Figure 26.3. Why is it allowed to take a greater load in summer than in winter?

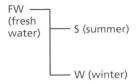

Figure 26.3 Simplified Plimsoll mark

b) Submarines

A submarine, Figure 26.4, sinks by taking water into its buoyancy tanks. Once submerged, the upthrust is unchanged but the weight of the submarine increases with the inflow of water and it sinks faster. To surface, compressed air is used to blow the water out of the tanks.

In the 'Cartesian diver', Figure 26.5, pressure on the cork forces more water into the bulb. The diver's weight increases and it sinks. Decreasing the pressure causes it to rise.

Figure 26.4 A submarine

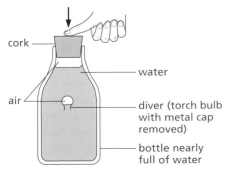

Figure 26.5 Demonstrating the Cartesian diver

c) Balloons

A balloon filled with hydrogen or hot air weighs less than the weight of air it displaces. The upthrust is therefore greater than its weight and the resultant upward force on the balloon causes it to rise.

Meteorological balloons, Figure 26.6, carrying scientific instruments called **radiosondes** are sent into the upper atmosphere. A small radio transmitter sends signals back to Earth which contain information about the temperature, pressure and humidity. They are tracked by radar to give data on wind direction and speed.

Figure 26.6 A meteorological balloon being launched

Figure 26.7 An airship

d) Airships

An airship such as that in Figure 26.7 has a plastic gas bag filled with non-flammable helium (a gas less dense than air). It is powered by two car engines which drive swivelling propellers to provide vertical thrust for take-off and landing and horizontal thrust for forward motion. It can climb at about 700 metres a minute, cruise sedately at 80 km per hour (50 mph) and travel 1600 km (1000 miles) on just 550 litres (120 gallons) of fuel.

Airships are used for aerial surveys, photography and advertising.

Practical work

Investigating fluid flow in tubes

(a) The rate of flow of water through tubes of different lengths and diameters can be investigated using apparatus like that in Figure 27.3. The level of water in the tank never changes so the pressure causing the flow remains constant.

(b) The viscosity of different liquids can be compared with the apparatus in Figure 27.4. The time taken for the liquid level to drop from the upper to the lower rubber band is measured for different liquids.

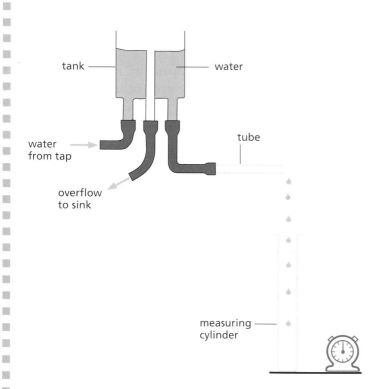

Figure 27.3

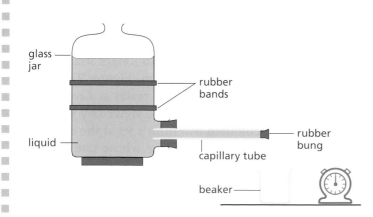

Figure 27.4

Motion of an object in a fluid

The behaviour of a fluid when an object is moving in it is similar to what occurs when a fluid flows in a pipe.

If the object, e.g. a small sphere, moves slowly, **streamlines** like those in Figure 27.5a can be drawn to show how the apparent motion of the fluid would appear to someone on the moving sphere. They represent **steady** flow.

If the speed of the sphere increases, a **critical speed** is reached when the flow breaks up and eddies are formed behind the sphere as in Figure 27.5b. The flow becomes **turbulent** and the viscous drag on the sphere increases sharply.

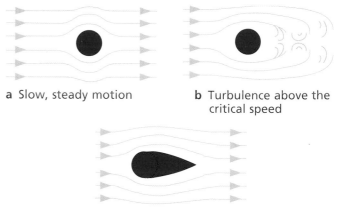

a Slow, steady motion

b Turbulence above the critical speed

c A streamlined object has a greater critical speed

Figure 27.5

The critical speed can be raised by changing the shape of the object, so reducing the drag and causing steady flow to replace turbulent flow. This is called **streamlining** the object and Figure 27.5c shows how it is done for a sphere. The pointed tail can be regarded as filling the region where eddies occur in turbulent motion, thus ensuring that the streamlines merge again behind the sphere. The shape of a speed cyclist's helmet, Figure 27.6, is a clear application of this.

Figure 27.6 A cyclist's helmet is designed with streamlining in mind

Streamlining is especially important in the design of high-speed aircraft and other fast-moving vehicles such as racing cars. A wind tunnel is often used to improve aerodynamic design, Figure 27.7. In nature, streamlining enables fish, dolphins and other creatures to move faster through water or the air, Figure 27.8, sometimes a matter of importance to their survival.

Figure 27.7 Prototype racing car in a wind tunnel

Figure 27.8 Dolphins have streamlined bodies to assist their movement in water

Practical work

Objects falling in a liquid

The time of fall of small ball-bearings dropped into a viscous liquid such as glycerine, engine oil or clear syrup can be studied using the apparatus of Figure 27.9. The glass jar should be as wide and as tall as possible. To reduce the risk of air bubbles sticking to the ball-bearings, they should be dipped in some of the liquid and thereby coated with it, before dropping. The investigations that can be carried out include:

(i) measurement of the speed of a ball-bearing over a distance at different depths,
(ii) the effect of using balls of different diameters, and
(iii) a comparison of the viscosities of different liquids.

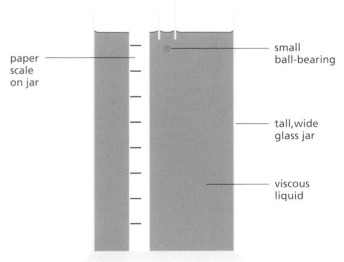

Figure 27.9

At the start the weight W of the ball causing it to fall vertically downwards under gravity is greater than the upwards viscous force V opposing its motion. As the ball speeds up, V increases until eventually V (+ upthrust) equals W. The forces acting on the ball then balance and it moves with constant speed (see Chapter 31), called the **terminal speed**.

Bernoulli's principle

As we have seen (Chapter 25), the pressure is the same at all points on the same level in a fluid at rest; this is not so when the fluid is moving.

a) Liquids

When a liquid flows steadily through a uniform tube the pressure falls steadily due to viscosity, as shown by the decreasing height of liquid in the three vertical tubes (manometers, Chapter 25) in Figure 27.10a.

In Figure 27.10b the pressure falls in the narrow part B but rises again in the wider part C. Since the same volume of liquid passes through B in a certain time as enters A, the liquid must be moving faster in B than in A. Therefore a decrease of pressure occurs when the speed of the liquid increases. Conversely an increase of pressure accompanies a fall in speed. This effect, called **Bernoulli's principle**, is stated as follows.

> When the speed of a fluid in smooth flow increases, the pressure in the fluid decreases and vice versa.

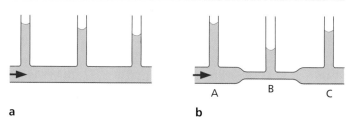

a   **b**

Figure 27.10

b) Gases

Bernoulli effects in air streams can be shown as in Figures 27.11a, b, c. In all cases the pressure is lower in the fast-moving air stream.

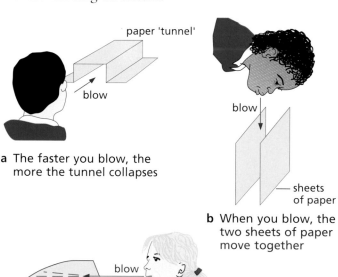

a The faster you blow, the more the tunnel collapses

b When you blow, the two sheets of paper move together

c The paper rises when you blow

Figure 27.11 Demonstrating Bernoulli's principle

Applications of Bernoulli

a) Jets and nozzles

A slow stream of water from a tap can be changed into a fast jet by narrowing the outlet with a finger. The more it is narrowed, the greater is the increase of speed and so the greater is the pressure drop. Several devices with jets or nozzles use this effect. Figures 27.12a, b show the action of a Bunsen burner and a paint spray. It is also used in carburettors and filter pumps.

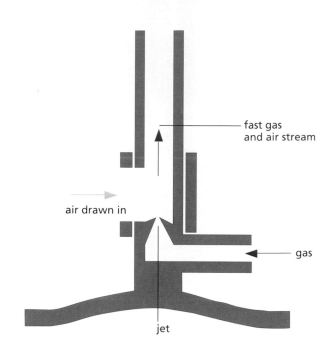

a A Bunsen burner

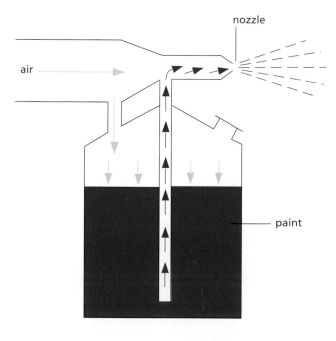

b A paint spray

Figure 27.12 Jets and nozzles make use of the Bernoulli effect

b) Spinning ball

If a tennis ball is 'cut' or a golf ball 'sliced', it spins as it travels through the air and experiences a sideways force which causes it to curve in flight. This is due to air being dragged round by the spinning ball, thereby increasing the air flow on one side and decreasing it on the other. A pressure difference is created, Figure 27.13. The swing of a spinning cricket ball is complicated by its raised seam.

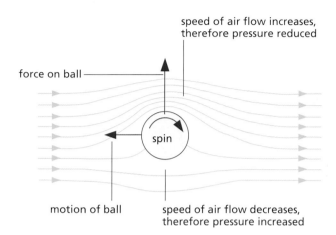

Figure 27.13 A spinning ball follows a curved path

c) Aerofoil

This is a device which is shaped so that when a fluid flows over it, a force is produced at right angles to the direction of flow. Examples of aerofoils are aircraft wings, propellers and turbine blades.

An aircraft wing has a curved upper surface and a flat under surface so that air passing over the top of the wing has to travel further and so faster than over the bottom. This is represented by the streamlines being closer together above than below the aerofoil, Figure 27.14. As a result the pressure underneath is increased and that above reduced, and the resultant upward force on the wing provides the 'lift' for the aircraft.

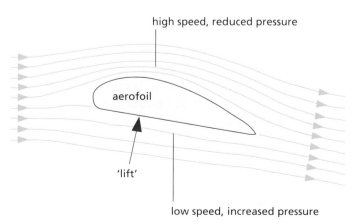

Figure 27.14 When fluid flows over an aerofoil a force is produced at right angles to the flow

The lift increases when the angle between the wing and the air flow, called the 'angle of attack', increases by raising the nose of the aircraft. At a certain angle (about 16°) the flow separates from the upper surface, i.e. the air stops flowing smoothly over the top of the wing. Lift is lost almost completely, the flow downstream becomes very turbulent, drag increases sharply, the aircraft stalls and starts to dive. The pilot loses control until lift can be obtained again.

The sail of a yacht 'tacking into the wind', Figure 27.15a, is another example of an aerofoil. The air flow over the sail results in a pressure increase on the windward side and a decrease on the leeward side. The resultant force is at right angles to the sail, Figure 27.15b.

Figure 27.15a A yacht 'tacking into the wind'

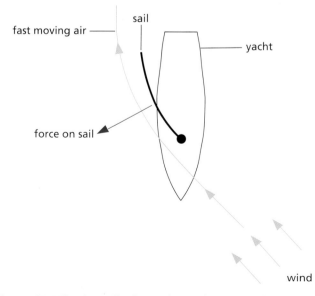

Figure 27.15b The sail of a yacht 'tacking into the wind' acts as an aerofoil

Questions

1 In a model yacht race on a river five identical yachts A, B, C, D, E are released at the same time from different points along the starting line, as in Figure 27.16.
   a Why should C win?
   b As they approach the finishing line they all slow down. Why?

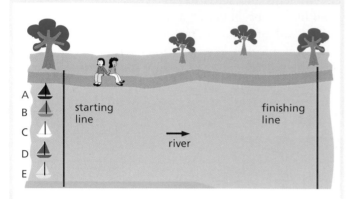

Figure 27.16

2 How would you expect the rate of flow of a liquid through a pipe to depend on
   a the width of the tube,
   b the length of the tube, and
   c the viscosity of the liquid?

3 Explain the following tricks which you could also try.
   a If you blow hard over the top of a small glass containing a table tennis ball, it jumps out of the glass, Figure 27.17a.
   b A table tennis ball can be supported in the stream of air coming from the nozzle of a hairdryer, Figure 27.17b.
   c If you suddenly blow really hard just above a small coin near a saucer, Figure 27.17c, it may jump into the saucer.

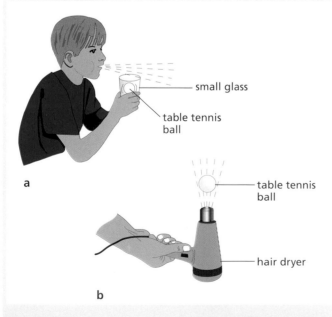

Figure 27.17a, b

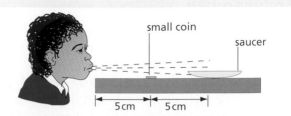

Figure 27.17c

4 a Why does streamlining improve the efficiency of a car?
   b Wind tunnel tests are carried out on models of new cars, in which the air moves past the car rather than the car moving through the air. The air flow pattern is revealed by fine smoke trails. For the car in Figure 27.18 eddies occur at the foot of the windscreen and the flow becomes unsteady behind the car. Make a sketch to show how you would improve the streamlining.

Figure 27.18

▪ *Checklist*

After studying this chapter you should be able to

- ▪ recall that viscosity is a kind of frictional force in fluids and give a molecular explanation,
- ▪ explain the difference between **steady** and **turbulent** flow, using the term **critical speed**,
- ▪ describe experiments to investigate the flow of liquids in tubes,
- ▪ explain using a sketch the purpose of **streamlining** an object,
- ▪ describe an experiment to investigate the time of fall of small objects in liquids,
- ▪ state Bernoulli's principle and describe some effects and applications,
- ▪ explain the action of an aerofoil.

Forces and pressure
Additional questions

Moments and levers

1 The diagram shows a reading lamp. The light is at the end of an arm which can be rotated about the pivot. A cord is fixed near the right-hand end of the arm. The cord passes vertically down to a stretched spring.

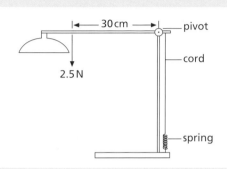

The weight of the lamp and arm is 2.5 N. The weight acts at a distance of 30 cm from the pivot.
a Write down the equation used to calculate the **moment** of a force.
b Calculate the moment of the 2.5 N force about the pivot.
c The arm is now raised as shown in the diagram below.

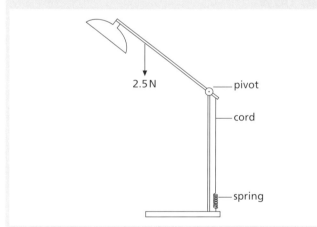

Complete the following sentences by choosing the correct ending.
(i) The moment of the 2.5 N force about the pivot has *increased/stayed the same/decreased.*
(ii) The length of the spring has *increased/stayed the same/decreased.*

(AQA (NEAB) Foundation, June 99)

2 The uniform plank shown weighs 200 N, rests on two trestles A and B, and supports a boy of weight 500 N in the position shown. *P* and *Q* are the reaction forces at A and B.
a Write down the moment of each force about A.
b Use the principle of moments to find *Q*.
c What is the total upward force *P+Q*?
d What is the value of *P*?

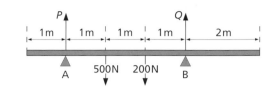

Adding forces

3 Find the size of the resultant of two forces of 5 N and 12 N acting
a in opposite directions to each other,
b at 90° to each other.

Energy transfer

4 State what energy transfers occur in
a a hairdryer,
b a refrigerator,
c an audio system.

5 The diagram shows an experimental solar-powered bike.

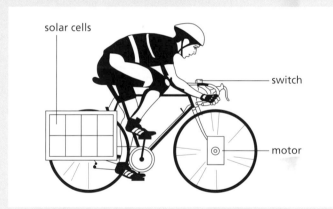

A battery is connected to the solar cells. The solar cells charge up the battery. There is a switch on the handlebars. When the switch is closed, the battery drives a motor attached to the front wheel.

Use words from the list to complete the following sentences. Words may be used once, more than once, or not at all.

> **chemical   electrical   heat (thermal)**
>
> **kinetic   light   potential   sound**

a (i) The solar cells transfer ...... energy to ...... energy.
(ii) When the battery is being charged up, ... energy is transferred to ...... energy.
(iii) The motor is designed to transfer ...... energy to ...... energy.
b Name *one* form of wasted energy which is produced when the motor is running.

(AQA (NEAB) Foundation, June 99)

6 An escalator carries 60 people of average mass 70 kg to a height of 5 m in one minute. Find the power needed to do this.

7 Some students carry out an experiment to measure their own power. One of the students measures her mass. She then runs up the stairs. Her friend uses a stopwatch to time her.

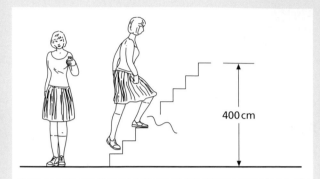

400 cm

Mass of student = 45 kg

a (i) Complete the following sentence.

On Earth the gravitational field strength is about ...... N/kg.

(ii) Use the equation below to calculate the weight of the student.

> **weight** = **mass** × **gravitational field**
> **(newton, N)** **(kilogram, kg)** **strength**
> **(newton/kilogram, N/kg)**

b (i) Write down the relationship used to calculate **work done**.
(ii) Calculate the work done by the student in climbing the stairs.
c It takes the student 2.5 seconds to climb the stairs. Use the equation below to calculate her power.

$$\text{power (watt, W)} = \frac{\text{work done (joule, J)}}{\text{time taken (second, s)}}$$

(*NEAB Higher, June 98*)

Pressure

8 Which of the following will damage a wood-block floor that can withstand a pressure of 2000 kPa (2000 kN/m²)?

1 A block weighing 2000 kN standing on an area of 2 m².
2 An elephant weighing 200 kN standing on an area of 0.2 m².
3 A girl of weight 0.5 kN wearing stiletto-heeled shoes standing on an area of 0.0002 m².

Use the answer code:

A 1, 2, 3   **B** 1, 2   **C** 2, 3   **D** 1   **E** 3

9 The pressure at a point in a liquid

1 increases as the depth increases
2 increases if the density of the liquid increases
3 is greater vertically than horizontally.

Which statement(s) is (are) correct?

A 1, 2, 3   **B** 1, 2   **C** 2, 3   **D** 1   **E** 3

Floating, sinking and flying

10 An object weighs 200 N in air and 120 N when totally submerged in a liquid of density 800 kg/m³. What is
a the upthrust on the object,
b the weight of liquid displaced,
c the mass of liquid displaced,
d the volume of liquid displaced,
e the volume of the object?

Fluid flow

11 Give molecular explanations of the following.
a The viscosity of a fluid affects how it flows through a pipe.
b The viscosity of some fluids decreases if the temperature increases.

Motion and energy

28 Velocity and acceleration

■ Speed

If a car travels 300 km from Liverpool to London in five hours, its **average speed** is 300 km/5 h = 60 km/h. The speedometer would certainly not read 60 km/h for the whole journey but might vary considerably from this value. That is why we state the average speed. If a car could travel at a constant speed of 60 km/h for five hours, the distance covered would still be 300 km. It is *always* true that

$$\text{average speed} = \frac{\text{distance moved}}{\text{time taken}}$$

To find the actual speed at any instant we would need to know the distance moved in a very short interval of time. This can be done by multiflash photography. In Figure 28.1 the golfer is photographed while a flashing lamp illuminates him 100 times a second. The speed of the club-head as it hits the ball is about 200 km/h.

■ Velocity

Speed is the distance travelled in unit time; **velocity is the distance travelled in unit time in a stated direction**. If two cars travel due north at 20 m/s, they have the same speed of 20 m/s and the same velocity of 20 m/s *due north*. If one travels north and the other south, their speeds are the same but not their velocities since their directions of motion are different. Speed is a scalar and velocity a vector quantity (see Chapter 23).

$$\text{velocity} = \frac{\text{distance moved in a stated direction}}{\text{time taken}}$$

The velocity of a body is **uniform** or **constant** if it moves with a steady speed in a straight line. It is not uniform if it moves in a curved path. Why?

The units of speed and velocity are the same, e.g. km/h, m/s, and

$$60 \text{ km/h} = 60\,000 \text{ m}/3600 \text{ s} = 17 \text{ m/s}$$

Distance moved in a stated direction is called the **displacement**. It is a vector, unlike distance which is a scalar. Velocity may also be defined as

$$\text{velocity} = \frac{\text{displacement}}{\text{time taken}}$$

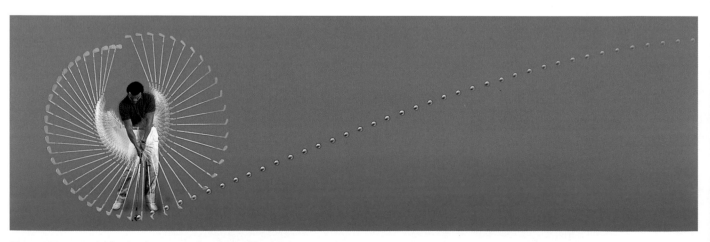

118 ■ **Figure 28.1** Multiflash photograph of a golf swing

■ *Acceleration*

When the velocity of a body changes we say the body **accelerates**. If a car starting from rest and moving due north has velocity 2 m/s after 1 second, its velocity has increased by 2 m/s in 1 s and its acceleration is 2 m/s per second due north. We write this as $2\,\text{m/s}^2$.

Acceleration is the change of velocity in unit time, or

$$\text{acceleration} = \frac{\text{change of velocity}}{\text{time taken for change}}$$

For a steady increase of velocity from 20 m/s to 50 m/s in 5 s

$$\text{acceleration} = \frac{(50-20)\,\text{m/s}}{5\,\text{s}} = 6\,\text{m/s}^2$$

Acceleration is also a vector and both its magnitude and direction should be stated. However, at present we will consider only motion in a straight line and so the magnitude of the velocity will equal the speed, and the magnitude of the acceleration will equal the change of speed in unit time.

The speeds of a car accelerating on a straight road are shown below.

Time/s	0	1	2	3	4	5	6
Speed/m/s	0	5	10	15	20	25	30

The speed increases by 5 m/s every second and the acceleration of $5\,\text{m/s}^2$ is said to be **uniform**.

An acceleration is positive if the velocity increases and negative if it decreases. A negative acceleration is also called a **deceleration** or **retardation**.

■ *Timers*

A number of different devices are useful for analysing motion in the laboratory.

a) Motion sensors

Motion sensors use the ultrasonic echo technique (see p. 57) to determine the distance of an object from the sensor (see Figure 28.6). Connection of a datalogger and computer to the motion sensor then enables a distance–time graph to be plotted directly. Further data analysis by the computer allows a velocity–time graph to be obtained, as in Figures 29.1 and 29.2, p. 122.

b) Tickertape timer: tape charts

A tickertape timer also enables us to measure speeds and hence accelerations. One type, Figure 28.2, has a marker which vibrates 50 times a second and makes dots at $\frac{1}{50}$ s intervals on the paper tape being pulled through it. $\frac{1}{50}$ s is called a 'tick'.

The distance between successive dots equals the average speed of whatever is pulling the tape in, say, cm per $\frac{1}{50}$ s, i.e. cm per tick. The 'tentick' ($\frac{1}{5}$ s) is also used as a unit of time. Since ticks and tenticks are small we drop the 'average' and just refer to the 'speed'.

Tape charts are made by sticking successive strips of tape, usually tentick lengths, side by side. That in Figure 28.3a (overleaf) represents a body moving with **uniform speed** since equal distances have been moved in each tentick interval.

The chart in Figure 28.3b is for **uniform acceleration**: the 'steps' are of equal size showing that the speed increased by the same amount in every tentick ($\frac{1}{5}$ s). The acceleration (average) can be found from the chart as follows.

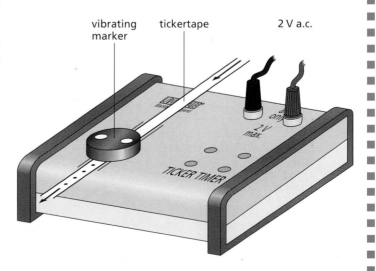

Figure 28.2 Tickertape timer

The speed during the *first* tentick is $2\,\text{cm}/\frac{1}{5}\,\text{s}$ or $10\,\text{cm/s}$. During the *sixth* tentick it is $12\,\text{cm}/\frac{1}{5}\,\text{s}$ or $60\,\text{cm/s}$. And so during this interval of 5 tenticks, i.e. 1 second, the change of speed is $(60 - 10)\,\text{cm/s} = 50\,\text{cm/s}$.

$$\text{acceleration} = \frac{\text{change of speed}}{\text{time taken}}$$

$$= \frac{50\,\text{cm/s}}{1\,\text{s}}$$

$$= 50\,\text{cm/s}^2$$

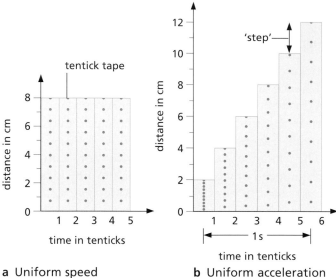

a Uniform speed **b** Uniform acceleration

Figure 28.3 Tape charts

c) Photogate timer

Photogate timers may be used to record the time taken for a trolley to pass through the gate, Figure 28.4. If the length of the 'interrupt card' on the trolley is measured, the velocity of the trolley can then be calculated. Photogates are most useful in experiments where the velocity at only one or two positions is needed, as for example in measurements of momentum (Chapter 32).

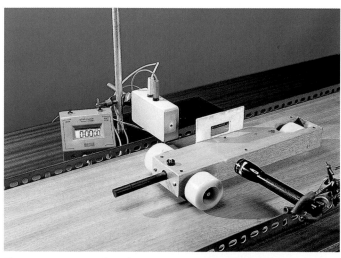

Figure 28.4 Use of a photogate timer

Practical work

Analysing motion

a) Your own motion

Pull a 2 m length of tape through a tickertape timer as you walk away from it quickly, then slowly, then speeding up again and finally stopping.

Cut the tape into tentick lengths and make a tape chart. Write labels on it to show where you speeded up, slowed down, etc.

b) Trolley on a sloping runway

Attach a length of tape to a trolley and release it at the top of a runway, Figure 28.5. The dots will be very crowded at the start – ignore those; but beyond them cut the tape into tentick lengths.

Make a tape chart. Is the acceleration uniform? What is its average value?

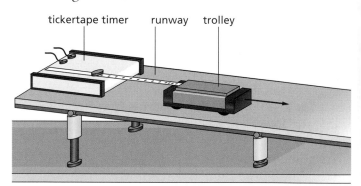

Figure 28.5

c) Datalogging

Replace the tickertape timer with a motion sensor connected to a datalogger and computer, Figure 28.6. Repeat the experiments in **(a)** and **(b)** and obtain distance–time and velocity–time graphs for each case; identify regions where you think the acceleration changes or remains uniform.

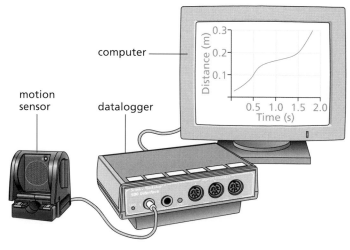

Figure 28.6 Use of a motion sensor

Questions

1 What is the average speed of
a a car which travels 400 m in 20 s,
b an athlete who runs 1500 m in 4 minutes?

2 Four runners compete in a 200 m race. The table shows the time it took for each runner to finish the race.

Runner	Time (s)
Adam	25
Ahmed	28
Neil	26
Naveed	24

a (i) Who came last in the race?
   (ii) Explain how you could tell.
b (i) What is the formula for working out speed?
   (ii) What is Adam's speed during the race?

(London Foundation, June 99)

3 A train increases its speed **steadily** from 10 m/s to 20 m/s in 1 minute.
a What is its average speed during this time in m/s?
b How far does it travel while increasing its speed?

4 A motor cyclist starts from rest and reaches a speed of 6 m/s after travelling with uniform acceleration for 3 s. What is his acceleration?

5 The tape in Figure 28.7 was pulled through a timer by a trolley travelling down a runway. It was marked off in tentick lengths.
a What can you say about the trolley's motion?
b Find its acceleration in cm/s^2.

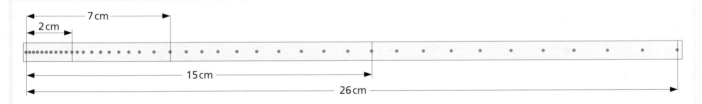

Figure 28.7

6 An aircraft travelling at 600 km/h accelerates steadily at 10 km/h per second. Taking the speed of sound as 1100 km/h at the aircraft's altitude, how long will it take to reach the 'sound barrier'?

7 A vehicle moving with a uniform acceleration of 2 m/s^2 has a velocity of 4 m/s at a certain time. What will its velocity be
a 1 s later,
b 5 s later?

8 If a bus travelling at 20 m/s is subject to a steady deceleration of 5 m/s^2, how long will it take to come to rest?

9 Each strip in the tape chart of Figure 28.8 is for a time interval of 1 tentick.
a If the timer makes 50 dots per second, what time intervals are represented by OA and AB?
b What is the acceleration between O and A in
   (i) cm/tentick^2,
   (ii) cm/s per tentick,
   (iii) cm/s^2?
c What is the acceleration between A and B?

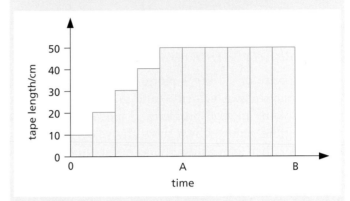

Figure 28.8

■ *Checklist*

After studying this chapter you should be able to

■ explain the meaning of the terms **speed, velocity, acceleration, displacement,**

■ describe how speed and acceleration may be found using tape charts and motion sensors.

29 Graphs and equations

Velocity–time graphs

If the velocity of a body is plotted against the time, the graph obtained is a velocity–time graph. It provides a way of solving motion problems. Tape charts are crude velocity–time graphs which show the velocity changing in jumps rather than smoothly, as occurs in practice. A motion sensor gives a smoother plot.

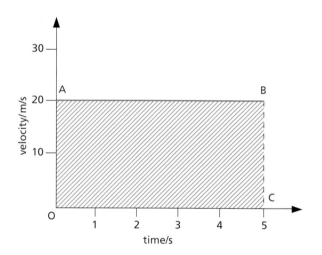

Figure 29.1 Uniform velocity

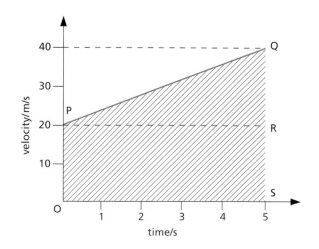

Figure 29.2 Uniform acceleration

The area under a velocity–time graph measures the distance travelled.

In Figure 29.1, AB is the velocity–time graph for a body moving with a **uniform velocity** of 20 m/s. Since distance = average velocity × time, after 5 s it will have moved 20 m/s × 5 s = 100 m. This is the shaded area under the graph, i.e. rectangle OABC.

In Figure 29.2, PQ is the velocity–time graph for a body moving with **uniform acceleration**. At the start of the timing the velocity is 20 m/s but it increases steadily to 40 m/s after 5 s. If the distance covered equals the area under PQ, i.e. the shaded area OPQS, then

$$\text{distance} = \text{area of rectangle OPRS} + \text{area of triangle PQR}$$

$$= \text{OP} \times \text{OS} + \tfrac{1}{2} \times \text{PR} \times \text{QR}$$
(area of a triangle = $\tfrac{1}{2}$ base × height)

$$= 20\,\text{m/s} \times 5\,\text{s} + \tfrac{1}{2} \times 5\,\text{s} \times 20\,\text{m/s}$$

$$= 100\,\text{m} + 50\,\text{m} = 150\,\text{m}$$

Notes
1 When calculating the area from the graph the unit of time must be the same on both axes.
2 This rule for finding distances travelled is true even if the acceleration is not uniform.

The slope or gradient of a velocity–time graph represents the acceleration of the body.

In Figure 29.1 the slope of AB is zero, as is the acceleration. In Figure 29.2 the slope of PQ is QR/PR = 20/5 = 4: the acceleration is 4 m/s².

■ Distance–time graphs

A body travelling with uniform velocity covers equal distances in equal times. Its distance–time graph is a straight line, like OL in Figure 29.3 for a velocity of 10 m/s. The slope of the graph is LM/OM = 40 m/4 s = 10 m/s, which is the value of the velocity. The following statement is true in general.

> The slope or gradient of a distance–time graph represents the velocity of the body.

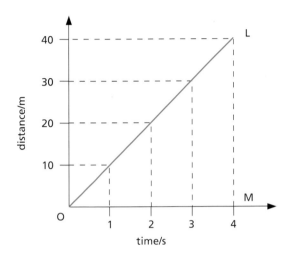

Figure 29.3 Uniform velocity

When the velocity of the body is changing, the slope of the distance–time graph varies, Figure 29.4, and at any point equals the slope of the tangent. For example, the slope of the tangent at T is AB/BC = 40 m/2 s = 20 m/s. The velocity at the instant corresponding to T is therefore 20 m/s.

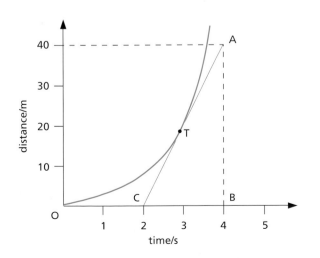

Figure 29.4 Non-uniform velocity

■ Equations for uniform acceleration

Problems involving bodies moving with **uniform acceleration** can often be solved quickly using the **equations of motion**.

First equation

If a body is moving with uniform acceleration a and its velocity increases from u to v in time t, then

$$a = \frac{\text{change of velocity}}{\text{time taken}} = \frac{v - u}{t}$$

$\therefore \qquad at = v - u$

or

$$v = u + at \qquad (1)$$

Note that the initial velocity u and the final velocity v refer to the start and the finish of the *timing* and do not necessarily mean the start and finish of the motion.

Second equation

The velocity of a body moving with uniform acceleration increases steadily. Its average velocity therefore equals half the sum of its initial and final velocities, that is,

$$\text{average velocity} = \frac{u + v}{2}$$

If s is the distance moved in time t, then since average velocity = distance/time = s/t,

$$\frac{s}{t} = \frac{u + v}{2}$$

or

$$s = \frac{(u + v)}{2} t \qquad (2)$$

Third equation

From equation (1), $v = u + at$
From equation (2),

$$\frac{s}{t} = \frac{u + v}{2}$$

$$= \frac{u + u + at}{2} = \frac{2u + at}{2}$$

$$= u + \tfrac{1}{2}at$$

and so

$$s = ut + \tfrac{1}{2}at^2 \qquad (3)$$

Fourth equation

This is obtained by eliminating t from equations (1) and (3). We have

$$v = u + at$$

$$\therefore \qquad v^2 = u^2 + 2uat + a^2t^2$$

$$= u^2 + 2a(ut + \tfrac{1}{2}at^2)$$

But $\qquad s = ut + \tfrac{1}{2}at^2$

$$\therefore \qquad\qquad v^2 = u^2 + 2as \qquad\qquad (4)$$

If we know any *three* of u, v, a, s and t, the others can be found from the equations.

■ *Worked example*

A sprint cyclist starts from rest and accelerates at $1\,\text{m/s}^2$ for 20 seconds. He then travels at a constant speed for 1 minute and finally decelerates at $2\,\text{m/s}^2$ until he stops. Find his maximum speed in km/h and the total distance covered in metres.

First stage

$$u = 0 \quad a = 1\,\text{m/s}^2 \quad t = 20\,\text{s}$$

We have $\quad v = u + at = 0 + 1\,\text{m/s}^2 \times 20\,\text{s}$

$$= 20\,\text{m/s}$$

$$= \frac{20}{1000} \times 60 \times 60 = 72\,\text{km/h}$$

The distance s moved in the first stage is given by

$$s = ut + \tfrac{1}{2}at^2 = 0 \times 20\,\text{s} + \tfrac{1}{2} \times 1\,\text{m/s}^2 \times 20^2\,\text{s}^2$$

$$= \tfrac{1}{2} \times 1\,\frac{\text{m}}{\text{s}^2} \times 400\,\text{s}^2 = 200\,\text{m}$$

Second stage

$$u = 20\,\text{m/s (constant)} \quad t = 60\,\text{s}$$

$$\text{distance moved} = \text{speed} \times \text{time} = 20\,\text{m/s} \times 60\,\text{s}$$

$$= 1200\,\text{m}$$

Third stage

$$u = 20\,\text{m/s} \quad v = 0 \quad a = -2\,\text{m/s}^2 \text{ (a deceleration)}$$

We have

$$v^2 = u^2 + 2as$$

$$\therefore \quad s = \frac{v^2 - u^2}{2a} = \frac{0 - (20)^2\,\text{m}^2/\text{s}^2}{2 \times (-2)\,\text{m/s}^2} = \frac{-400\,\text{m}^2/\text{s}^2}{-4\,\text{m/s}^2}$$

$$= 100\,\text{m}$$

Answers

Maximum speed $= 72\,\text{km/h}$

Total distance covered $= 200\,\text{m} + 1200\,\text{m} + 100\,\text{m}$
$$= 1500\,\text{m}$$

Questions

1 The distance–time graph for a girl on a cycle ride is shown in Figure 29.5.
   a How far did she travel?
   b How long did she take?
   c What was her average speed in km/h?
   d How many stops did she make?
   e How long did she stop for altogether?
   f What was her average speed *excluding* stops?
   g How can you tell from the shape of the graph when she travelled fastest? Over which stage did this happen?

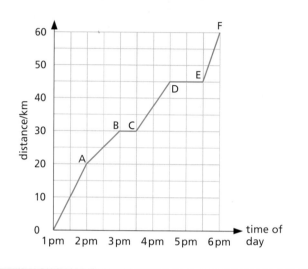

Figure 29.5

2 The graph in Figure 29.6 represents the distance travelled by a car plotted against time.
   a How far has the car travelled at the end of 5 seconds?
   b What is the speed of the car during the first 5 seconds?
   c What has happened to the car after A?
   d Draw a graph showing the speed of the car plotted against time during the first 5 seconds.

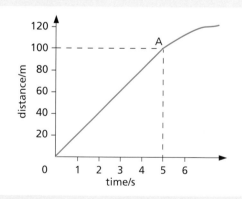

Figure 29.6

3 Figure 29.7 shows an incomplete velocity–time graph for a boy running a distance of 100 m.
   a What is his acceleration during the first 4 seconds?
   b How far does the boy travel during (i) the first 4 seconds, (ii) the next 9 seconds?

c Copy and complete the graph showing clearly at what time he has covered the distance of 100 m. Assume his speed remains constant at the value shown by the horizontal portion of the graph.

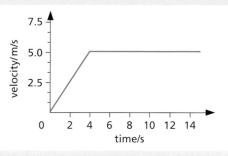

Figure 29.7

4 A car accelerates from 4 m/s to 20 m/s in 8 s. How far does it travel in this time?

5 A motor cyclist travelling at 12 m/s decelerates at 3 m/s².
a How long does he take to come to rest?
b How far does he travel in coming to rest?

6 A hot air balloon called Global Challenger was used to try to break the record for travelling round the world. The graph in Figure 29.8 shows how the height of the balloon changed during the flight.

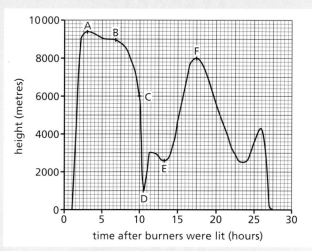

Figure 29.8

The balloon took off from Marrakesh one hour after the burners were lit and climbed rapidly.
a Use the graph to find
  (i) the maximum height reached,
  (ii) the total time of the flight.
b Describe the vertical motion of the balloon
  (i) just before point B on the graph,
  (ii) immediately after point C on the graph.
c Several important moments during the flight are labelled on the graph with the letters A, B, C, D, E and F. At which of these moments did the following happen?
  (i) The balloon began a slow controlled descent to 2500 metres.
  (ii) The crew threw out all the cargo on board in order to stop a very rapid descent.
  (iii) The balloon started to descend from 9000 metres.

(NEAB Higher, June 98)

7 For this question these equations may be useful.

$$v = u + at$$
$$s = ut + \tfrac{1}{2}at^2$$

Figure 29.9 shows the Shuttle spacecraft as it is launched into space.

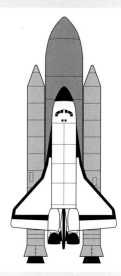

Figure 29.9

During the first eight minutes of the launch the average acceleration of the Shuttle is 17.5 m/s².
a Calculate the speed of the Shuttle after the first 8 minutes. You *must* show how you work out your answer.
b Calculate how far the Shuttle travels in the first 8 minutes. You *must* show how you work out your answer.
c In fact the acceleration of the Shuttle is increasing. Suggest a reason for this. Explain your answer.

(OCR Higher, June 99)

■ *Checklist*

After studying this chapter you should be able to

■ draw, interpret and use velocity–time and distance–time graphs to solve problems,
■ recall and use the four equations of motion for solving problems.

30 Falling bodies

Acceleration of free fall
Measuring g
Distance–time graphs

Projectiles
Practical work
Motion of a falling body.

In air, a coin falls faster than a small piece of paper. In a vacuum they fall at the same rate as may be shown with the apparatus of Figure 30.1. The difference in air is due to **air resistance** having a greater effect on light bodies than on heavy bodies. The air resistance to a light body is big when compared with the body's weight. With a dense piece of metal the resistance is negligible at low speeds.

There is a story, untrue we now think, that in the 16th century the Italian scientist Galileo dropped a small iron ball and a large cannon ball ten times heavier from the top of the Leaning Tower of Pisa, Figure 30.2. And we are told that, to the surprise of onlookers who expected the cannon ball to arrive first, they reached the ground almost simultaneously. You will learn more about air resistance in the next chapter.

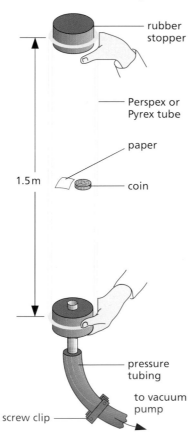

Figure 30.1 A coin and a piece of paper fall at the same rate in a vacuum

Figure 30.2 The Leaning Tower of Pisa, where Galileo is said to have experimented with falling objects

126

Practical work

Motion of a falling body

Arrange things as in Figure 30.3 and investigate the motion of a 100 g mass falling from a height of about 2 m.

Construct a tape chart using *one tick* lengths. Choose as dot '0' the first one you can distinguish clearly. What does the tape chart tell you about the motion of the falling mass? Repeat the experiment with a 200 g mass; what do you note?

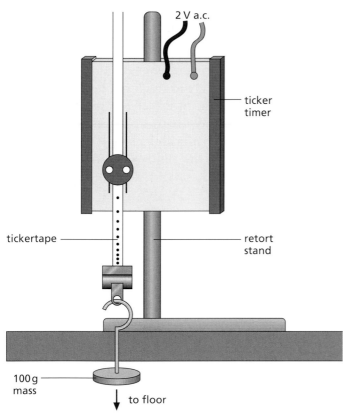

Figure 30.3

■ *Acceleration of free fall*

All bodies falling freely under the force of gravity do so with **uniform acceleration** if air resistance is negligible (i.e. the 'steps' in the previous tape chart should all be equal).

This acceleration, called the **acceleration of free fall**, is denoted by the italic letter *g*. Its value varies over the Earth. In Britain it is about 9.8 m/s^2 or near enough 10 m/s^2. The velocity of a free-falling body therefore increases by 10 m/s every second. A ball shot straight upwards with a velocity of 30 m/s decelerates by 10 m/s every second and reaches its highest point after 3 s.

In calculations using the equations of motion, *g* replaces *a*. It is given a positive sign for falling bodies (i.e. $a = +10$ m/s^2) and a negative sign for rising bodies since they are decelerating (i.e. $a = -10$ m/s^2).

■ *Measuring g*

Using the arrangement in Figure 30.4 the time for a steel ball-bearing to fall a known distance is measured by an electronic timer.

When the two-way switch is changed to the 'down' position, the electromagnet releases the ball and simultaneously the clock starts. At the end of its fall the ball opens the 'trap-door' on the impact switch and the clock stops.

The result is found from the third equation of motion $s = ut + \frac{1}{2}at^2$, where *s* is the distance fallen (in m), *t* is the time taken (in s), $u = 0$ (the ball starts from rest) and $a = g$ (in m/s^2). Hence

$$s = \tfrac{1}{2}gt^2$$

or

$$g = 2s/t^2$$

Air resistance is negligible for a dense object such as a steel ball-bearing falling a short distance.

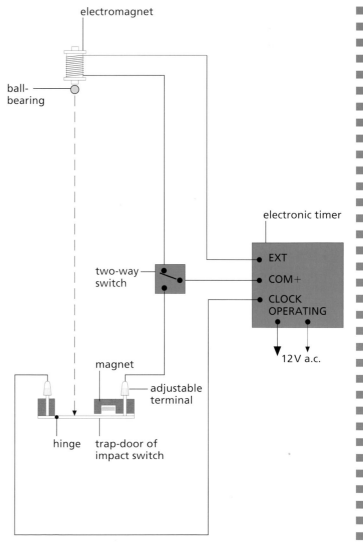

Figure 30.4

■ *Worked example*

A ball is projected vertically upwards with an initial velocity of 30 m/s. Find **a** its maximum height, **b** the time taken to return to its starting point. Neglect air resistance and take $g = 10 \, m/s^2$.

a We have $u = 30 \, m/s$, $a = -10 \, m/s^2$ (a deceleration) and $v = 0$ since the ball is momentarily at rest at its highest point. Substituting in $v^2 = u^2 + 2as$,

$$0 = 30^2 \, m^2/s^2 + 2(-10 \, m/s^2) \times s$$

or

$$-900 \, m^2/s^2 = -s \times 20 \, m/s^2$$

$\therefore$
$$s = \frac{-900 \, m^2/s^2}{-20 \, m/s^2} = 45 \, m$$

b If t is the time to reach the highest point, we have, from $v = u + at$,

$$0 = 30 \, m/s + (-10 \, m/s^2) \times t$$

or

$$-30 \, m/s = -t \times 10 \, m/s^2$$

$\therefore$
$$t = \frac{-30 \, m/s}{-10 \, m/s^2} = 3 \, s$$

The downward trip takes exactly the same time as the upward one and so the answer is 6 s.

■ *Distance–time graphs*

For a body falling freely from rest we have

$$s = \tfrac{1}{2} g t^2$$

A graph of s against t is shown in Figure 30.5a and for s against t^2 in Figure 30.5b. The second graph is a straight line through the origin since $s \propto t^2$ (g being constant at one place).

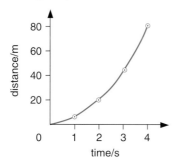

a

b

Figure 30.5 Graphs for a body falling freely from rest

■ *Projectiles*

The photograph in Figure 30.6 was taken while a lamp emitted regular flashes of light. One ball was **dropped** from rest and the other, a 'projectile', was **thrown sideways** at the same time. Their vertical accelerations (due to gravity) are equal, showing that a projectile falls like a body which is dropped from rest. Its horizontal velocity does not affect its vertical motion.

> The horizontal and vertical motions of a body are independent and can be treated separately.

For example if a ball is thrown horizontally from the top of a cliff and takes 3 s to reach the beach below we can calculate the height of the cliff by considering the vertical motion only. We have $u = 0$ (since the ball has no vertical velocity initially), $a = g = +10 \, m/s^2$ and $t = 3 \, s$. The height s of the cliff is given by

$$s = ut + \tfrac{1}{2} a t^2$$

$$= 0 \times 3 \, s + \tfrac{1}{2}(+10 \, m/s^2) 3^2 \, s^2$$

$$= 45 \, m$$

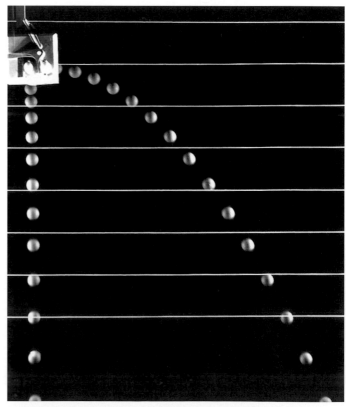

Figure 30.6 Comparing free fall and projectile motion using multiflash photography

Projectiles such as cricket balls and explosive shells are projected from near ground level and at an angle. The horizontal distance they travel, i.e. their **range**, depends on

(i) the **speed** of projection – the greater this is, the greater the range, and

(ii) the **angle** of projection – it can be shown that, neglecting air resistance, the range is a maximum when the angle is 45°, Figure 30.7.

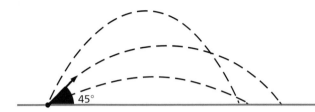

Figure 30.7 The range is greatest for an angle of projection of 45°

Questions

1 A stone falls from rest from the top of a high tower. Ignoring air resistance and taking $g = 10 \, \text{m/s}^2$,
   a what is its velocity after
   (i) 1 s,
   (ii) 2 s,
   (iii) 3 s,
   (iv) 5 s?
   b how far has it fallen after
   (i) 1 s,
   (ii) 2 s,
   (iii) 3 s,
   (iv) 5 s?

2 An object is dropped from a helicopter at a height of 45 m above the ground. If the helicopter is at rest, how long does the object take to reach the ground and what is its velocity on arrival? ($g = 10 \, \text{m/s}^2$)

3 An object is released from an aircraft travelling horizontally with a constant velocity of 200 m/s at a height of 500 m. Ignoring air resistance and taking $g = 10 \, \text{m/s}^2$ find
   a how long it takes the object to reach the ground,
   b the horizontal distance covered by the object between leaving the aircraft and reaching the ground.

4 A gun pointing vertically upwards is fired from an open car B moving with uniform velocity. When the bullet returns to the level of the gun, car B has travelled to C and another car A has reached the position occupied by B when the gun was fired, Figure 30.8. Are the occupants of A or B in danger?

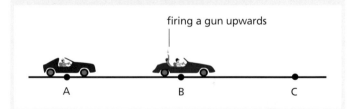

firing a gun upwards

A   B   C

Figure 30.8

■ *Checklist*

After studying this chapter you should be able to

■ describe the behaviour of falling objects and solve problems on them,

■ describe an experiment to find the acceleration of free fall.

31 Newton's laws of motion

■ First law

Friction and air resistance cause a car to come to rest when the engine is switched off. If these forces were absent we believe that a body, once set in motion, would go on moving forever with a constant speed in a straight line. That is, force is not needed to keep a body moving with uniform velocity so long as no opposing forces act on it.

This idea was proposed by Galileo and is summed up in Newton's first law:

> A body stays at rest, or if moving it continues to move with uniform velocity, unless an external force makes it behave differently.

It seems that the question we should ask about a moving body is not 'what keeps it moving' but 'what changes or stops its motion'.

The smaller the external forces opposing a moving body, the smaller is the force needed to keep it moving with uniform velocity. An 'airboard', which is supported by a cushion of air, Figure 31.1, can skim across the ground with little frictional opposition, so that relatively little power is needed to maintain motion.

Figure 31.1 Friction is much reduced for an airboard

■ Mass and inertia

The first law is another way of saying that all matter has a built-in opposition to being moved if it is at rest or, if it is moving, to having its motion changed. This property of matter is called **inertia** (from the Latin word for laziness).

Its effect is evident on the occupants of a car which stops suddenly; they lurch forwards in an attempt to continue moving, and this is why seat belts are needed. The reluctance of a stationary object to move can be shown by placing a large coin on a piece of card on your finger, Figure 31.2. If the card is flicked *sharply* the coin stays where it is while the card flies off.

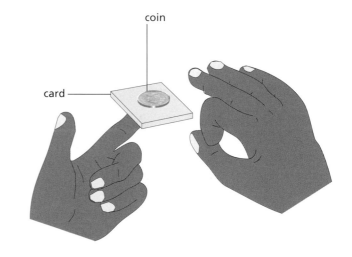
coin
card

Figure 31.2 Flick the card sharply

The larger the mass of a body the greater is its inertia, i.e. the more difficult it is to move it when at rest and to stop it when in motion. Because of this we consider that **the mass of a body measures its inertia**. This is a better definition of mass than the one given earlier (Chapter 17) in which it was stated to be the 'amount of matter' in a body.

Practical work

Effect of force and mass on acceleration

The apparatus consists of a trolley to which a force is applied by a stretched length of elastic, Figure 31.3. The velocity of the trolley is found from a tickertape timer or a motion sensor, datalogger and computer (see Figure 28.6, p. 120).

First compensate the runway for friction by raising one end until the trolley runs down with uniform velocity when given a push. The dots on the tickertape should be equally spaced, or a horizontal trace obtained on a velocity–time graph. There is now no resultant force on the trolley and any acceleration produced later will be due only to the force caused by the stretched elastic.

a) Force and acceleration (mass constant)

Fix one end of a short length of elastic to the rod at the back of the trolley and stretch it until the other end is level with the front of the trolley. Practise pulling the trolley down the runway, keeping the same stretch on the elastic. After a few trials you should be able to produce a steady accelerating force.

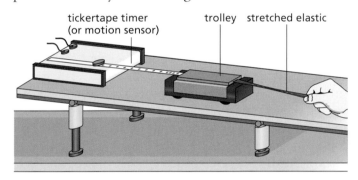

tickertape timer (or motion sensor)   trolley   stretched elastic

Figure 31.3

Repeat using first two and then three *identical* pieces of elastic, stretched side by side by the same amount, to give two and three units of force.

If you are using tickertape make a tape chart for each force and use it to find the acceleration produced in cm/tentick² (see Chapter 28). Ignore the start of the tape (where the dots are too close) and the end (where the force may not be steady). If you use a motion sensor and computer to plot a velocity–time graph, the acceleration can be obtained in m/s² from the slope of the graph (Chapter 29).

Does a steady force cause a steady acceleration? Put the results in a table. Do they suggest any relationship between *a* and *F*?

Force (F) (no. of pieces of elastic)	1	2	3
Acceleration (a) /cm/tentick² or m/s²			

b) Mass and acceleration (force constant)

Do the experiment as in **(a)** using two pieces of elastic (i.e. constant *F*) to accelerate first one trolley, then two (stacked one above the other) and finally three. Check the friction compensation of the runway each time.

Find the accelerations from the tape charts or computer plots and tabulate the results. Do they suggest any relationship between *a* and *m*?

Mass (m) (no. of trolleys)	1	2	3
Acceleration (a) /cm/tentick² or m/s²)			

■ *Second law*

The previous experiment should show roughly that the acceleration *a* is

(i) directly proportional to the applied force *F* for a fixed mass, i.e. $a \propto F$, and
(ii) inversely proportional to the mass *m* for a fixed force, i.e $a \propto 1/m$.

Combining the results into one equation, we get

$$a \propto F/m \quad \text{or} \quad F \propto ma$$

Therefore $\qquad\qquad F = kma$

where *k* is the constant of proportionality.

> One newton is defined as the force which gives a mass of 1 kg an acceleration of 1 m/s², i.e. $1\,\text{N} = 1\,\text{kg m/s}^2$.

So if $m = 1\,\text{kg}$ and $a = 1\,\text{m/s}^2$ then $F = 1\,\text{N}$. Substituting in $F = kma$ we get $k = 1$ and so we can write

$$F = ma$$

This is Newton's second law of motion. When using it two points should be noted. First, *F* is the resultant (or unbalanced) force causing the acceleration *a*. Second, *F* must be in newtons, *m* in kilograms and *a* in metres per second squared, otherwise *k* is not 1. The law shows that *a* will be largest when *F* is large and *m* small.

You should now appreciate that when the forces acting on a body do not balance there is a net (resultant) force which causes a **change** of motion, i.e. the body accelerates or decelerates. If they balance, there is no change of motion. However, there may be a change of shape, in which case internal forces in the body (i.e. forces between neighbouring atoms) balance the external forces.

■ Worked examples

1 A block of mass 2 kg is pushed along a table with a constant velocity by a force of 5 N. When the push is increased to 9 N what is

a the resultant force,

b the acceleration?

When the block moves with constant velocity the forces acting on it are balanced. The force of friction opposing its motion must therefore be 5 N.

a When the push is increased to 9 N the resultant (unbalanced) force F on the block is $(9 - 5)\,\text{N} = 4\,\text{N}$ (since the frictional force is still 5 N).

b The acceleration a is obtained from $F = ma$ where $F = 4\,\text{N}$ and $m = 2\,\text{kg}$.

$$\therefore \quad a = F/m = 4\,\text{N}/2\,\text{kg} = \frac{4\,\text{kg}\,\text{m}/\text{s}^2}{2\,\text{kg}} = 2\,\text{m}/\text{s}^2$$

2 A car of mass 1200 kg travelling at 72 km/h is brought to rest in 4 s. Find

a the average deceleration,

b the average braking force,

c the distance moved during the deceleration.

a The deceleration is found from $v = u + at$ where $v = 0$.

$$u = 72\,\text{km/h} = \frac{72 \times 1000}{60 \times 60} = 20\,\text{m/s}$$

since 1 km = 1000 m and 1 hour = 3600 s

and $\qquad\qquad\qquad t = 4\,\text{s}$

Hence $\qquad\qquad 0 = 20\,\text{m/s} + a \times 4\,\text{s}$

or $\qquad\qquad -20\,\text{m/s} = a \times 4\,\text{s}$

$$\therefore \quad a = -20\,\frac{\text{m}}{\text{s}} \times \frac{1}{4\,\text{s}} = -5\,\text{m/s}^2$$

b The average braking force F is given by $F = ma$ where $m = 1200\,\text{kg}$ and $a = -5\,\text{m/s}^2$. Therefore

$$F = 1200\,\text{kg} \times (-5)\,\text{m/s}^2 = -6000\,\text{kg}\,\text{m/s}^2$$
$$= 6000\,\text{N}$$

c To find the distance moved s we use

$$s = \frac{(u + v)}{2}\,t$$

$$= \frac{(20 + 0)}{2}\,\text{m/s} \times 4\,\text{s}$$

$$= 40\,\text{m}$$

■ Weight and gravity

The weight W of a body is the force of gravity acting on it which gives it an acceleration g when it is falling freely near the Earth's surface. If the body has mass m, then W can be calculated from $F = ma$ if we put $F = W$ and $a = g$ to give

$$W = mg$$

Taking $g = 9.8\,\text{m/s}^2$ and $m = 1\,\text{kg}$, this gives $W = 9.8\,\text{N}$, i.e. a body of mass 1 kg has weight 9.8 N, or near enough 10 N. Similarly a body of mass 2 kg has weight of about 20 N and so on. While the mass of a body is always the same, its weight varies depending on the value of g. On the Moon the acceleration of free fall is only about $1.6\,\text{m/s}^2$, and so a mass of 1 kg has a weight of just 1.6 N there.

The weight of a body is directly proportional to its mass, which explains why g is the same for all bodies. The greater the mass of a body, the greater is the force of gravity on it but it does not accelerate faster when falling because of its greater inertia (i.e. its greater resistance to acceleration).

■ Gravitational field

The force of gravity acts through space and can cause a body, not in contact with the Earth, to fall to the ground. It is an invisible, action-at-a-distance force. We try to 'explain' its existence by saying that the Earth is surrounded by a **gravitational field** which exerts a force on any body in the field. Later, magnetic and electric fields will be considered.

> The strength of a gravitational field is defined as the force acting on unit mass in the field.

Measurement shows that on the Earth's surface a mass of 1 kg experiences a force of 9.8 N, i.e. its weight is 9.8 N. The strength of the Earth's field is therefore 9.8 N/kg (near enough 10 N/kg). It is denoted by g, the letter also used to denote the acceleration of free fall. Hence

$$g = 9.8\,\text{N/kg} = 9.8\,\text{m/s}^2$$

We now have two ways of regarding g. When considering bodies **falling freely** we can think of it as an acceleration of $9.8\,\text{m/s}^2$, but when a body of known mass is **at rest** and we wish to know the force of gravity (in N) acting on it we think of g as the Earth's gravitational field strength of 9.8 N/kg.

■ *Third law*

> If a body A exerts a force on body B, then body B exerts an equal but opposite force on body A.

The law states that forces never occur singly but always in pairs as a result of the action between two bodies. For example, when you step forwards from rest your foot pushes backwards on the Earth, and the Earth exerts an equal and opposite force forward on you. Two bodies and two forces are involved. The small force you exert on the large mass of the Earth gives no noticeable acceleration to the Earth but the equal force it exerts on your very much smaller mass causes you to accelerate.

Note that the pair of equal and opposite forces **do not act on the same body**; if they did, there could never be any resultant forces and acceleration would be impossible. For a book resting on a table, the book exerts a downward force on the table and the table exerts an equal and opposite upward force on the book; this *pair* of forces act on *different* objects and are represented by the red arrows in Figure 31.4. The weight of the book (blue arrow) *does not* form a pair with the upward force on the book (although they are equal numerically) as these two forces act on the *same* body.

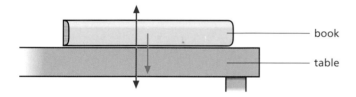

Figure 31.4 Forces between book and table

An appreciation of the third law and the effect of friction is desirable when stepping from a rowing boat, Figure 31.5. You push backwards on the boat and, although the boat pushes you forwards with an equal force, it is itself now moving backwards (because friction with the water is slight). This reduces your forwards motion by the same amount – so you may fall in!

Figure 31.5 The boat moves backwards when you step forwards!

■ *Air resistance: terminal velocity*

When an object falls in air, the air resistance (fluid friction) opposing its motion **increases as its speed rises**, so reducing its acceleration. Eventually, air resistance acting upwards equals the weight of the object acting downwards. The resultant force on the object is then zero since the gravitational force balances the frictional force. The object falls at a constant velocity, called its **terminal velocity**, whose value depends on the size, shape and weight of the object.

A small dense object, e.g. a steel ball-bearing, has a high terminal velocity and falls a considerable distance with a constant acceleration of 9.8 m/s^2 before air resistance equals its weight. A light object, e.g. a raindrop, or one with a large surface area, e.g. a parachute, has a low terminal velocity and only accelerates over a comparatively short distance before air resistance equals its weight. A sky diver, Figure 31.6, has a terminal velocity of more than 50 m/s (180 km/h).

Figure 31.6 Synchronized sky divers

Objects falling in liquids behave similarly (see Chapter 27).

■ *Explanation of Bernoulli's principle*

In Figure 27.10b (p. 112) the liquid speeds up going from the wide part A of the tube to the narrower part B, i.e. it is accelerated. Therefore, since $F = ma$, the force at A, and so also the pressure at A, must be greater than the force and pressure at B. Between B and C the liquid slows down due to the pressure at C being greater than that at B.

Questions

1 Which one of the diagrams in Figure 31.7 shows the arrangement of forces that gives the block of mass *M* the greatest acceleration?

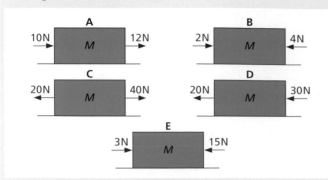

Figure 31.7

2 In Figure 31.8 if *P* is a force of 20 N and the object moves with **constant velocity** what is the value of the opposing force *F*?

Figure 31.8

3 a What resultant force produces an acceleration of 5 m/s² in a car of mass 1000 kg?
  b What acceleration is produced in a mass of 2 kg by a resultant force of 30 N?

4 A block of mass 500 g is pulled from rest on a horizontal frictionless bench by a steady force *F* and travels 8 m in 2 s. Find
  a the acceleration,
  b the value of *F*.

5 Starting from rest on a level road a girl can reach a speed of 5 m/s in 10 s on her bicycle. Find
  a the acceleration,
  b the average speed during the 10 s,
  c the distance she travels in 10 s.
  Eventually, even though she is still pedalling as fast as she can, she stops accelerating and her speed reaches a maximum value. Explain in terms of the forces acting why this happens.

6 What does an astronaut of mass 100 kg weigh
  a on Earth where the gravitational field strength is 10 N/kg,
  b on the Moon where the gravitational field strength is 1.6 N/kg?

7 A rocket has a mass of 500 kg.
  a What is its weight on Earth where *g* = 10 N/kg?
  b At lift-off the rocket engine exerts an upward force of 25 000 N. What is the resultant force on the rocket? What is its initial acceleration?

8 Figure 31.9 shows the forces acting on a raindrop which is falling to the ground.
  a (i) *A* is the force which causes the raindrop to fall. What is this force called?
     (ii) *B* is the total force opposing the motion of the drop. State *one* possible cause of this force.

b What happens to the drop when force *A* = force *B*?

raindrop

Figure 31.9

9 Explain the following using *F* = *ma*.
  a A racing car has a powerful engine and is made of strong but lightweight material.
  b A car with a small engine can still accelerate rapidly.

10 A hot-air balloon is tied to the ground by two ropes to stop it from taking off. Figure 31.10 shows the forces acting on the balloon.

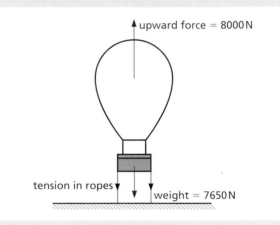

upward force = 8000 N

tension in ropes

weight = 7650 N

Figure 31.10

The ropes are untied and the balloon starts to move upwards.
  a Calculate the size of the unbalanced force acting on the balloon. State the direction of this force.
  b The mass of the balloon is 765 kg. Calculate the initial acceleration of the balloon.
  c Explain how the acceleration of the balloon changes during the first ten seconds of its flight.
  d When the balloon is still accelerating, the balloonist throws some bags of sand over the side. Explain how this affects the acceleration of the balloon.
  (London Higher, June 99)

■ *Checklist*

After studying this chapter you should be able to

■ describe an experiment to investigate the relationship between force, mass and acceleration,
■ define the unit of force,
■ state Newton's three laws of motion and use them to solve problems,
■ define the strength of the Earth's gravitational field,
■ describe the motion of an object falling in air.

32 *Momentum*

Conservation of momentum	***Sport: impulse and collision time***
Explosions	***Discovery of the neutron***
Rockets and jets	***Practical work***
Force and momentum	Collisions and momentum. Explosions and momentum.

Momentum is a useful quantity to consider when bodies are involved in collisions and explosions. It is defined as the **mass of the body multiplied by its velocity** and is measured in kilogram metre per second (kg m/s) or newton second (N s).

$$\text{momentum} = \text{mass} \times \text{velocity}$$

A 2 kg mass moving at 10 m/s has momentum 20 kg m/s, the same as the momentum of a 5 kg mass moving at 4 m/s.

Practical work

Collisions and momentum

Figure 32.1 shows an arrangement which can be used to find the velocity of a trolley before and after a collision. If a trolley of length l takes time t to pass through a photogate its velocity = distance/time = l/t. Two photogates are needed, placed each side of the collision point, to find the velocities before and after the collision. Set them up so that they will record the time taken for the passage of a trolley.

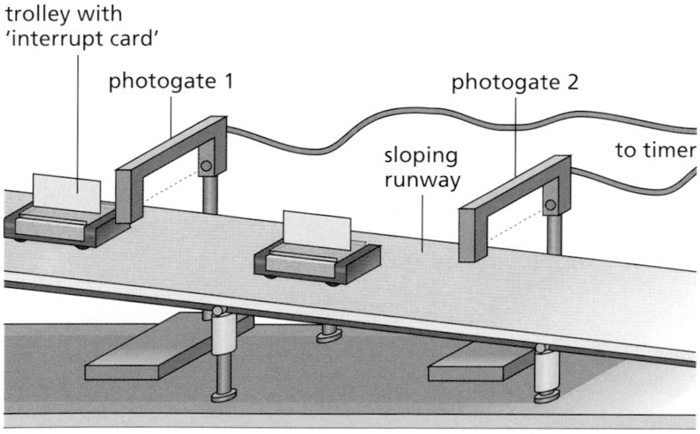

Figure 32.1

Two tickertape timers or two motion sensors placed at each end of the runway could be used instead of the photogates if preferred.

Attach a strip of Velcro to each trolley so that they 'stick' to each other on collision and compensate the runway for friction (see Chapter 31). Place one trolley at rest halfway down the runway and another at the top; give the top trolley a push. It will move forwards with uniform velocity and should hit the second trolley so that they travel on as one. Using the times recorded by the photogate timer, calculate the velocity of the moving trolley before the collision and the common velocity of both trolleys after the collision.

Repeat the experiment with another trolley stacked on top of the one to be pushed so that two are moving before the collision and three after.

Copy and complete the tables of results.

Before collision (m_2 at rest)		
Mass m_1 (no. of trolleys)	**Velocity** v /m/s	**Momentum** m_1v
1		
2		

After collision (m_1 and m_2 together)		
Mass $m_1 + m_2$ (no. of trolleys)	**Velocity** v_1 /m/s	**Momentum** $(m_1 + m_2)v_1$
2		
3		

Do the results suggest any connection between the momentum before the collision and after it?

Conservation of momentum

When two or more bodies act on one another, as in a collision, the total momentum of the bodies remains constant, provided no external forces act (e.g. friction).

This statement is called the **principle of conservation of momentum**. Experiments like those on the previous page show that it is true for all types of collisions.

As an example, suppose a truck of mass 60 kg moving with velocity 3 m/s collides and couples with a stationary truck of mass 30 kg, Figure 32.2a. The two move off together with the same velocity v which we can find as follows, Figure 32.2b.

Total momentum before is

$$(60\,\text{kg} \times 3\,\text{m/s} + 30\,\text{kg} \times 0\,\text{m/s}) = 180\,\text{kg\,m/s}$$

Total momentum after is

$$(60\,\text{kg} + 30\,\text{kg})v = 90\,\text{kg} \times v$$

Since momentum is not lost

$$90\,\text{kg} \times v = 180\,\text{kg\,m/s} \quad \text{or} \quad v = 2\,\text{m/s}$$

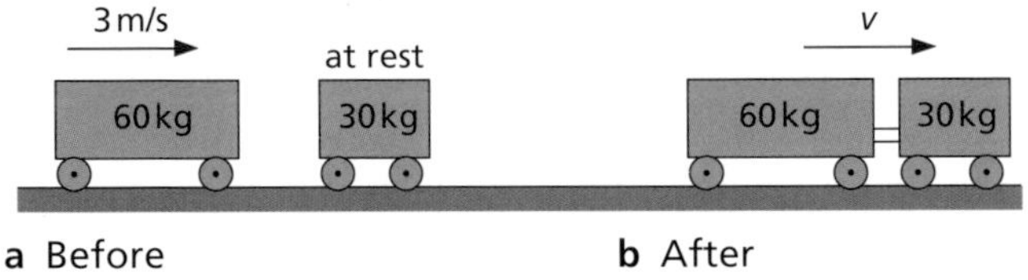

a Before   **b** After
Figure 32.2

Explosions

Momentum, like velocity, is a vector since it has both magnitude and direction. Vectors cannot be added by ordinary addition unless they act in the same direction. If they act in exactly opposite directions, e.g. east and west, the smaller subtracts from the greater, or if the same they cancel out.

Momentum is conserved in an explosion such as occurs when a rifle is fired. Before firing, the total momentum is zero since both rifle and bullet are at rest. During the firing the rifle and bullet receive **equal** but **opposite** amounts of momentum so that the **total** momentum after firing is zero. For example, if a rifle fires a bullet of mass 0.01 kg with a velocity of 300 m/s,

$$\text{forward momentum of bullet} = 0.01\,\text{kg} \times 300\,\text{m/s}$$
$$= 3\,\text{kg\,m/s}$$

$$\therefore \quad \text{backward momentum of rifle} = 3\,\text{kg\,m/s}$$

If the rifle has mass m, it recoils (kicks back) with a velocity v such that

$$mv = 3\,\text{kg\,m/s}$$

Taking $m = 6\,\text{kg}$ gives $v = 3/6\,\text{m/s} = 0.5\,\text{m/s}$.

Practical work

Explosions and momentum

The principle of conservation of momentum can be tested experimentally for 'explosions' with the apparatus in Figure 32.3 arranged as shown. (Two tickertape timers or two motion sensors placed at each end of the runway could be used instead of the photogates if preferred).

Tap one of the buffer rods to release the spring inside. The trolleys fly apart. Work out the velocities v_1 and v_2 of each from the times recorded by the photogate timer.

Since the trolleys are initially at rest

$$\text{total momentum before explosion} = 0$$

If the trolleys have masses m_1 and m_2 then

$$\text{total momentum after explosion} = m_1 v_1 - m_2 v_2$$

If the principle holds for explosions, $m_1 v_1 - m_2 v_2$ should be zero. (For trolleys of equal mass it simplifies matters to take '1 trolley' as the unit of mass (as before), then $m_1 = m_2 = 1$.)

Repeat the experiment with another trolley stacked on top of one of the trolleys so that, for example, $m_1 = 1$ and $m_2 = 2$.

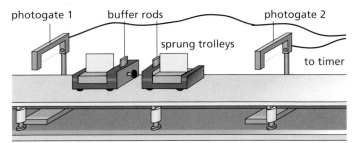

Figure 32.3

Rockets and jets

If you release an inflated balloon with its neck open, it flies off in the opposite direction to that of the escaping air. In Figure 32.4 the air has momentum to the left and the balloon moves to the right with equal momentum.

This is the principle of rockets and jet engines. In both, a high-velocity stream of hot gas is produced by burning fuel and leaves the exhaust with large momentum. The rocket or jet engine itself acquires an equal forward momentum. Space rockets carry their own oxygen supply; jet engines use the surrounding air.

Figure 32.4 A deflating balloon demonstrates the principle of a rocket or a jet engine

Force and momentum

If a steady force F acting on a body of mass m increases its velocity from u to v in time t, the acceleration a is given by

$$a = (v - u)/t \quad \text{(from } v = u + at\text{)}$$

Substituting for a in $F = ma$,

$$F = \frac{m(v - u)}{t} = \frac{mv - mu}{t}$$

Therefore

$$\text{force} = \frac{\text{change of momentum}}{\text{time}} = \frac{\text{rate of change of}}{\text{momentum}}$$

This is another version of the second law of motion. For some problems it is more useful than $F = ma$.

We also have

$$Ft = mv - mu$$

where mv is the final momentum, mu the initial momentum and Ft is called the **impulse**.

Sport: impulse and collision time

The good cricketer or tennis player 'follows through' with the bat or racket when striking the ball, Figure 32.5a. The force applied then acts for a longer time, the impulse is greater and so also is the gain of momentum (and velocity) of the ball.

When we want to stop a moving ball such as a cricket ball, however, its momentum has to be reduced to zero. An impulse is then required in the form of an opposing force acting for a certain time. While any number of combinations of force and time will give a particular impulse, the 'sting' can be removed from the catch by drawing back the hands as the ball is caught, Figure 32.5b. A smaller average force is then applied for a longer time.

Figure 32.5a Batsman 'following through' after hitting the ball

Figure 32.5b Cricketer drawing back the hands on catching the ball

The use of sand gives a softer landing for long-jumpers, Figure 32.6, as a smaller stopping force is applied over a longer time. In a car crash the car's momentum is reduced to zero in a very short time. If the time of impact can be extended by using crumple zones (see Figure 33.6, p. 141) and extensible seat belts, the average force needed to stop the car is reduced so the injury to passengers should also be less.

Figure 32.6 Sand reduces the athlete's momentum more gently

Discovery of the neutron

The principle of conservation of momentum was responsible for leading to the discovery of the neutron in 1932 by Chadwick. He found that in collisions between alpha particles (Chapter 59) and the element beryllium, the principle only held if it was assumed that a particle was produced which had about the same mass as the proton, but no electric charge. It was named the **neutron**.

Questions

1 What is the momentum in kg m/s of a 10 kg truck travelling at
   a 5 m/s,
   b 20 cm/s,
   c 36 km/h?

2 A ball X mass 1 kg travelling at 2 m/s has a head-on collision with an identical ball Y at rest. X stops and Y moves off. What is Y's velocity?

3 A boy with mass 50 kg running at 5 m/s jumps on to a 20 kg trolley travelling in the same direction at 1.5 m/s. What is their common velocity?

4 A girl of mass 50 kg jumps out of a rowing boat of mass 300 kg on to the bank, with a horizontal velocity of 3 m/s. With what velocity does the boat begin to move backwards?

5 A truck of mass 500 kg moving at 4 m/s collides with another truck of mass 1500 kg moving in the same direction at 2 m/s. What is their common velocity just after the collision if they move off together?

6 The velocity of a body of mass 10 kg increases from 4 m/s to 8 m/s when a force acts on it for 2 s.
   a What is the momentum before the force acts?
   b What is the momentum after the force acts?
   c What is the momentum gain per second?
   d What is the value of the force?

7 A rocket of mass 10 000 kg uses 5.0 kg of fuel and oxygen to produce exhaust gases ejected at 5000 m/s. Calculate the increase in its velocity.

8 A rocket launched vertically sends out 50 kg of exhaust gases every second with a velocity of 200 m/s.
   a What is the upward force on the rocket?
   b If the mass of the rocket is 500 kg, what is its initial upward acceleration?

9 a Figure 32.7 shows a golfer about to strike a stationary golf ball of mass 0.045 kg.

Figure 32.7

(i) What is the momentum of the golf ball before it is struck?
(ii) What is the unit of momentum?

b When the golf club strikes the ball it is in contact for 0.001 s and exerts a force of 3600 N on the ball.
   (i) Use the following equation

$$\text{force} = \frac{\text{change in momentum}}{\text{time}}$$

to calculate the velocity at which the ball leaves the club. Show clearly how you work out your final answer and give the unit.
   (ii) Give *two* ways in which the impulse acting on the golf ball can be increased.
   (iii) What is the effect of increasing the impulse on the velocity of the ball as it leaves the club?

c A modern car with a rigid passenger safety cage has zones at the front and rear which are designed to crumple in a crash, Figure 32.8.

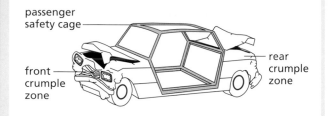

passenger safety cage

front crumple zone

rear crumple zone

Figure 32.8

Use the idea of momentum to explain why 'crumple zones' should reduce passenger injury in a car crash.

(SEG Higher, Summer 98)

■ *Checklist*

After studying this chapter you should be able to

■ define **momentum**,

■ describe experiments to demonstrate the principle of conservation of momentum,

■ state and use the principle of conservation of momentum to solve problems,

■ understand the action of rocket and jet engines,

■ state the relationship between force and rate of change of momentum and use it to solve problems,

■ use the definition of **impulse** to explain how the time of impact affects the force acting in a collision.

33 Kinetic and potential energy

Kinetic energy	Elastic and inelastic collisions	Practical work
Potential energy	Driving and car safety	Change of p.e. to k.e.
Conservation of energy		

Energy and its different forms were discussed earlier (Chapter 24). Here we will consider kinetic energy (k.e.) and potential energy (p.e.) in more detail.

■ Kinetic energy

Kinetic energy is the energy a body has because of its motion.

We can obtain an expression for k.e. Suppose a body of mass m starts from rest and is acted on by a steady force F which gives it a uniform acceleration a. If the velocity of the body is v when it has travelled a distance s, then, using $v^2 = u^2 + 2as$, we get, since $u = 0$,

$$v^2 = 2as \quad \text{or} \quad a = v^2/2s$$

Substituting in $F = ma$,

$$F = m\left(\frac{v^2}{2s}\right) \quad \text{or} \quad Fs = \tfrac{1}{2}mv^2$$

Fs is the work done on the body to give it velocity v and therefore equals its k.e.

$$\therefore \quad \text{kinetic energy} = E_k = \tfrac{1}{2}mv^2$$

If m is in kg and v in m/s, then k.e. is in J. For example, a football of mass 0.4 kg (400 g) moving with velocity 20 m/s has

$$\begin{aligned} \text{k.e.} &= \tfrac{1}{2}mv^2 = \tfrac{1}{2} \times 0.4\,\text{kg} \times (20)^2\,\text{m}^2/\text{s}^2 \\ &= 0.2 \times 400\,\text{kg}\,\text{m/s}^2 \times \text{m} \\ &= 80\,\text{N}\,\text{m} = 80\,\text{J} \end{aligned}$$

Since k.e. depends on v^2, a high-speed vehicle travelling at 1000 km/h, Figure 33.1, has one hundred times the k.e. it has at 100 km/h.

■ Potential energy

Potential energy is the energy a body has because of its position or condition.

A body above the Earth's surface is considered to have an amount of gravitational p.e. equal to the work that has been done against gravity by the force used to raise it. To lift a body of mass m through a vertical height h at a place where the Earth's gravitational field strength is g, needs a force equal and opposite to the weight mg of the body. Hence

$$\begin{aligned} \text{work done by force} &= \text{force} \times \text{vertical height} \\ &= mg \times h \end{aligned}$$

$$\therefore \quad \text{potential energy} = E_p = mgh$$

When m is in kg, g in N/kg (or m/s^2) and h in m, the p.e. is in J. For example, if $g = 10$ N/kg, the p.e. gained by a 0.1 kg (100 g) mass raised vertically by 1 m is $0.1\,\text{kg} \times 10\,\text{N/kg} \times 1\,\text{m} = 1\,\text{N}\,\text{m} = 1\,\text{J}$.

Note Strictly speaking we are concerned with *changes* in p.e. from that which a body has at the Earth's surface, rather than with actual values. The expression for p.e. is therefore more correctly written

$$\Delta E_p = mgh$$

where Δ (pronounced delta) stands for 'a change in'.

Figure 33.1 Kinetic energy depends on the square of the velocity

Practical work

Change of p.e. to k.e.

Friction-compensate a runway and arrange the apparatus as in Figure 33.2 with the bottom of the 0.1 kg (100 g) mass 0.5 m from the floor.

Start the timer and release the trolley. It will accelerate until the falling mass reaches the floor; after that it moves with *constant* velocity v.

From your results calculate v in m/s (on the tickertape 50 ticks = 1 s). Find the mass of the trolley in kg. Work out:

k.e. gained by trolley and 0.1 kg mass =   J
p.e. lost by 0.1 kg mass =   J

Compare and comment on the results.

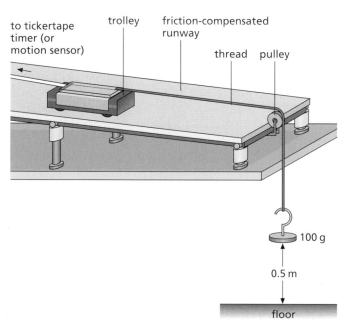

Figure 33.2

■ *Conservation of energy*

A mass m at height h above the ground has p.e. = mgh, Figure 33.3. When it falls, its velocity increases and it gains k.e. at the expense of its p.e. If it starts from rest and air resistance is negligible, its velocity v on reaching the ground is given by

$$v^2 = u^2 + 2as = 0 + 2gh = 2gh$$

Also, as it reaches the ground, its k.e. is

$$\tfrac{1}{2}mv^2 = \tfrac{1}{2}m \times 2gh = mgh$$

∴   loss of p.e. = gain of k.e.

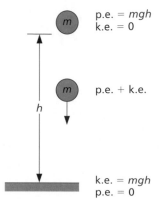

Figure 33.3 Loss of p.e. = gain of k.e.

This is an example of the **principle of conservation of energy** which was discussed in Chapter 24.

In the case of a pendulum k.e. and p.e. are interchanged continually. The energy of the bob is all p.e. at the end of the swing and all k.e. as it passes through its central position. In other positions it has both p.e. and k.e., Figure 33.4. Eventually all the energy is changed to heat as a result of overcoming air resistance.

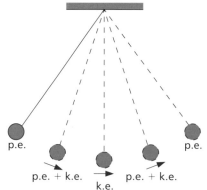

Figure 33.4 Interchange of p.e. and k.e. for a simple pendulum

■ *Elastic and inelastic collisions*

In all collisions (where no external force acts) momentum is conserved but there is normally a loss of k.e., usually to heat energy and to a small extent to sound energy. The greater the proportion of k.e. lost, the less **elastic** is the collision, i.e. the more **inelastic** it is. In a perfectly elastic collision k.e. is conserved.

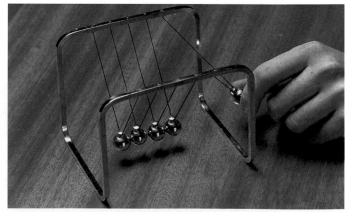

Figure 33.5 Newton's cradle is an instructive toy for studying collisions and conservation

Driving and car safety

a) Braking distance and speed

In the equation derived earlier

$$Fs = \tfrac{1}{2}mv^2$$

if F is the **steady** braking force applied to a car of mass m moving with speed v, then s is the braking distance needed to stop it. Since F and m are constant we can say

$$s \propto v^2$$

That is, the **braking distance is directly proportional to the square of the speed**, i.e. if v is doubled, s is quadrupled. The **thinking distance** (i.e. the distance travelled while the driver is reacting before applying the brakes) has to be added to the **braking distance** to obtain the **overall stopping distance**, i.e.

$$\frac{\text{stopping}}{\text{distance}} = \frac{\text{thinking}}{\text{distance}} + \frac{\text{braking}}{\text{distance}}$$

Typical values taken from the Highway Code are given in Table 33.1 for different speeds. The greater the speed the greater the stopping distance for a given braking force. (To stop the car in a given distance a greater braking force would be needed for higher speeds.)

Thinking distance depends on the driver's reaction time – this will vary with factors such as the driver's degree of tiredness, use of alcohol or drugs, eyesight and the visibility of the hazard. Braking distance varies with both the road conditions and the state of the car; it is longer when the road is wet or icy, when friction between the tyres and the road is low, than when conditions are dry. Efficient brakes and high tyre tread help to reduce the braking distance.

Table 33.1

Speed/mph	20	40	60	80
Thinking distance/metres	6	12	18	24
Braking distance/metres	6	24	54	96
Total stopping distance/metres	12	36	72	120

b) Car design and safety

When a car stops rapidly in a collision, large forces are produced on the car and its passengers, and their k.e. has to be dissipated.

Crumple zones at the front and rear collapse in such a way that the k.e. is absorbed gradually, Figure 33.6. As we saw in Chapter 32 this extends the collision time and reduces the decelerating force and hence the potential for injury to the passengers.

Extensible seat belts exert a backwards force (of 10 000 N or so) over about 0.5 m, which is roughly the distance between the front seat occupants and the windscreen. In a car travelling at 15 m/s (34 mph), the effect felt by anyone *not* using a seat belt is the same as that produced by jumping off a building 12 m high!

Air bags in some cars inflate and protect the driver from injury by the steering wheel.

Head restraints ensure that if the car is hit from behind, the head goes forwards with the body and this prevents damage to the top of the spine.

All these are **secondary** safety devices which aid **survival** in the event of an accident. **Primary** safety factors help to **prevent** accidents and depend on the car's roadholding, brakes, steering, handling and above all on the driver since most accidents are due to driver error.

The chance of being killed in an accident is about **five times less** if seat belts are worn and head restraints are installed.

Figure 33.6 Car in an impact test showing the collapse of the front crumple zone

Worked example

A boulder of mass 4 kg rolls over a cliff and reaches the beach below with a velocity of 20 m/s.

a What is the k.e. of the boulder just before it lands?
b What is its p.e. on the cliff?
c How high is the cliff?

a Mass of boulder $= m = 4\,\text{kg}$
Velocity of boulder as it lands $= v = 20\,\text{m/s}$

$$\therefore \quad \text{k.e. of boulder as it lands} = E_k = \tfrac{1}{2}mv^2$$
$$= \tfrac{1}{2} \times 4\,\text{kg} \times (20)^2\,\text{m}^2/\text{s}^2$$
$$= 800\,\text{kg m/s}^2 \times \text{m} = 800\,\text{N m}$$
$$= 800\,\text{J}$$

b Applying the principle of conservation of energy (and neglecting energy lost in overcoming air resistance),

$$\text{p.e. of boulder on cliff} = \text{k.e. as it lands}$$
$$\therefore \quad \Delta E_p = E_k = 800\,\text{J}$$

c If h is the height of the cliff,

$$\Delta E_p = mgh$$
$$\therefore \quad h = \frac{\Delta E_p}{mg} = \frac{800\,\text{J}}{4\,\text{kg} \times 10\,\text{m/s}^2} = \frac{800\,\text{N m}}{40\,\text{kg m/s}^2}$$
$$= \frac{800\,\text{kg m/s}^2 \times \text{m}}{40\,\text{kg m/s}^2} = 20\,\text{m}$$

Questions

1 Calculate the k.e. of
   a a 1 kg trolley travelling at 2 m/s,
   b a 2 g (0.002 kg) bullet travelling at 400 m/s,
   c a 500 kg car travelling at 72 km/h.

2 **a** What is the velocity of an object of mass 1 kg which has 200 J of k.e.?
   b Calculate the p.e. of a 5 kg mass when it is (i) 3 m, (ii) 6 m, above the ground. ($g = 10$ N/kg)

3 A 100 g steel ball falls from a height of 1.8 m onto a metal plate and rebounds to a height of 1.25 m. Find
   a the p.e. of the ball before the fall ($g = 10$ m/s^2),
   b its k.e. as it hits the plate,
   c its velocity on hitting the plate,
   d its k.e. as it leaves the plate on the rebound,
   e its velocity of rebound.

4 It is estimated that 7×10^6 kg of water pours over the Niagara Falls every second. If the Falls are 50 m high, and if all the energy of the falling water could be harnessed, what power would be available? ($g = 10$ N/kg)

5 **a** A student claims that

> When the velocity of a 500 kg car changes from 5 m/s to 10 m/s its momentum will double...

Check the student's claim by using this equation.

momentum = mass × velocity

momentum at 5 m/s = ... kgm/s
momentum at 10 m/s = ... kgm/s

   b Another student has written this.

> A 500 kg car travelling at 10 m/s has twice the kinetic energy of a 500 kg car travelling at 5 m/s.

Use the equation below to show that this claim is **incorrect**.

kinetic energy = $\frac{1}{2}mv^2$

   c Describe the relationship between the speed of the car and its braking distance.
   d Su-Ling drives along a straight road at a speed of 30 m/s. She sees a broken-down lorry completely blocking the road 350 m away. Her car stops 50 m before the lorry. Use the equation shown below to calculate the deceleration of the car. You *must* show how you work out your answer.

$$v^2 = u^2 + 2as$$
(*OCR Higher, June 99*)

6 The Highway Code gives tables of the shortest stopping distances for cars travelling at various speeds. An extract from the Highway Code is given in Figure 33.7.

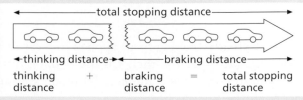

Figure 33.7

   a A driver's reaction time is 0.7 s. Calculate the 'thinking distance' when travelling at 10 m/s.
   b (i) Write down *two* factors which could increase a driver's reaction time.
   (ii) What effect does an increase in reaction time have on
   A thinking distance,
   B braking distance,
   C total stopping distance?
   c Explain why the braking distance would change on a wet road.
   d A car was travelling at 30 m/s. The driver braked. The graph in Figure 33.8 is a velocity–time graph showing the velocity of the car during braking.

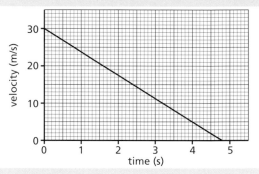

Figure 33.8

Calculate
(i) the rate at which the velocity decreases (deceleration),
(ii) the braking force, if the mass of the car is 900 kg,
(iii) the braking distance.
(*NEAB Higher, June 98*)

■ *Checklist*

After studying this chapter you should be able to

- define **kinetic energy** (k.e.),
- perform calculations using $E_k = \frac{1}{2}mv^2$,
- define **potential energy** (p.e.),
- calculate changes in p.e. using $\Delta E_p = mgh$,
- apply the principle of conservation of energy to simple mechanical systems, e.g. a swinging pendulum,
- recall the meaning of **elastic** and **inelastic** collisions,
- recall that the braking distance of a vehicle is proportional to the square of the speed,
- describe secondary safety devices in cars.

34 Circular motion

Centripetal force	Looping the loop	Practical work
Rounding a bend	Satellites	Investigating circular motion.

There are many examples of bodies moving in circular paths – rides at a fun fair, clothes being spun dry in a washing machine, the planets going round the Sun and the Moon circling the Earth. When a car turns a corner it may follow an arc of a circle. 'Throwing the hammer' is a sport practised at Highland Games in Scotland, Figure 34.1, in which the hammer is whirled round and round before it is released.

Figure 34.1 'Throwing the hammer'

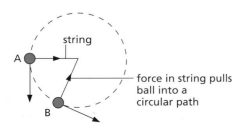

Figure 34.2

■ Centripetal force

In Figure 34.2 a ball attached to a string is being whirled round in a horizontal circle. Its direction of motion is constantly changing. At A it is along the tangent at A; shortly afterwards, at B, it is along the tangent at B; and so on.

Velocity has both size and direction; speed has only size. Velocity is speed in a stated direction and if the direction of a moving body changes, even if its speed does not, then its velocity has changed. A change of velocity is an acceleration and so during its whirling motion the ball is accelerating.

It follows from Newton's first law of motion that if we consider a body moving in a circle to be accelerating then there must be a force acting on it to cause the acceleration. In the case of the whirling ball it is reasonable to say the force is provided by the string pulling inwards on the ball. Like the acceleration, the force acts towards the centre of the circle and keeps the body at a fixed distance from the centre.

A larger force is needed if

(i) the speed v of the ball is increased,
(ii) the radius r of the circle is decreased,
(iii) the mass m of the ball is increased.

The rate of change of direction, and so the acceleration a, is increased by (i) and (ii). It can be shown that $a = v^2/r$ and so, from $F = ma$, we can write

$$F = \frac{mv^2}{r}$$

This force which acts **towards the centre** and keeps a body moving in a circular path is called the **centripetal force** (centre-seeking force).

Should the force be greater than the string can bear, the string breaks and the ball flies off with steady speed in a straight line **along the tangent**, i.e. in the direction of travel when the string broke (as the first law of motion predicts). It is not thrown outwards.

Whenever a body moves in a circle (or circular arc) there must be a centripetal force acting on it. In throwing the hammer it is the pull of the athlete's arms acting on the hammer towards the centre of the whirling path. When a car rounds a bend a frictional force is exerted inwards by the road on the car's tyres.

Practical work

Investigating circular motion

Use the apparatus in Figure 34.3 to investigate the various factors which affect circular motion. Make sure the rubber bung is **tied securely** to the string and that **the area around you is clear of other students**. The paper clip acts as an indicator to aid keeping the radius of the circular motion constant.

Spin the rubber bung at a constant speed while adding more weights to the holder; it will be found that the radius of the orbit decreases. Show that if the rubber bung is spun faster, more weights must be added to the holder to keep the radius constant. Are these findings in agreement with the formula given above for the centripetal force $F = mv^2/r$?

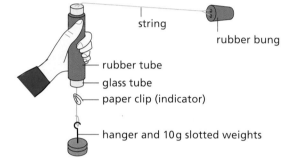

string
rubber bung
rubber tube
glass tube
paper clip (indicator)
hanger and 10 g slotted weights

Figure 34.3

■ *Rounding a bend*

When a car rounds a bend a frictional force is exerted **inwards** by the road on the car's tyres so providing the centripetal force needed to keep it in its curved path, Figure 34.4a. Here friction acts as an accelerating force (towards the centre of the circle) rather than a retarding force (p.95, p.130). The successful negotiation of a bend on a flat road therefore depends on the tyres and the road surface being in a condition that enables them to provide a sufficiently large frictional force – otherwise skidding occurs.

Safe cornering that does not rely entirely on friction is achieved by 'banking' the road as in Figure 34.4b. Some of the centripetal force is then supplied by the part of the contact force N from the road surface on the car which acts horizontally. A bend in a railway track is banked so that the outer rail is not strained by having to supply the centripetal force, by pushing inwards on the wheel flanges.

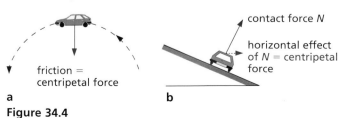

friction = centripetal force

contact force N

horizontal effect of N = centripetal force

a
b

Figure 34.4

■ *Looping the loop*

A pilot who is not strapped into his aircraft can loop the loop without falling downwards at the top of the loop. A bucket of water can be swung round in a vertical circle without spilling. Some amusement park rides, Figure 34.5, give similar effects. Can you suggest what provides the centripetal force for each of these three cases (i) at the top of the loop and (ii) at the bottom of the loop?

Figure 34.5 Looping the loop at an amusement park

■ *Satellites*

For a satellite of mass m orbiting the Earth at radius r with orbital speed v, the centripetal force $F = mv^2/r$ is provided by gravity.

To put an artificial satellite in orbit at a certain height above the Earth it must enter the orbit at the correct speed. If it does not, the force of gravity, which decreases with height, will not be equal to the centripetal force needed for the orbit.

This can be seen by imagining a shell fired horizontally from the top of a very high mountain, Figure 34.6. If gravity did not pull it towards the centre of the Earth it would continue to travel horizontally, taking path A. In practice it might take path B. A second shell fired faster might take path C and travel farther. If a third shell is fired even faster, it might never catch up with the rate at which the Earth's surface is falling away. It would remain at the same height above the Earth (path D) and return to the mountain top, behaving like a satellite.

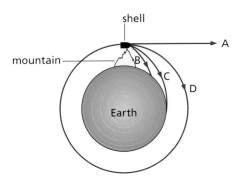

Figure 34.6

The orbital period T (the time for one orbit) of a satellite = distance/velocity. So for a circular orbit

$$T = \frac{2\pi r}{v}$$

Satellites in high orbits have longer periods than those in low orbits.

The Moon is kept in a circular orbit round the Earth by the force of gravity between it and the Earth. It has an orbital period of 27 days.

a) Communication satellites

These circle the Earth in orbits above the equator. **Geostationary** satellites (see Figure 2c, p. viii) have an orbit high above the equator (36 000 km); they travel with the same speed as the Earth so appear to be stationary at a particular point above the Earth's surface – their orbital period is 24 hours. They are used for transmitting television, intercontinental telephone and data signals (Chapter 61). Geostationary satellites need to be well separated so that they do not interfere with each other; there is room for about 400.

Mobile phone networks use many satellites in much lower equatorial orbits; they are slowed by the Earth's atmosphere and their orbit has to be regularly adjusted by firing a rocket engine. Eventually they run out of fuel and burn up in the atmosphere as they fall to Earth.

b) Monitoring satellites

These circle the Earth rapidly in low **polar** orbits, i.e. passing over both poles; at a height of 850 km the orbital period is only 100 minutes. The Earth rotates below them so they scan the whole surface at short range in a 24 hour period and can be used to map or monitor regions of the Earth's surface which may be inaccessible by other means. They are widely used in weather forecasting to transmit infrared pictures of cloud patterns continuously down to Earth, Figure 34.7, which are picked up in turn by receiving stations around the world.

Figure 34.7 Satellite image of cloud over Europe

c) Orbital relationships

For a geostationary satellite at height 36 000 km above the Earth's surface, $r = (36\,000 + 6000)\,\text{km} = 42\,000\,\text{km} = 42 \times 10^6\,\text{m}$ (taking the Earth's radius as 6000 km) and $T = 24\,\text{hours} = 24 \times 3600\,\text{s}$. Hence the orbital velocity v is

$$v = \frac{2\pi r}{T} = \frac{2\pi \times 42 \times 10^6\,\text{m}}{24 \times 3600\,\text{s}}$$

$$= 3.1 \times 10^3\,\text{m/s} = 3.1\,\text{km/s}$$

If F is the gravitational force per kg to keep the satellite in orbit, then

$$F = \frac{mv^2}{r} = \frac{1.0\,\text{kg}}{42 \times 10^6\,\text{m}} \times (3.1 \times 10^3)^2\,\frac{\text{m}^2}{\text{s}^2}$$

$$= \frac{9.6}{42}\,\text{kg m/s}^2 = 0.23\,\text{N}$$

Questions

1 An apple is whirled round in a horizontal circle on the end of a string which is tied to the stalk. It is whirled faster and faster and at a certain speed the apple is torn from the stalk. Why?

2 A car rounding a bend travels in an arc of a circle.
   a What provides the centripetal force?
   b Is a larger or a smaller centripetal force required if
      (i) the car travels faster,
      (ii) the bend is less curved,
      (iii) the car has more passengers?

3 Racing cars are fitted with tyres called 'slicks' (which have no tread pattern) for dry tracks and 'treads' for wet tracks. Why?

4 A satellite close to the Earth (at a height of about 200 km) has an orbital speed of 8 km/s. Taking the radius of the orbit as approximately equal to the Earth's radius of 6400 km, calculate the time it takes to make one orbit.

5 Figure 34.8 shows a satellite in orbit around the Earth.

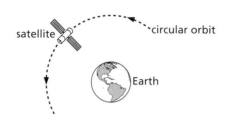

Figure 34.8

   a Copy the diagram and draw an arrow to show the direction of the centripetal force which acts on the satellite.
   b Use words from the following list to complete the sentences.

   greater   less   unchanged

   (i) If the mass of the satellite decreases then the centripetal force needed is ......
   (ii) If the speed of the satellite increases then the centripetal force needed is ......
   (iii) If the radius of the orbit increases then the centripetal force needed is ......
   (AQA (NEAB) Foundation, June 99)

6 Figure 34.9 shows **circular** orbits for two satellites around the Earth.

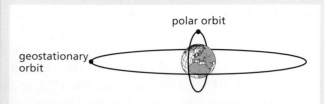

Figure 34.9

   a A centripetal force is needed to make an object travel in a circular path. What provides the centripetal force which keeps the satellites in orbit around the Earth?
   b If the two satellites have the same mass, explain why the centripetal forces are different.
   c (i) How long does it take a geostationary satellite to complete one orbit?
      (ii) How is the orbital time of the polar satellite shown in the diagram different to that of the geostationary one? Explain your answer.
   d Suggest *one* use of a satellite in
      (i) a geostationary orbit,
      (ii) a polar orbit.
   e The Hubble Telescope is in orbit round the Earth. What is the advantage of this telescope over telescopes on Earth?
   (NEAB Higher, June 98)

■ *Checklist*

After studying this chapter you should be able to

■ explain circular motion in terms of an unbalanced **centripetal force**,

■ describe an experiment to investigate the factors affecting circular motion,

■ explain how the centripetal force arises for a car rounding a bend,

■ understand satellite motion.

Motion and energy
Additional questions

Velocity and acceleration; graphs and equations

1 The approximate velocity–time graph for a car on a 5-hour journey is shown below. (There is a very quick driver change midway to prevent driving fatigue!)
  a State in which of the regions OA, AB, BC, CD, DE the car is (i) accelerating, (ii) decelerating, (iii) travelling with uniform velocity.
  b Calculate the value of the acceleration, deceleration or constant velocity in each region.
  c What is the distance travelled over each region?
  d What is the total distance travelled?
  e Calculate the average velocity for the whole journey.

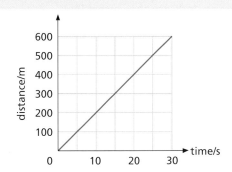

2 The distance–time graph for a motor cyclist riding off from rest is shown below.
  a Describe the motion.
  b How far does the motorbike move in 30 seconds?
  c Calculate the speed.

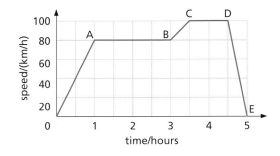

3 The speeds of a car travelling on a straight road are given below at successive intervals of 1 second.

Time/s	0	1	2	3	4
Speed/m/s	0	2	4	6	8

The car travels
1 with an average velocity of 4 m/s
2 16 m in 4 s
3 with a uniform acceleration of 2 m/s².
Which statement(s) is (are) correct?

A 1, 2, 3   **B** 1, 2   **C** 2, 3   **D** 1   **E** 3

4 If a train travelling at 10 m/s starts to accelerate at 1 m/s² for 15 s on a straight track, its final velocity in m/s is

A 5   **B** 10   **C** 15   **D** 20   **E** 25

5 A small pebble is thrown **horizontally** at 20 m/s from the top of a cliff 80 m high. Ignoring air resistance and taking $g = 10$ m/s², calculate
  a how long the pebble takes to reach the ground,
  b the distance from the foot of the cliff to where the pebble strikes the ground,
  c the vertical velocity of the pebble as it hits the ground, and
  d the horizontal velocity of the pebble as it hits the ground.

6 Two students Anna and Graham took part in a sponsored run. The distance–time graph for Graham's run is shown. Four points have been labelled A, B, C and D.

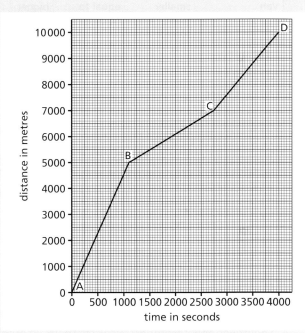

  a Between which pair of points was Graham running the slowest?
  b Anna did not start the run until 10 minutes after Graham. She completed the whole run at a constant speed of 4 m/s.
  (i) Write down the equation that links distance, speed and time.
  (ii) Calculate, in seconds, how long it took Anna to complete the run. Show clearly how you work out your answer.
  (iii) Copy the graph above and draw a line to show Anna's run.
  (iv) How far had Graham run when he was overtaken by Anna?

(SEG Higher, Summer 98)

b Some communications satellites are geostationary. They remain above the same point on the Earth's surface.
(i) Write down the orbit time of a goestationary satellite.
(ii) Use the graph to find the height above the Earth's surface of a geostationary satellite.

c Some weather satellites occupy very low orbits, close to the surface of the Earth. Describe *two* advantages of a low orbit for a weather satellite.

(London Foundation, June 98)

13 The diagram shows an astronaut and a space telescope. He is stationary relative to the telescope. He is about to use the thrusters on his backpack to move towards the telescope.

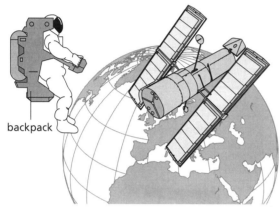

backpack

not to scale

a He switches on the thrusters. In a short burst, 0.020 kg of gas is ejected backwards at 300 m/s.
(i) Calculate the momentum gained by the gas. Use the equation below. You *must* show how you work out your answer.

momentum = mass × velocity

(ii) What is the momentum gained by the fully equipped astronaut in this time? Explain your answer.
(iii) The thrusters give the astronaut a velocity of 0.06 m/s towards the telescope. Calculate the mass of the fully equipped astronaut. You *must* show how you work out your answer.

b The astronaut arrives at the telescope. He uses a long spanner to undo a nut. On Earth he can do this easily. In space it is more difficult. Use your ideas about forces to explain why.

(OCR Higher, June 99)

14 The data in the table below shows how the stopping distance of a car depends on its speed.
a Write down *two* factors, apart from speed, that affect the stopping distance of a car.
b Use the data to draw a graph of stopping distance against speed.

Stopping distance (m)	0	4	12	22	36	52	72
Speed (m/s)	0	5	10	15	20	25	30

c The speed limit in a supermarket car park is 7.5 m/s. Use your graph to estimate the stopping distance of a car travelling at this speed.
d Describe how the stopping distance changes as the speed of a car increases.
e Explain why the stopping distance is reduced if a driver is made aware of possible hazards.
f The speed limit on roads in towns is 15 m/s. Some road safety campaigners are asking the government to change this to 10 m/s. Suggest why this may be a good idea.

(London Higher, June 99)

15 The diagram shows an orbiter, the reusable part of a space shuttle. The data refers to a typical flight.

Orbiter data	
Mass	78 000 kg
Orbital speed	7.5 km/s
Orbital altitude	200 km
Landing speed	100 m/s
Flight time	7 days

a (i) What name is given to the force which keeps the orbiter in orbit around the Earth?
(ii) Use the following equation to calculate the kinetic energy, in joules, of the orbiter while it is in orbit.

$$\text{kinetic energy} = \tfrac{1}{2}mv^2$$

(iii) What happens to most of this kinetic energy as the orbiter re-enters the Earth's atmosphere?

b After touchdown the orbiter decelerates uniformly coming to a halt in 50 s.
(i) Give the equation that links acceleration, time and velocity.
(ii) Calculate the deceleration of the orbiter. Show clearly how you work out your answer and give the unit.

c (i) Give the equation that links acceleration, force and mass.
(ii) Calculate, in newtons, the force needed to bring the orbiter to a halt. Show clearly how you work out your answer.

(SEG Higher, Summer 98)

Heat and energy

35 Thermometers

The **temperature** of a body tells us how hot the body is. It is measured by a thermometer, usually in **degrees Celsius** (°C). The kinetic theory (Chapter 20) regards temperature as a measure of the average k.e. of the molecules of the body. The greater this is, the faster the molecules move and the higher the temperature of the body.

There are different kinds of thermometer, each type being more suitable than another for a certain job. Figure 35.1 shows the temperature of a lava flow being measured.

Figure 35.1 Use of a thermocouple probe thermometer to measure a temperature of about 1160 °C

Liquid-in-glass thermometer

In this type the liquid in a glass bulb expands up a capillary tube when the bulb is heated. The liquid must be easily seen and must expand (or contract) rapidly and by a large amount over a wide range of temperature. It must not stick to the inside of the tube or the reading will be too high when the temperature is falling.

Mercury and coloured alcohol are in common use. Mercury freezes at −39 °C and boils at 357 °C; alcohol freezes at −115 °C and boils at 78 °C and is therefore more suitable for low temperatures.

Scale of temperature

A scale and unit of temperature are obtained by choosing two temperatures, called the **fixed points**, and dividing the range between them into a number of equal divisions or **degrees**.

On the Celsius scale (named after the Swedish scientist who suggested it), **the lower fixed point is the temperature of pure melting ice** and is taken as 0 °C. **The upper fixed point is the temperature of the steam above water boiling at normal atmospheric pressure**, 10^5 Pa (or N/m^2), and is taken as 100 °C.

When the fixed points have been marked on the thermometer, the distance between them is divided into 100 equal degrees, Figure 35.2. The thermometer now has a scale, i.e. it has been calibrated or graduated.

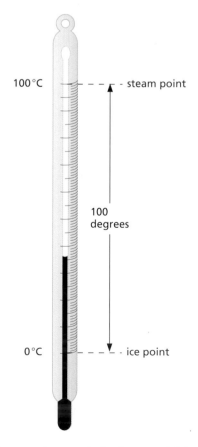

Figure 35.2 A temperature scale in degrees Celsius

Clinical thermometer

A clinical thermometer is a special type of mercury-in-glass thermometer used by doctors and nurses. Its scale only extends over a few degrees on either side of the normal body temperature of 37°C, Figure 35.3.

The tube has a constriction (i.e. a narrower part) just beyond the bulb. When the thermometer is placed under the tongue the mercury expands, forcing its way past the constriction. When the thermometer is removed (after 1 minute) from the mouth, the mercury in the bulb cools and contracts, breaking the mercury thread at the constriction. The mercury beyond the constriction stays in the tube and shows the body temperature. After use the mercury is returned to the bulb by a flick of the wrist.

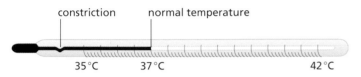

Figure 35.3 A clinical thermometer

Thermocouple thermometer

A thermocouple consists of two wires of different materials, e.g. copper and iron, joined together, Figure 35.4. When one junction is at a higher temperature than the other an electric current flows and produces a reading on a sensitive meter which depends on the temperature difference.

Thermocouples are used in industry to measure a wide range of temperatures from −250°C up to about 1500°C, especially rapidly changing ones and those of small objects.

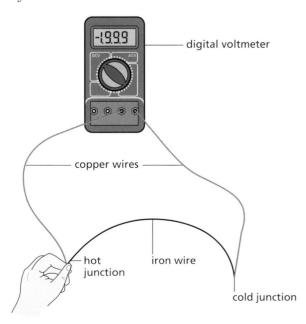

Figure 35.4 A simple thermocouple thermometer

Other thermometers

One type of **resistance thermometer** uses the fact that the electrical resistance (Chapter 46) of a platinum wire increases with temperature according to the equation

$$\frac{\theta}{100} = \frac{R_\theta - R_0}{R_{100} - R_0}$$

θ is the required temperature, R_{100}, R_0 and R_θ are the resistances of the wire at 100°C, 0°C and θ. It can measure temperatures accurately in the range −200°C to 1200°C but it is bulky and best for steady temperatures. A **thermistor** (Chapter 60) can also be used but over a small range, e.g. −5°C to 70°C; its resistance decreases with temperature.

The **constant-volume gas thermometer** uses the change in pressure of a gas to measure temperatures over a wide range. It is an accurate but bulky instrument, basically similar to the apparatus of Figure 37.3.

Thermochromic liquids which change colour with temperature have a limited range around room temperatures.

Heat and temperature

It is important not to confuse the temperature of a body with the heat energy that can be obtained from it. For example, a red-hot spark from a fire is at a higher temperature than the boiling water in a saucepan. In the boiling water the average k.e. of the molecules is lower than in the spark; but since there are many more water molecules, their total energy is greater, and therefore more heat energy can be supplied by the water than by the spark.

Heat passes from a body at a higher temperature to one at a lower temperature. This is due to the average k.e. (and speed) of the molecules in the 'hot' body falling as a result of having collisions with molecules of the 'cold' body whose average k.e., and therefore temperature, increases. When the average k.e. of the molecules is the same in both bodies, they are at the same temperature. For example, if the red-hot spark landed in the boiling water, heat would pass from it to the water even though much more heat energy could be obtained from the water.

Heat is also called **thermal** or **internal** energy; it is the energy a body has because of the kinetic energy *and* the potential energy of its molecules. Increasing the temperature of a body increases its heat energy due to the k.e. of its molecules increasing. But as we will see later (Chapter 39), the internal energy of a body can also be increased by increasing the p.e. of its molecules.

153

Questions

1 1530°C   120°C   55°C   37°C   19°C   0°C
   −12°C   −50°C

From the above list of temperatures choose the most likely value for *each* of the following:
a the melting point of iron,
b the temperature of a room that is comfortably warm,
c the melting point of pure ice at normal pressure,
d the lowest outdoor temperature recorded in London in winter,
e the normal body temperature of a healthy person.

2 In order to make a mercury thermometer which will measure small changes in temperature accurately

A decrease the volume of the mercury bulb
B put the degree markings farther apart
C decrease the diameter of the capillary tube
D put the degree markings closer together
E leave the capillary tube open to the air.

3 a How must a property behave to measure temperature?
   b Name three properties that qualify.
   c Name a suitable thermometer for measuring
      (i) a steady temperature of 1000°C,
      (ii) the changing temperature of a small object,
      (iii) a winter temperature at the North Pole.

4 Describe the main features of a clinical thermometer.

Checklist

After studying this chapter you should be able to

■ define the fixed points on the Celsius scale,
■ recall the properties of mercury and alcohol as liquids suitable for use in thermometers,
■ describe clinical and thermocouple thermometers,
■ recall some other types of thermometer and the physical properties on which they depend,
■ distinguish between heat and temperature and recall that temperature decides the direction of heat flow.

36 Expansion of solids and liquids

Uses of expansion
Precautions against expansion
Bimetallic strip

Linear expansivity
Unusual expansion of water

In general, when matter is heated it expands and when cooled it contracts. If the changes are resisted large forces are created which are sometimes useful but at other times are a nuisance.

According to the kinetic theory (Chapter 20) the molecules of solids and liquids are in constant vibration. When heated they vibrate faster and force each other a little farther apart. Expansion results, and this is greater for liquids. The linear (length) expansion of solids is small and for the effect to be noticed the solid must be long and/or the temperature change large.

■ Uses of expansion

In Figure 36.1 the axles have been shrunk by cooling in liquid nitrogen at −196 °C until the gear wheels can be slipped on to them. On regaining normal temperature the axles expand to give a very tight fit.

Figure 36.1 'Shrink-fitting' of axles into gear wheels

In the kitchen, a tight metal lid can be removed from a glass jar by immersing the lid in hot water so that it expands.

■ *Precautions against expansion*

Gaps used to be left between lengths of railway lines to allow for expansion in summer. They caused a familiar 'clickety-click' sound as the train passed over them. These days rails are welded into lengths of about 1 km and are held by concrete 'sleepers' that can withstand the large forces created without buckling. Also, at the joints the ends are tapered and overlap, Figure 36.2a. This gives a smoother journey and allows some expansion near the ends of each length of rail.

For similar reasons slight gaps are left between lengths of aluminium guttering. In central heating pipes 'expansion joints' are used to join lengths of pipe, Figure 36.2b; these allow the copper pipes to expand in length inside the joints when carrying very hot water.

Figure 36.2a Tapered overlap of rails

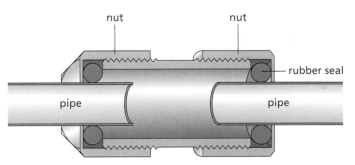

Figure 36.2b Expansion joint

155

■ *Bimetallic strip*

If equal lengths of two different metals, e.g. copper and iron, are riveted together so that they cannot move separately, they form a bimetallic strip, Figure 36.3a. When heated, copper expands more than iron and to allow this the strip bends with copper on the outside, Figure 36.3b. If they had expanded equally the strip would have stayed straight.

Bimetallic strips have many uses.

a Before heating

b After heating

Figure 36.3 A bimetallic strip

a) Fire alarm

Heat from the fire makes the bimetallic strip bend and complete the electrical circuit, so ringing the alarm bell, Figure 36.4a.

A bimetallic strip is also used in this way to work the flashing direction indicator lamps in a car, being warmed by an electric heating coil wound round it.

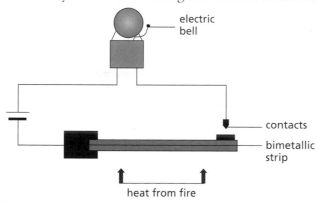

a A fire alarm

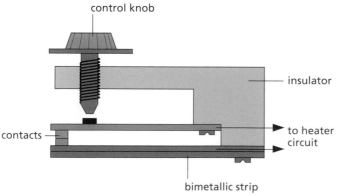

b A thermostat in an iron

Figure 36.4 Uses of a bimetallic strip

b) Thermostat

A thermostat keeps the temperature of a room or an appliance constant. The one in Figure 36.4b uses a bimetallic strip in the electrical heating circuit of, for example, an iron.

When the iron reaches the required temperature the strip bends down, breaks the circuit at the contacts and switches off the heater. After cooling a little the strip remakes contact and turns the heater on again. A near-steady temperature results.

If the control knob is screwed down, the strip has to bend more to break the heating circuit and this needs a higher temperature.

■ *Linear expansivity*

An engineer has to allow for the linear expansion of a bridge when designing it. The expansion can be calculated if

(i) the length of the bridge,
(ii) the range of temperature it will experience, and
(iii) the **linear expansivity** of the material to be used,

are known.

> The linear expansivity α of a substance is the increase in length of 1 m for a 1 °C rise in temperature.

The linear expansivity of a material is found by experiment. For steel it is 0.000 012 per °C. This means that 1 m will become 1.000 012 m for a temperature rise of 1 °C. A steel bridge 100 m long will expand by 0.000 012 × 100 m for each 1 °C rise in temperature. If the maximum temperature **change** $\Delta\theta$ expected is 60 °C (e.g. from −15 °C to +45 °C), the expansion Δl will be 0.000 012 °C × 100 m × 60 °C = 0.072 m = 7.2 cm. In general,

> expansion = linear expansivity × original length
> × temperature rise

In symbols

$$\Delta l = \alpha l_0 \Delta\theta$$

where l_0 is the original length.

Unusual expansion of water

As water is cooled to 4°C it contracts, as we would expect. However between 4°C and 0°C it expands, surprisingly. **Water has a maximum density at 4°C**, Figure 36.5.

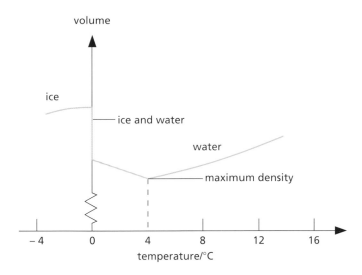

Figure 36.5 Water expands on cooling below 4°C

At 0°C, when it freezes, a considerable expansion occurs and every 100 cm³ of water becomes 109 cm³ of ice. This accounts for the bursting of unlagged water pipes in very cold weather and for the fact that ice is less dense than cold water and so floats. Figure 36.6 shows a bottle of frozen milk, the main constituent of which is water.

The unusual expansion of water between 4°C and 0°C explains why fish survive in a frozen pond. The water at the top of the pond cools first, contracts and being denser sinks to the bottom. Warmer less dense water rises to the surface to be cooled. When all the water is at 4°C the circulation stops. If the temperature of the surface water falls below 4°C, it becomes less dense and *remains at the top*, eventually forming a layer of ice at 0°C. Temperatures in the pond are then as in Figure 36.7.

The expansion of water between 4°C and 0°C is due to the breaking up of the groups which water molecules form above 4°C. The new arrangement requires a larger volume and more than cancels out the contraction due to the fall in temperature.

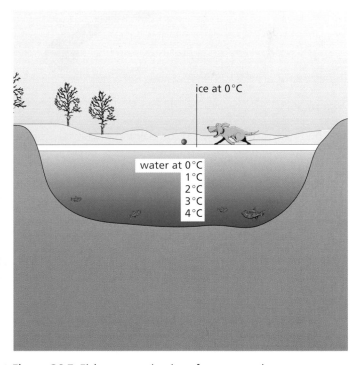

Figure 36.7 Fish can survive in a frozen pond

Figure 36.6 Result of the expansion of water on freezing

Questions

1 Explain why
  a the metal lid on a glass jam jar can be unscrewed easily if the jar is inverted for a few seconds with the *lid* in very hot water,
  b furniture may creak at night after a warm day,
  c concrete roads are laid in sections with pitch between them.

2 A bimetallic strip is made from aluminium and copper. When heated it bends in the direction shown in Figure 36.8.

  Which metal expands more for the same rise in temperature?

  Draw a diagram to show how the bimetallic strip would appear if it were cooled to below room temperature.

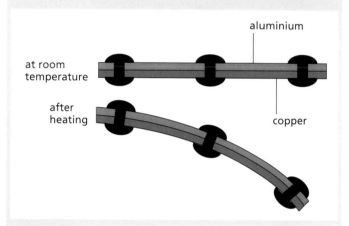

Figure 36.8

3 The linear expansivities of some common substances are given below.

aluminium	0.000 03 per °C	glass	0.000 009 per °C
concrete	0.000 01 per °C	platinum	0.000 009 per °C
copper	0.000 02 per °C	steel	0.000 01 per °C

  a Rods of aluminium, copper and steel are the same length at a temperature of 0°C. Which will be the longest at 300°C?
  b If the aluminium rod is 1.000 m long at 0°C, what will be its length at 300°C?
  c Why is steel a suitable material for reinforcing concrete?
  d Which of the substances listed above would be most suitable for carrying a current of electricity through the walls of a glass vessel?
  e How might a steel cylinder liner be fitted tightly inside a cylinder block made of aluminium?

4 Using the values for linear expansivities in question 3 calculate the expansion of
  a 100 m of copper pipe heated through 50°C,
  b 50 cm of steel pipe heated through 80°C,
  c 200 m of aluminium pipe heated from 10°C to 60°C.

5 Discuss what is unusual about the expansion of water.

Checklist

After studying this chapter you should be able to

■ describe uses of expansion, including the bimetallic strip,
■ describe precautions taken against expansion,
■ define linear expansivity,
■ recall that water has its maximum density at 4°C and explain why a pond freezes at the top first.

37 The gas laws

Pressure of a gas
Absolute zero
The gas laws
Gases and the kinetic theory

Practical work
Effect on volume of temperature.
Effect on pressure of temperature.
Effect on volume of pressure.

◾ *Pressure of a gas*

The air forming the Earth's atmosphere stretches up-wards a long way. Air has weight; the air in a normal room weighs about the same as you do, e.g. 500 N. Because of its weight the atmosphere exerts a large pressure at sea-level, about $100\,000\,\text{N/m}^2 = 10^5\,\text{Pa} = 100\,\text{kPa}$. This pressure acts equally in all directions.

A gas in a container exerts a pressure on the walls of the container. If air is removed from a can by a vacuum pump, Figure 37.1, the can collapses because the air pressure outside is greater than that inside. A space from which all the air has been removed is a **vacuum**. Alternatively the pressure in a container can be increased, for example by pumping more gas into the can; a Bourdon gauge (p. 104) is used for measuring fluid pressures.

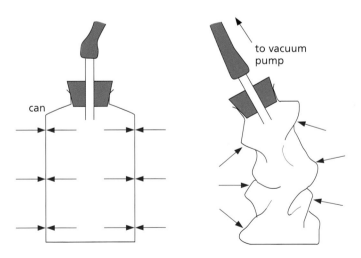

Figure 37.1 Atmospheric pressure collapses the evacuated can

When a gas is heated, as air is in a jet engine, its pressure as well as its volume may change. To study the effect of temperature on these two quantities we must keep one fixed while the other is changed.

Practical work

Effect on volume of temperature (pressure constant) – Charles' law

Arrange the apparatus as in Figure 37.2. The index of concentrated sulphuric acid traps the air column to be investigated and also dries it. Adjust the capillary tube so that the bottom of the air column is opposite a convenient mark on the ruler.

Note the length of the air column (to the *lower* end of the index) at different temperatures but, before taking a reading, stop heating and stir well to make sure that the air has reached the temperature of the water. Put the results in a table.

Plot a graph of volume (in cm, since the length of the air column is a measure of it) on the *y*-axis and temperature (in °C) on the *x*-axis.

The pressure of (and on) the air column is constant and equals atmospheric pressure plus the pressure of the acid index.

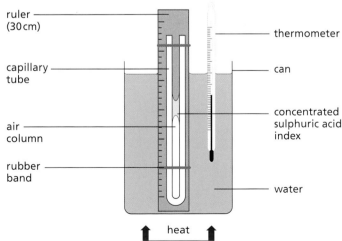

Figure 37.2

Practical work

Effect on pressure of temperature (volume constant) – the Pressure law

The apparatus is shown in Figure 37.3. The rubber tubing from the flask to the pressure gauge should be as short as possible. The flask must be in water almost to the top of its neck and be securely clamped to keep it off the bottom of the can.

Record the pressure over a wide range of temperatures but before taking a reading, stop heating, stir and allow time for the gauge reading to become steady; the air in the flask will then be at the temperature of the water. Tabulate the results.

Plot a graph of pressure on the y-axis and temperature on the x-axis.

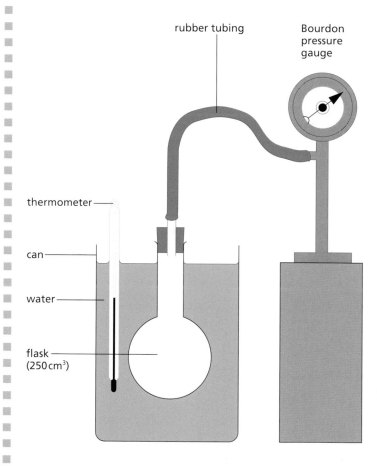

Figure 37.3

Absolute zero

The volume–temperature and pressure–temperature graphs for a gas are straight lines, Figure 37.4. They show that gases expand **uniformly** with temperature as measured on a mercury thermometer, i.e. equal temperature increases cause equal volume or pressure increases.

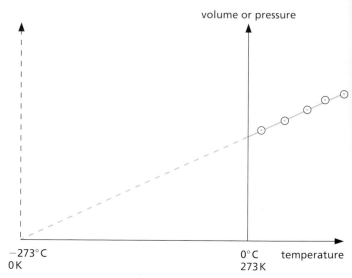

Figure 37.4

The graphs do not pass through the Celsius temperature origin (0 °C). If they are produced backwards they cut the temperature axis at about −273 °C. This temperature is called **absolute zero** because we believe it is the lowest temperature possible. It is the zero of the **absolute** or **Kelvin scale of temperature**. At absolute zero molecular motion ceases and a substance has no internal energy.

Degrees on this scale are called **kelvins** and are denoted by K. They are exactly the same size as Celsius degrees. Since −273 °C = 0 K, conversions from °C to K are made by adding 273. For example

$$0\,°C = 273\,K$$
$$15\,°C = 273 + 15 = 288\,K$$
$$100\,°C = 273 + 100 = 373\,K$$

Kelvin or absolute temperatures are represented by the letter T and if θ stands for a Celsius scale temperature then, in general

$$T = 273 + \theta$$

Near absolute zero strange things occur. Liquid helium becomes a 'superfluid'. It cannot be kept in an open vessel because it flows up the inside of the vessel, over the edge and down the outside. Some metals and compounds become 'superconductors' of electricity and a current once started in them flows for ever without a battery. Figure 37.5 shows research equipment that is being used to create materials which are superconductors at very much higher temperatures, e.g. $-23\,°C$.

Figure 37.5 Equipment that is being used to make films of complex composite materials which are superconducting at temperatures far above absolute zero

Practical work

Effect on volume of pressure (temperature constant) – Boyle's law

Changes in the volume of a gas due to pressure changes can be studied using the apparatus in Figure 37.6. The volume V of air trapped in the glass tube is read off on the scale behind. The pressure is altered by pumping air from a foot pump into the space above the oil reservoir. This forces more oil into the glass tube and increases the pressure p on the air in it; p is measured by the Bourdon gauge.

If a graph of pressure against volume is plotted using the results, a curve like that in Figure 37.7a is obtained. Close examination of it shows that if p is doubled, V is halved. That is, p **is inversely proportional to** V. In symbols

$$p \propto \frac{1}{V} \quad \text{or} \quad p = \text{constant} \times \frac{1}{V}$$

$$\therefore \qquad pV = \text{constant}$$

If several pairs of readings p_1V_1, p_2V_2, etc. are taken, then it can be confirmed that $p_1V_1 = p_2V_2 = \text{constant}$. This is **Boyle's law** which is stated as follows.

The pressure of a fixed mass of gas is inversely proportional to its volume if its temperature is kept constant.

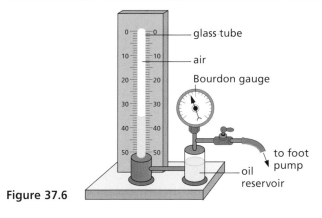

Figure 37.6

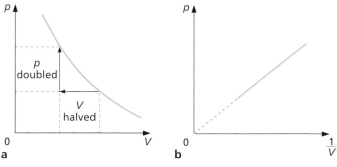

Figure 37.7

Since p is inversely proportional to V, then p is directly proportional to $1/V$. A graph of p against $1/V$ is therefore a straight line through the origin, Figure 37.7b.

The gas laws

Using absolute temperatures the gas laws can be stated in a convenient form for calculations.

a) Charles' law

In Figure 37.4 (p.160) the volume–temperature graph passes through the origin if temperatures are measured on the Kelvin scale, that is, if we take $0\,K$ as the origin. We can then say that the volume V is directly proportional to the absolute temperature T, i.e. doubling T doubles V, etc. Therefore

$$V \propto T \quad \text{or} \quad V = \text{constant} \times T$$

or $\qquad\qquad V/T = \text{constant} \qquad\qquad$ (1)

Charles' law may be stated as follows.

The volume of a fixed mass of gas is directly proportional to its absolute temperature if the pressure is kept constant.

b) Pressure law

From Figure 37.4 we can say similarly for the pressure p that

$$p \propto T \quad \text{or} \quad p = \text{constant} \times T$$

or $\qquad\qquad p/T = \text{constant} \qquad\qquad$ (2)

The Pressure law may be stated as follows.

The pressure of a fixed mass of gas is directly proportional to its absolute temperature if the volume is kept constant.

c) Boyle's law

For a fixed mass of gas at constant temperature

$$pV = \text{constant} \qquad\qquad (3)$$

These three equations can be combined giving

$$\frac{pV}{T} = \text{constant}$$

For cases in which p, V and T all change from, say, p_1, V_1 and T_1 to p_2, V_2 and T_2, then

$$\frac{p_1 V_1}{T_1} = \frac{p_2 V_2}{T_2} \qquad\qquad (4)$$

Worked example

A bicycle pump contains $50\,cm^3$ of air at $17\,°C$ and at 1.0 atmosphere pressure. Find the pressure when the air is compressed to $10\,cm^3$ and its temperature rises to $27\,°C$.

We have

$$p_1 = 1.0 \text{ atmosphere} \qquad p_2 = ?$$
$$V_1 = 50\,cm^3 \qquad\qquad V_2 = 10\,cm^3$$
$$T_1 = 273 + 17 = 290\,K \qquad T_2 = 273 + 27 = 300\,K$$

From equation (4) we get

$$p_2 = p_1 \times \frac{V_1}{V_2} \times \frac{T_2}{T_1}$$

$$= 1 \times \frac{50}{10} \times \frac{300}{290} = 5.2 \text{ atmosphere}$$

Notes

1 All temperatures must be in K.
2 Any units can be used for p and V so long as they are the same on both sides of the equation.
3 In some calculations the volume of the gas has to be found at s.t.p. (standard temperature and pressure). This is $0\,°C$ and $10^5\,Pa$ pressure.

Gases and the kinetic theory

The kinetic theory can explain the behaviour of gases.

a) Cause of gas pressure

All the molecules in a gas are in rapid motion, with a wide range of speeds, and repeatedly hit the walls of the container in huge numbers per second. The average force and hence the pressure they exert on the walls is constant since pressure is force on unit area.

b) Boyle's law

If the volume of a fixed mass of gas is halved by halving the volume of the container, Figure 37.8, the number of molecules per cm^3 will be doubled. There will be twice as many collisions per second with the walls, i.e. the pressure is doubled. This is Boyle's law.

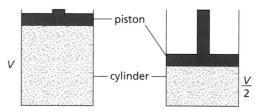

Figure 37.8 Halving the volume doubles the pressure

c) Temperature

When a gas is heated and its temperature rises, the average speed of its molecules increases. If the volume of the gas is to remain constant, its pressure increases due to more frequent and more violent collisions of the molecules with the walls. If the pressure of the gas is to remain constant, the volume must increase so that the frequency of collisions does not.

Questions

1 A gas of volume $2\,m^3$ at $27\,^{\circ}C$ is
  a heated to $327\,^{\circ}C$,
  b cooled to $-123\,^{\circ}C$,
  at constant pressure. What is its new volume in each case?

2 A container holds gas at $27\,^{\circ}C$. To what temperature must it be heated for its pressure to double? Assume the volume of the gas remains unchanged and the container does not leak.

3 a Figure 37.9 of a thermometer shows both the Celsius and Kelvin temperature scales.

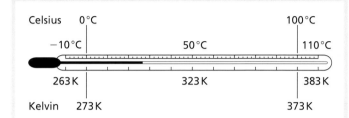

Figure 37.9

  (i) What temperature, in degrees Celsius, is shown by the thermometer?
  (ii) How do you change a temperature given in degrees Celsius to a temperature in kelvins?
  b The graph in Figure 37.10 shows how the pressure of a gas trapped inside a sealed container changes with temperature. The pressure is caused by the gas particles continually hitting the sides of the container.

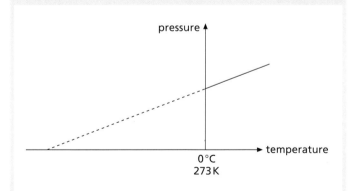

Figure 37.10

  (i) What name is given to the temperature at which the gas particles stop hitting the sides of the container?
  (ii) What is the momentum of the gas particles at this temperature? Give a reason for your answer.
  (iii) What is this temperature in kelvins?
                    (AQA (SEG) Foundation, Summer 99)

4 If a certain quantity of gas has a volume of $30\,cm^3$ at a pressure of $1 \times 10^5\,Pa$, what is its volume when the pressure is
  a $2 \times 10^5\,Pa$,
  b $5 \times 10^5\,Pa$?

5 a Figure 37.11 shows a sealed gas syringe in a heated water bath.

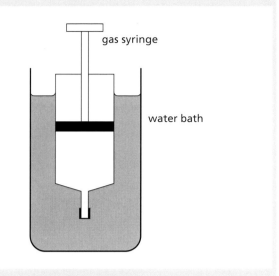

Figure 37.11

  Explain in terms of the gas molecules, why, if the pressure inside the syringe is to remain constant the gas must expand as its temperature rises.
  b A student was asked to investigate the relationship between the volume and pressure of a gas. The student used a fixed mass of oxygen kept at a constant temperature of 300 kelvins. The results of the investigation were used to plot a graph, Figure 37.12.

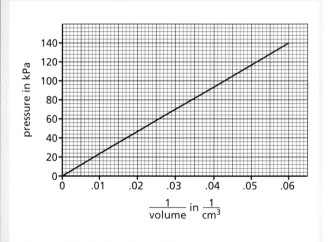

Figure 37.12

  (i) Write down the relationship between the volume and pressure of a gas as found by this investigation.
  (ii) Copy the graph and draw the line that would be obtained if it were possible to repeat the investigation at a temperature of 600 kelvins.

c A gas cylinder contains 0.05 m³ of helium at a pressure of 16 000 kPa. The gas is used to blow up balloons. Each balloon when fully inflated contains 0.04 m³ of helium at a pressure of 102 kPa. The temperature of the helium inside both the cylinder and the balloons is the same.

Use the following equation to calculate the maximum number of balloons that can be fully inflated using this gas cylinder. Show clearly how you work out your final answer.

$$\frac{pV}{T} = \text{constant}$$

(AQA (SEG) Higher, Summer 99)

■ *Checklist*

After studying this chapter you should be able to

■ describe experiments to study the relationships between the pressure, volume and temperature of a gas,

■ explain the establishment of the Kelvin (absolute) temperature scale from graphs of pressure or volume against temperature and recall the equation connecting the Kelvin and Celsius scales, i.e. $T = 273 + \theta$,

■ state Charles' law, the Pressure law and Boyle's law and use them to solve problems,

■ recall that $pV/T = $ constant and use it to solve problems,

■ explain the behaviour of gases using the kinetic theory.

38 *Specific heat capacity*

The heat equation
Importance of the high specific heat capacity of water

Practical work
Finding specific heat capacities: water, aluminium.

If 1 kg of water and 1 kg of paraffin are heated in turn for the same time by the same heater, the temperature rise of the paraffin is about *twice* that of the water. Since the heater gives equal amounts of heat energy to each liquid, it seems that different substances require different amounts of heat to cause the same temperature rise in the same mass, say 1 °C in 1 kg.

The 'thirst' of a substance for heat is measured by its **specific heat capacity** (symbol *c*).

The specific heat capacity of a substance is the heat required to produce a 1 °C rise in 1 kg.

Heat, like other forms of energy, is measured in joules (J) and the unit of specific heat capacity is the joule per kilogram per °C, i.e. J/(kg °C).

In physics the word 'specific' means 'unit mass' is being considered.

■ *The heat equation*

If a substance has a specific heat capacity of 1000 J/(kg °C) then

1000 J raise the temp. of 1 kg by 1 °C
∴   2 × 1000 J raise the temp. of 2 kg by 1 °C
∴   3 × 2 × 1000 J raise the temp. of 2 kg by 3 °C

That is 6000 J will raise the temperature of 2 kg of this substance by 3 °C. We have obtained this answer by multiplying together:

(i) the **mass** in kg,
(ii) the **temperature rise** in °C, and
(iii) the **specific heat capacity** in J/(kg °C).

If the temperature of the substance fell by 3 °C, the heat given out would also be 6000 J. In general, we can write the 'heat equation' as

heat received or given out
   = mass × temp. change × sp. heat capacity

In symbols

$$Q = m \times \Delta\theta \times c$$

For example, if the temperature of a 5 kg mass of copper of specific heat capacity 400 J/(kg °C) rises from 15 °C to 25 °C, the heat received Q

$$= 5\,kg \times (25 - 15)\,°C \times 400\,J/(kg\,°C)$$
$$= 5\,kg \times 10\,°C \times 400\,J/(kg\,°C) = 20\,000\,J$$

Practical work

Finding specific heat capacities

You need to know the power of the 12-volt electric immersion heater to be used. (**Precaution.** Do not use one with a cracked seal.) A 40-watt heater converts 40 joules of electrical energy into heat energy per second. If the power is not marked on the heater ask about it.[1]

a) Water

Weigh out 1 kg of water into a container, e.g. an aluminium saucepan. Note the temperature of the water, insert the heater, Figure 38.1, and switch on the 12 V supply. Stir the water and after 5 minutes switch off, but continue stirring and note the *highest* temperature reached.

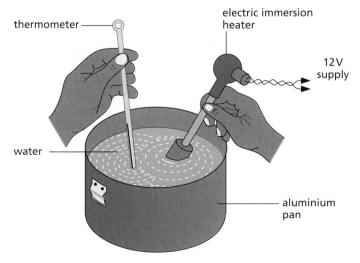

Figure 38.1

[1] The power is found by immersing the heater in water, connecting it to a 12-volt d.c. supply and measuring the current taken (usually 3–4 amperes). Then power in watts = volts × amperes.

Assuming that the heat supplied by the heater equals the heat received by the water, work out the specific heat capacity of water in J/(kg°C), as shown below.

heat received by water (J)
= power of heater (J/s) × time heater on (s)

Rearranging the 'heat equation' we get

$$\text{sp. heat cap. of water} = \frac{\text{heat received by water (J)}}{\text{mass (kg)} \times \text{temp. rise (°C)}}$$

Suggest causes of error in this experiment.

b) Aluminium

An aluminium cylinder weighing 1 kg and having two holes drilled in it is used. Place the immersion heater in the central hole and a thermometer in the other hole, Figure 38.2.

Note the temperature, connect the heater to a 12 V supply and switch it on for 5 minutes. When the temperature stops rising record its highest value.

Calculate the specific heat capacity as before.

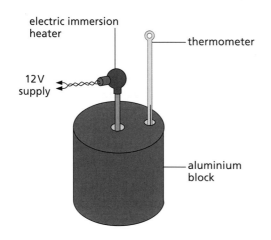

Figure 38.2

Importance of the high specific heat capacity of water

The specific heat capacity of water is 4200 J/(kg°C) and that of soil is about 800 J/(kg°C). As a result, the temperature of the sea rises and falls more slowly than that of the land. A certain mass of water needs five times more heat than the same mass of soil for its temperature to rise by 1 °C. Water also has to give out more heat to fall 1 °C. Since islands are surrounded by water they experience much smaller changes of temperature from summer to winter than large land masses such as Central Asia.

The high specific heat capacity of water (as well as its cheapness and availability) accounts for its use in cooling engines and in the radiators of central heating systems.

Worked examples

1 A tank holding 60 kg of water is heated by a 3 kW electric immersion heater. If the specific heat capacity of water is 4200 J/(kg°C), estimate the time for the temperature to rise from 10 °C to 60 °C.

A 3 kW (3000 W) heater supplies 3000 J of heat energy per second.

Let t = time taken in seconds to raise the temperature of the water by $(60 - 10) = 50$ °C,

$\therefore$ heat supplied to water in time $t = 3000 \times t$ J

From the 'heat equation', we can say

heat received by water
$$= 60 \text{ kg} \times 4200 \text{ J/(kg°C)} \times 50 \text{ °C}$$

Assuming heat supplied = heat received

$$3000 \text{ J/s} \times t = (60 \times 4200 \times 50) \text{ J}$$

$\therefore \quad t = \dfrac{(60 \times 4200 \times 50) \text{ J}}{3000 \text{ J/s}} = 4200 \text{ s (70 mins)}$

2 A piece of aluminium of mass 0.5 kg is heated to 100 °C and then placed in 0.4 kg of water at 10 °C. If the resulting temperature of the mixture is 30 °C, what is the specific heat capacity of aluminium if that of water is 4200 J/(kg°C)?

When two substances at different temperatures are mixed, heat flows from the one at the higher temperature to the one at the lower temperature until both are at the same temperature – the temperature of the mixture. If there is no loss of heat, then in this case

heat given out by aluminium
$$= \text{heat taken in by water}$$

Using the 'heat equation' and letting c be the specific heat capacity of aluminium in J/(kg°C), we have

heat given out $= 0.5 \text{ kg} \times c \times (100 - 30)$ °C

heat taken in
$$= 0.4 \text{ kg} \times 4200 \text{ J/(kg°C)} \times (30 - 10) \text{ °C}$$

$\therefore \quad 0.5 \text{ kg} \times c \times 70 \text{ °C} = 0.4 \text{ kg} \times 4200 \text{ J/(kg°C)} \times 20 \text{ °C}$

$\therefore \quad c = \dfrac{(4200 \times 8) \text{ J}}{35 \text{ kg°C}} = 960 \text{ J/(kg°C)}$

Questions

1 How much heat is needed to raise the temperature by 10°C of 5 kg of a substance of specific heat capacity 300 J/(kg °C)?

2 The same quantity of heat was given to different masses of three substances A, B and C. The temperature rise in each case is shown in the table. Calculate the specific heat capacities of A, B and C.

Material	Mass /kg	Heat given /J	Temp. rise /°C
A	1.0	2000	1.0
B	2.0	2000	5.0
C	0.5	2000	4.0

3 a The table shows how much energy is needed to make the temperature of 1 kg of different substances go up by 1 °C.

Substance	Energy in joules
copper	390
mercury	140
silver	240
steel	450

Which *one* of these four substances has the highest specific heat capacity? Give a reason for your answer.

b A microwave oven is used to heat a cold mug of coffee. In a few seconds the temperature of the coffee goes up by 70°C. Use the following equation

energy exchanged = mass × specific heat
capacity × temperature change

to calculate the energy in joules gained by the coffee. Show clearly how you work out your final answer.
(Take the mass of the coffee to be 0.2 kg and the specific heat capacity of the coffee to be 4000 J/kg °C.)

(*SEG Foundation, Summer 98*)

4 The jam in a hot pop tart always seems hotter than the pastry. Why?

■ *Checklist*

After studying this chapter you should be able to

■ define **specific heat capacity**, c,
■ solve problems on specific heat capacity using the heat equation $Q = m \times \Delta\theta \times c$,
■ describe experiments to measure the specific heat capacity of metals and liquids by electrical heating,
■ explain the importance of the high specific heat capacity of water.

39 Latent heat

When a solid is heated, it may melt and **change its state** from solid to liquid. If ice is heated it becomes water. The opposite process, freezing, occurs when a liquid solidifies.

A pure substance melts at a definite temperature, called the **melting point**; it solidifies at the same temperature – sometimes then called the **freezing point**.

Practical work

Cooling curve of ethanamide

Half-fill a test-tube with ethanamide (acetamide) and place it in a beaker of water, Figure 39.1a. Heat the water until all the ethanamide has melted and its temperature reaches about 90 °C.

Remove the test-tube and arrange it as in Figure 39.1b with a thermometer in the liquid ethanamide. Record the temperature every minute until it has fallen to 70 °C.

Plot a cooling curve of temperature against time. What is the freezing (melting) point of ethanamide?

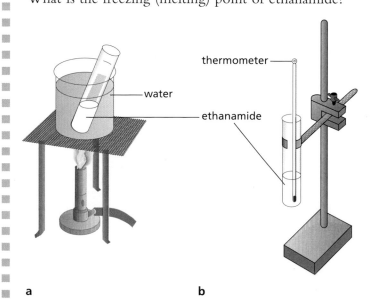

a   **b**

Figure 39.1

Latent heat of fusion

The previous experiment shows that the temperature of liquid ethanamide falls until it starts to solidify (at 82 °C) and remains constant until it has all solidified. The cooling curve in Figure 39.2 is for a pure substance; the flat part AB occurs at the melting point when the substance is solidifying.

During solidification a substance loses heat to its surroundings but its temperature does not fall. Conversely when a solid is melting, the heat supplied does not cause a temperature rise; heat is added but the substance does not get hotter. For example, the temperature of a well-stirred ice–water mixture remains at 0 °C until all the ice is melted.

Heat which is **absorbed** by a solid during melting or **given out** by a liquid during solidification is called **latent heat of fusion**. 'Latent' means hidden and 'fusion' means melting. Latent heat does not cause a temperature change; it seems to disappear.

The specific latent heat of fusion (l_f) of a substance is the quantity of heat needed to change unit mass from solid to liquid without temperature change.

It is measured in J/kg or J/g. In general, the quantity of heat Q to change a mass m from solid to liquid is given by

$$Q = m \times l_f$$

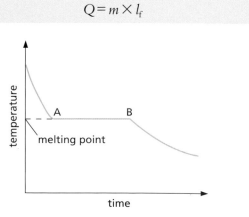

Figure 39.2 Cooling curve

168

Latent heat of vaporization

Latent heat is also needed to change a liquid into a vapour. The reading of a thermometer placed in water which is boiling remains constant at 100°C even though heat, called **latent heat of vaporization**, is still being absorbed by the water from whatever is heating it. When steam condenses to form water, latent heat is given out.

The specific latent heat of vaporization (l_v) of a substance is the quantity of heat needed to change unit mass from liquid to vapour without change of temperature.

It is measured in J/kg or J/g. In general, the quantity of heat Q to change a mass m from liquid to vapour is given by

$$Q = m \times l_v$$

Figure 39.3 Why is a scald from steam often more serious than one from boiling water?

Latent heat and the kinetic theory

a) Fusion

The kinetic theory explains latent heat of fusion as being the energy which enables the molecules of a solid to overcome the intermolecular forces that hold them in place and when it exceeds a certain value they break free. Their vibratory motion about fixed positions changes to the slightly greater range of movement they have as liquid molecules. Their p.e. increases but not their average k.e. as happens when the heat causes a temperature rise.

b) Vaporization

If liquid molecules are to overcome the forces holding them together and gain the freedom to move around independently as gas molecules, they need a large amount of energy. They receive this as latent heat of vaporization which, like latent heat of fusion, increases the p.e. of the molecules but not their k.e. It also gives the molecules the energy required to push back the surrounding atmosphere in the large expansion that occurs when a liquid vaporizes.

To change 1 kg of water at 100°C to steam at 100°C needs over *five* times as much heat as is needed to raise the temperature of 1 kg of water at 0°C to water at 100°C (see *Worked example* 1, p.170).

Evaporation and boiling

a) Evaporation

This occurs at all temperatures at the surface of the liquid. It happens more rapidly when

(i) the **temperature is higher**, since then more molecules in the liquid are moving fast enough to escape from the surface,
(ii) the **surface area** of the liquid is large so giving more molecules a chance to escape because more are near the surface, and
(iii) a **wind** or **draught** is blowing over the surface carrying vapour molecules away from the surface thus stopping them from returning to the liquid and making it easier for more liquid molecules to break free. (Evaporation into a vacuum occurs much more rapidly than into a region where there are air or other molecules.)

b) Boiling

This occurs for a pure liquid at a definite temperature called its **boiling point** and is accompanied by bubbles that form within the liquid, containing the gaseous or vapour form of the particular substance.

Latent heat is needed in both evaporation and boiling and is stored in the vapour, from which it is released when the vapour is cooled or compressed and changes to liquid again.

Cooling by evaporation

In evaporation latent heat is obtained by the liquid from its surroundings, as may be shown by the following demonstration, **done in a fume cupboard**.

a) Demonstration

Dichloromethane is a **volatile** liquid, i.e. it has a low boiling point and evaporates readily at room temperature, especially when air is blown through it, Figure 39.4. Latent heat is taken first from the liquid itself and then from the water below the can. The water soon freezes causing the block and can to stick together.

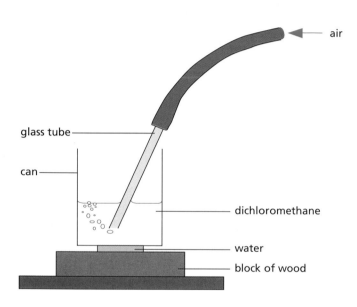

Figure 39.4 Demonstrating cooling by evaporation

b) Explanation

Evaporation occurs when faster-moving molecules escape from the surface of the liquid. The average speed and therefore the average k.e. of the molecules left behind decreases, i.e. the temperature of the liquid falls.

c) Uses

Water evaporates from the skin when we sweat. This is the body's way of using unwanted heat and keeping a constant temperature. After vigorous exercise there is a risk of the body being overcooled, especially in a draught; it is then less able to resist infection.

Ether acts as a local anaesthetic by chilling (as well as cleaning) your arm when you are having an injection. Refrigerators, freezers and air-conditioning systems use cooling by evaporation on a large scale.

Volatile liquids are used in perfumes.

Liquefaction of gases and vapours

A vapour can be liquefied if it is compressed enough. However, a gas must be cooled below a certain **critical temperature** T_c before liquefaction by pressure can occur. The critical temperatures for some gases are given in Table 39.1.

Table 39.1 Critical temperatures of some gases

Gas	carbon dioxide	oxygen	air	nitrogen	hydrogen	helium
T_c/K	304	154	132	126	33.3	5.3
T_c/°C	+31	−119	−141	−147	−239.7	−267.7

Low-temperature liquids have many uses. Liquid hydrogen and oxygen are used as the fuel and oxidant respectively in space rockets. Liquid nitrogen is used in industry as a coolant in, for example, shrink-fitting (see Figure 36.1, p.155). Materials which behave as super-conductors (see Chapter 37) when cooled by liquid nitrogen are increasingly being used in electrical power engineering and electronics.

Worked examples

The values in Table 39.2 are required.

Table 39.2

	water	ice	aluminium
sp. ht. cap./J/(g°C)	4.2	2.0	0.90
sp. lat. ht./J/g	2300	340	

1 How much heat is needed to change 20 g of ice at 0 °C to steam at 100 °C?

There are three stages in the change.

Heat to change 20 g **ice at 0 °C** to **water at 0 °C**
$$= \text{mass of ice} \times \text{sp. lat. ht. of ice}$$
$$= 20\,\text{g} \times 340\,\text{J/g} = 6800\,\text{J}$$

Heat to change 20 g **water at 0 °C** to **water at 100 °C**
$$= \text{mass of water} \times \text{sp. ht. cap. of water} \times \text{temp. rise}$$
$$= 20\,\text{g} \times 4.2\,\text{J/(g°C)} \times 100\,\text{°C} = 8400\,\text{J}$$

Heat to change 20 g **water at 100 °C** to **steam at 100 °C**
$$= \text{mass of water} \times \text{sp. lat. ht. of steam}$$
$$= 20\,\text{g} \times 2300\,\text{J/g} = 46\,000\,\text{J}$$

∴ Total heat supplied
$$= 6800 + 8400 + 46\,000 = 61\,200\,\text{J}$$

2 An aluminium can of mass 100 g contains 200 g of water. Both, initially at 15 °C, are placed in a freezer at −5.0 °C. Calculate the quantity of heat that has to be removed from the water and the can for their temperatures to fall to −5.0 °C.

Heat lost by **can** in falling from 15 °C to −5.0 °C
$$= \text{mass of can} \times \text{sp. ht. cap. of aluminium} \times \text{temp. fall}$$
$$= 100 \, \text{g} \times 0.90 \, \text{J/(g°C)} \times [15 - (-5)] \, °C$$
$$= 100 \, \text{g} \times 0.90 \, \text{J/(g°C)} \times 20 \, °C$$
$$= 1800 \, \text{J}$$

Heat lost by **water** in falling from 15 °C to 0 °C
$$= \text{mass of water} \times \text{sp. ht. cap. of water} \times \text{temp. fall}$$
$$= 200 \, \text{g} \times 4.2 \, \text{J/(g°C)} \times 15 \, °C = 12\,600 \, \text{J}$$

Heat lost by **water** at 0 °C freezing to ice at 0 °C
$$= \text{mass of water} \times \text{sp. lat. ht. of ice}$$
$$= 200 \, \text{g} \times 340 \, \text{J/g} = 68\,000 \, \text{J}$$

Heat lost by **ice** in falling from 0 °C to −5.0 °C
$$= \text{mass of ice} \times \text{sp. ht. cap. of ice} \times \text{temp. fall}$$
$$= 200 \, \text{g} \times 2.0 \, \text{J/(g°C)} \times 5.0 \, °C = 2000 \, \text{J}$$

∴ Total heat removed
$$= 1800 + 12\,600 + 68\,000 + 2000 = 84\,400 \, \text{J}$$

Questions

Use values given in Table 39.2.

1 a How much heat will change 10 g of ice at 0 °C to water at 0 °C?
   b What quantity of heat must be removed from 20 g of water at 0 °C to change it to ice at 0 °C?

2 a How much heat is needed to change 5 g of ice at 0 °C to water at 50 °C?
   b If a freezer cools 200 g of water from 20 °C to its freezing point in 10 minutes, how much heat is removed per minute from the water?

3 How long will it take a 50 W heater to melt 100 g of ice at 0 °C?

4 Some small aluminium rivets of total mass 170 g and at 100 °C are emptied into a hole in a large block of ice at 0 °C.
   a What will be the final temperature of the rivets?
   b How much ice will melt?

5 a How much heat is needed to change 4 g of water at 100 °C to steam at 100 °C?
   b Find the heat given out when 10 g of steam at 100 °C condenses and cools to water at 50 °C.

6 A 3 kW electric kettle is left on for 2 minutes after the water starts to boil. What mass of water is boiled off in this time?

7 a Why is ice good for cooling drinks?
   b Why do engineers often use superheated steam (steam above 100 °C) to transfer heat?

8 Some water is stored in a bag of porous material, e.g. canvas, which is hung where it is exposed to a draught of air. Explain why the temperature of the water is lower than that of the air.

9 Explain why a bottle of milk keeps better when it stands in water in a porous pot in a draught.

10 a Figure 39.5 shows one method of measuring the specific latent heat of fusion of ice. Two funnels, A and B, contain crushed ice at 0 °C. The mass of melted ice from each funnel is measured after 12 minutes. The joulemeter measures the energy supplied to the immersion heater.

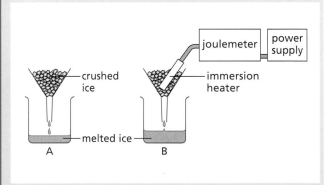

Figure 39.5

(i) What is the reason for setting up funnel A?

(ii) The measurements taken are listed below.

Mass of melted ice collected from funnel A	= 24 g
Mass of melted ice collected from funnel B	= 63 g
Joulemeter reading	= 17 160 J

Use these measurements and the following equation to calculate the specific latent heat of fusion of ice. Show clearly how you work out your answer and give the unit.

$$\text{energy exchanged} = \text{mass} \times \text{specific latent heat}$$

(iii) The measurements obtained in this experiment give a value for the specific latent heat which is higher than the accepted value. Give a reason why.

b (i) What happens to the average kinetic energy of the molecules as ice melts at 0 °C?

(ii) Energy must be given to ice if it is to change into water. Why?

(AQA (SEG) Higher, Summer 99)

11 a The table gives the melting and boiling points for lead and oxygen.

	Melting point in °C	Boiling point in °C
lead	327	1744
oxygen	−219	−183

(i) At 450 °C will the lead be a solid, a liquid or a gas?

(ii) At −200 °C will the oxygen be a solid, a liquid or a gas?

b The graph in Figure 39.6 shows how the temperature of a pure substance changes as it is heated.

(i) At what temperature does the substance boil?

(ii) Copy the graph and mark with an **X** any point where the substance exists as both a liquid and gas at the same time.

c (i) All substances consist of particles. What happens to the average kinetic energy of these particles as the substance changes from a liquid to a gas?

(ii) Explain, in terms of particles, why energy must be given to a liquid if it is to change to a gas.

(SEG Higher, Summer 98)

■ *Checklist*

After studying this chapter you should be able to

■ describe an experiment to show that during a change of state the temperature stays constant,

■ define **specific latent heat of fusion**, l_f,

■ define **specific latent heat of vaporization**, l_v,

■ explain latent heat using the kinetic theory,

■ solve problems on latent heat using $Q = ml$,

■ distinguish between evaporation and boiling,

■ describe an experiment to show that evaporation causes cooling,

■ explain cooling by evaporation using the kinetic theory,

■ explain the significance of the **critical temperature** in the liquefaction of gases.

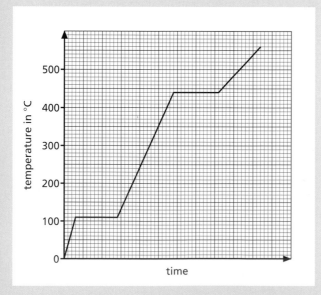

Figure 39.6

40 Conduction and convection

To keep a building or a house at a comfortable temperature in winter and in summer requires a knowledge of how heat travels, if it is to be done economically and efficiently.

Conduction

The handle of a metal spoon held in a hot drink soon gets warm. Heat passes along the spoon by **conduction**.

Conduction is the flow of heat through matter from places of higher temperature to places of lower temperature without movement of the matter as a whole.

A simple demonstration of the different conducting powers of various metals is shown in Figure 40.1. A match is fixed to one end of each rod using a little melted wax. The other ends of the rods are heated by a burner. When the temperatures of the far ends reach the melting point of wax, the matches drop off. The match on copper falls first showing it is the best conductor, followed by aluminium, brass and then iron.

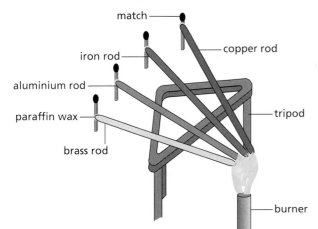

Figure 40.1 Comparing conducting powers

Heat is conducted faster through a rod if it has a large cross-section area, is short and has a large temperature difference between its ends.

Most metals are good conductors of heat; materials such as wood, glass, cork, plastics and fabrics are bad conductors. The arrangement in Figure 40.2 can be used to show the difference between brass and wood. If the rod is passed through a flame several times, the paper over the wood scorches but not over the brass. The brass conducts the heat away from the paper quickly and prevents it reaching the temperature at which it burns. The wood conducts it away only very slowly.

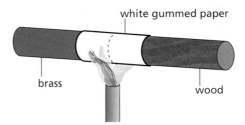

Figure 40.2 The paper over the brass does not burn

Metal objects below body temperature *feel* colder than those made of bad conductors because they carry heat away faster from the hand – even if all the objects are at exactly the same temperature.

Liquids and gases also conduct heat but only very slowly. Water is a very poor conductor as shown in Figure 40.3. The water at the top of the tube can be boiled before the ice at the bottom melts.

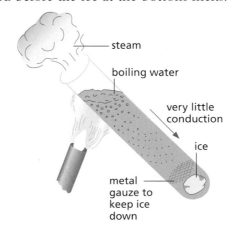

Figure 40.3 Water is a poor conductor of heat

Uses of conductors

a) Good conductors

These are used whenever heat is required to travel quickly through something. Saucepans, boilers and radiators are made of metals such as aluminium, iron and copper.

b) Bad conductors (insulators)

The handles of some saucepans are made of wood or plastic. Cork is used for table mats.

Air is one of the worst conductors, i.e. best insulators. This is why houses with cavity walls (i.e. two layers of bricks separated by an air space) and double-glazed windows keep warmer in winter and cooler in summer.

Materials which trap air, e.g. wool, felt, fur, feathers, polystyrene, fibreglass, are also very bad conductors. Some of these materials are used as 'lagging' to insulate water pipes, hot water cylinders, ovens, refrigerators and the walls and roofs of houses, Figures 40.4a, b. Others are used to make warm winter clothes like 'fleece' jackets, Figure 40.4c.

Figure 40.4a Lagging in a cavity wall provides extra insulation

Figure 40.4b Laying lagging in a house loft

Figure 40.4c Fleece jackets help you to retain your body warmth

'Wet suits' are worn by divers and water skiers to keep them warm. The suit gets wet and a layer of water gathers between the person's body and the suit. The water is warmed by body heat and stays warm because the suit is made of an insulating fabric, e.g. neoprene, a synthetic rubber.

Conduction and the kinetic theory

Two processes occur in metals. These have a large number of 'free' electrons (Chapter 44) which wander about inside them. When one part of a metal is heated, the electrons there move faster (i.e. their k.e. increases) and farther. As a result they 'jostle' atoms in cooler parts, so passing on their energy and raising the temperature of these parts. This process occurs quickly.

The second process is much slower. The atoms themselves at the hot part make 'colder' neighbouring atoms vibrate more vigorously. This is less important in metals but is the only way conduction occurs in non-metals since these do not have 'free' electrons; hence non-metals are poor conductors of heat.

Convection in liquids

Convection is the usual method by which heat travels through fluids, i.e. liquids and gases. It can be shown in water by dropping a few crystals of potassium permanganate down a tube to the bottom of a beaker or flask of water. When the tube is removed and the beaker heated just below the crystals by a *small* flame, Figure 40.5a, purple streaks of water rise upwards and fan outwards.

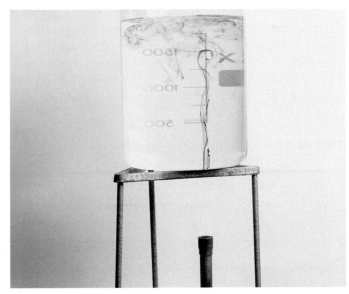

Figure 40.5a Convection currents shown by potassium permanganate in water

Figure 40.5b Lava lamps make use of convection

Streams of warm moving fluids are called **convection currents**. They arise when a fluid is heated because it expands, becomes less dense and is forced upwards by surrounding cooler, denser fluid which moves under it. We say 'hot water (or hot air) rises'. Warm fluid behaves like a cork released under water: being less dense it bobs up. Lava lamps, Figure 40.5b, use this principle.

Convection is the flow of heat through a fluid from places of higher temperature to places of lower temperature by movement of the fluid itself.

Convection in air

Black marks often appear on the wall or ceiling above a lamp or a radiator. They are caused by dust being carried upwards in air convection currents produced by the hot lamp or radiator.

A laboratory demonstration of convection currents in air can be given using the apparatus of Figure 40.6. The direction of the convection current created by the candle is made visible by the smoke from the touch paper (made by soaking brown paper in strong potassium nitrate solution and drying it).

Convection currents set up by electric, gas and oil heaters help to warm our homes. Many so-called 'radiators' are really convector heaters.

Where should the input and extraction ducts for cold/hot air be located in a room?

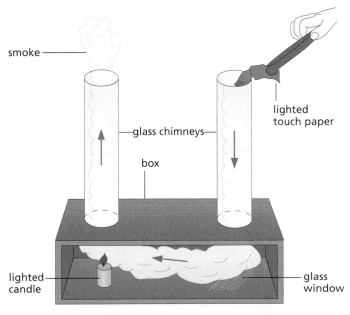

Figure 40.6 Demonstrating convection in air

Natural convection currents

a) Coastal breezes

During the day the temperature of the land increases more quickly than that of the sea (because the specific heat capacity of the land is much smaller, Chapter 38). The hot air above the land rises and is replaced by colder air from the sea. A breeze from the sea results, Figure 40.7a.

At night the opposite happens. The sea has more heat to lose and cools more slowly. The air above the sea is warmer than that over the land and a breeze blows from the land, Figure 40.7b.

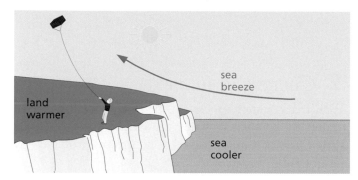

a Day

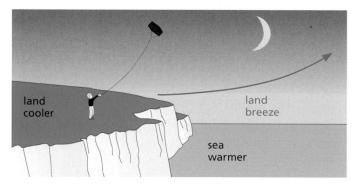

b Night

Figure 40.7 Coastal breezes are due to convection

b) Gliding

Gliders, including 'hang-gliders', Figure 40.8, depend on hot air currents, called **thermals**.

Figure 40.8 By flying from one thermal to another a hang-glider pilot can stay airborne for several hours

Energy losses from buildings

The inside of a building can only be kept at a steady temperature above that outside by heating it at a rate which equals the rate at which it is losing energy. The loss occurs mainly by conduction through the walls, roof, floors and windows. For a typical house in the UK where no special precautions have been taken, the contribution each of these makes to the total loss is shown in Table 40.1a.

As fuels (and electricity) become more expensive and the burning of fuels becomes of greater environmental concern (Chapter 42), more people are considering it worthwhile to reduce heat losses from their homes. The substantial reduction of this loss which can be achieved, especially by wall and roof insulation, is shown in Table 40.1b.

Table 40.1 Energy losses from a typical house

a

Percentage of total energy loss due to				
walls	roof	floors	windows	draughts
35	25	15	10	15

b

Percentage of each loss saved by				
insulating				
walls	roof	carpets on floors	double glazing	draught excluders
65	80	≈30	50	≈60
Percentage of total loss saved = 60				

U-value

This is a term used by heating engineers and is defined as follows.

> The *U*-value for a specified heat conductor is the heat energy lost per second through it per square metre when there is a temperature difference of 1 °C between its surfaces.

The rate of heat energy loss through a conductor can therefore be calculated knowing its *U*-value, from

$$\text{rate of energy loss} = U\text{-value} \times \text{surface area} \times \text{temp. difference}$$

Two *U*-values in joules per second per square metre per °C, i.e. $W/(m^2\,°C)$, are given below:

- tiled roof without insulation: 2.2
- tiled roof with 75 mm thick insulation: 0.45

Ventilation

In addition to supplying heat to compensate for the energy losses from a building, a heating system has also to warm the ventilated cold air, needed for comfort, which comes in to replace stale air.

If the rate of heat loss is, say, 6000 J/s, i.e. 6 kW, and the warming of ventilated air requires 2 kW, then the total power needed to maintain a certain temperature (e.g. 20 °C) in the building is 8 kW. Some of this is supplied by each person's 'body heat', estimated to be roughly equal to a 100 W heater.

Questions

1 Explain why
   a newspaper wrapping keeps hot things hot, e.g. fish and chips, and cold things cold, e.g. ice cream,
   b fur coats would keep their owners warmer if they were worn inside out,
   c a string vest helps to keep a person warm even though it is a collection of holes bounded by string.

2 Figure 40.9 illustrates three ways of reducing heat losses from a house.
   a As far as you can, explain how each of the three methods reduces heat losses. Draw diagrams where they will help your explanations.
   b Why are fibreglass and plastic foam good substances to use?
   c Air is one of the worst conductors of heat. What is the point of replacing it by the plastic foam in (ii)?
   d A vacuum is an even better heat insulator than air. Suggest one (scientific) reason why the double glazing should not have a vacuum between the sheets of glass.
   e The manufacturers of roof lagging suggest that two layers of fibreglass are more effective than one. Describe how you might set up an experiment in the laboratory to test whether this is true.

3 What is the advantage of placing an electric immersion heater
   a near the top,
   b near the bottom, of a tank of water?

4 a How much heat energy is lost in an hour through a window measuring 2.0 m by 2.5 m when the inside and outside temperatures are 18 °C and −2 °C respectively if the window is
   (i) single-glazed, U-value 6.0 W/(m²°C),
   (ii) double-glazed, U-value 3.0 W/(m²°C)?
   b What power of heater is required in case (ii) to maintain this temperature if there are no ventilation losses?

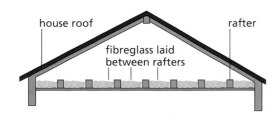

(i) Roof insulation

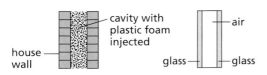

(ii) Cavity wall insulation   (iii) Double glazing

Figure 40.9

Checklist

After studying this chapter you should be able to

- describe experiments to show the different conducting powers of various substances,
- name good and bad conductors and state uses for each,
- explain conduction using the kinetic theory,
- describe experiments to show convection in fluids (liquids and gases),
- relate convection to phenomena such as land and sea breezes,
- explain the importance of insulating a building,
- recall the definition of *U*-value and use it to solve heat loss problems.

41 Radiation

Radiation is a third way in which heat can travel but whereas conduction and convection both need matter to be present, radiation can occur in a vacuum; particles of matter are not involved. It is the way heat reaches us from the Sun.

Radiation has all the properties of electromagnetic waves (Chapter 14), e.g. it travels at the speed of radio waves and gives interference effects. When it falls on an object, it is partly reflected, partly transmitted and partly absorbed; the absorbed part raises the temperature of the object.

Radiation is the flow of heat from one place to another by means of electromagnetic waves.

Radiation is emitted by all bodies above absolute zero and consists mostly of infrared radiation (Chapter 14) but light and ultraviolet are also present if the body is very hot (e.g. the Sun).

Figure 41.1 Why are buildings in hot countries often painted white?

■ *Good and bad absorbers*

Some surfaces absorb radiation better than others as may be shown using the apparatus in Figure 41.2. The inside surface of one lid is shiny and of the other dull black. The coins are stuck on the outside of each lid with candle wax. If the heater is midway between the lids they each receive the same amount of radiation. After a few minutes the wax on the black lid melts and the coin falls off. The shiny lids stay cool and the wax unmelted.

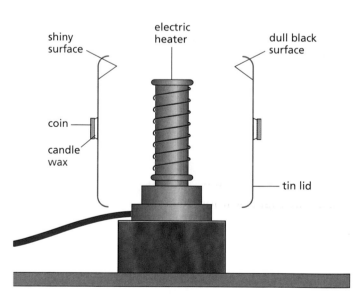

Figure 41.2 Comparing absorbers of radiation

Dull black surfaces are better absorbers of radiation than white shiny surfaces – the latter are good **reflectors** of radiation. Reflectors on electric fires are made of polished metal because of its good reflecting properties.

Good and bad emitters

Some surfaces also emit radiation better than others when they are hot. If you hold the backs of your hands on either side of a hot copper sheet which has one side polished and the other side blackened, Figure 41.3, it will be found that the **dull black surface is a better emitter of radiation than the shiny one**.

The cooling fins on the heat exchangers at the back of a refrigerator are painted black so that they lose heat more quickly. By contrast, saucepans etc. which are polished are poor emitters and keep their heat longer.

In general, surfaces that are good absorbers of radiation are good emitters when hot.

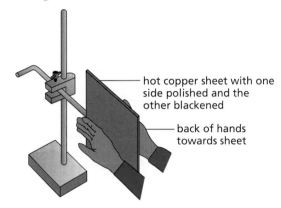

hot copper sheet with one side polished and the other blackened

back of hands towards sheet

Figure 41.3 Comparing emitters of radiation

Vacuum flask

A vacuum or Thermos flask keeps hot liquids hot or cold liquids cold. It is very difficult for heat to travel into or out of the flask.

Transfer by conduction and convection is minimized by making the flask a double-walled glass vessel with a vacuum between the walls, Figure 41.4. Radiation is reduced by silvering both walls on the vacuum side. Then if, for example, a hot liquid is stored, the small amount of radiation from the hot inside wall is reflected back across the vacuum by the silvering on the outer wall. The slight heat loss which does occur is by conduction up the thin glass walls and through the stopper.

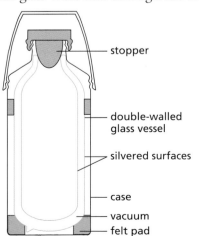

stopper

double-walled glass vessel

silvered surfaces

case

vacuum

felt pad

Figure 41.4 A vacuum flask

The greenhouse

The warmth from the Sun is not cut off by a sheet of glass but the warmth from a red-hot fire is. The radiation from very hot bodies like the Sun is mostly in the form of light and short-wavelength infrared. The radiation from less hot objects, e.g. a fire, is largely long-wavelength infrared which, unlike light and short-wavelength infrared, cannot pass through glass.

Light and short-wavelength infrared from the Sun penetrate the glass of a greenhouse and are absorbed by the soil, plants, etc., raising their temperature. These in turn emit infrared but, because of their relatively low temperature, this has a long wavelength and is not transmitted by the glass. The greenhouse thus acts as a 'heat-trap' and its temperature rises.

Carbon dioxide and other gases such as methane in the Earth's atmosphere act in a similar way to the glass of a greenhouse in trapping heat; this has serious implications for the global climate.

Rate of cooling of an object

The rate at which an object cools, i.e. at which its temperature falls, can be shown to be proportional to the ratio of its surface area A to its volume V.

For a cube of side l

$$A_1 / V_1 = 6 \times l^2 / l^3 = 6/l$$

For a cube of side $2l$

$$A_2 / V_2 = 6 \times 4l^2 / 8l^3 = 3/l = \tfrac{1}{2} \times 6/l = \tfrac{1}{2} A_1 / V_1$$

The larger cube has the smaller A/V ratio and so cools more slowly.

You could investigate this using two aluminium cubes, one having twice the length of side of the other. Each needs holes for a thermometer and an electric heater to raise them to the same starting temperature. Temperature against time graphs for both blocks can then be obtained.

Questions

1 a Choose words from the list below to complete the following sentences.

> **conduction   convection   diffusion**
> **evaporation   radiation**

(i) Thermal (heat) energy is transferred through a vacuum by......
(ii) Transfer of energy by hot liquids moving is called......
(iii) Transfer of energy by a substance, without the substance itself moving is called......

b Figure 41.5a shows a science toy. Figure 41.5b shows a close-up of one of its vanes.

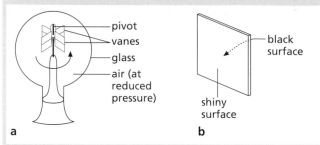

pivot
vanes
glass
air (at reduced pressure)

black surface
shiny surface

a   **b**

Figure 41.5

When the sun shines on the toy the vanes spin round. This happens because one side of each vane becomes warmer than the other.
(i) When the sun shines, which side becomes warmer, the shiny one or the black one?
(ii) Give *two* reasons for your answer.

(AQA (NEAB) Foundation, June 99)

2 The door canopy in Figure 41.6 shows in a striking way the difference between white and black surfaces when radiation falls on them. Explain why.

Figure 41.6

3 a The Earth has been warmed by the radiation from the Sun for millions of years yet we think its average temperature has remained fairly steady. Why is this?
b Why is frost less likely on a cloudy night than a clear one?

4 The diagram in Figure 41.7 comes from a leaflet about a 'fuel effect' gas fire. It shows how air circulates through the fire.

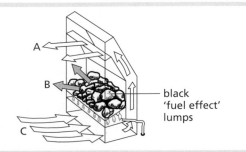

A
B
C
black 'fuel effect' lumps

Figure 41.7

a Explain in detail why the air travels from C to A.
b The black 'fuel effect' lumps become very hot.
(i) Name the process by which the lumps transfer thermal energy to the room, as shown at B.
(ii) Suggest a feature of the black 'fuel effect' lumps which make them efficient at transferring energy.

(AQA (NEAB) Higher, June 99)

5 The vacuum flask shown in Figure 41.8 has five features labelled, each one designed to reduce heat transfer.

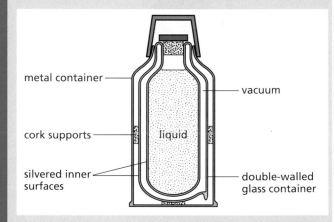

metal container
cork supports
silvered inner surfaces
vacuum
liquid
double-walled glass container

Figure 41.8

a (i) Which labelled feature of the vacuum flask reduces heat transfer by both conduction and convection?
(ii) Explain how this feature reduces heat transfer by **both** conduction and convection.
b (i) Which labelled feature of the vacuum flask reduces heat transfer by radiation?
(ii) Explain how this feature reduces heat transfer by radiation.

(SEG Higher, Summer 98)

■ Checklist

After studying this chapter you should be able to

■ describe experiments to study factors affecting the absorption and emission of radiation,
■ recall that good absorbers are also good emitters,
■ explain how a knowledge of heat transfer affects the design of a vacuum flask,
■ explain how a greenhouse acts as a 'heat-trap',
■ investigate how rate of cooling depends on the ratio of surface area to volume.

42 Energy sources

Energy is needed to heat buildings, to make cars move, to provide artificial light, to make computers work, and so on. The list is endless. This 'useful' energy needs to be produced in controllable energy transfers (Chapter 24). For example, in power stations a supply of useful energy in the form of electricity is produced. The 'raw materials' for energy production are **energy sources**. These may be **non-renewable** or **renewable**.

Non-renewable energy sources

Once used up these cannot be replaced.

a) Fossil fuels

These include coal, oil and natural gas, formed from the remains of plants and animals which lived millions of years ago and obtained energy originally from the Sun. At present they are our main energy source. Predictions vary as to how long they will last since this depends on what reserves are recoverable and on the future demands of a world population expected to increase from about 6000 million in 2000, to 7000 million by the year 2014. Some estimates say oil and gas will run low early in the present century but coal should last for 200 years or so.

Burning fossil fuels in power stations and in cars pollutes the atmosphere with harmful gases such as carbon dioxide and sulphur dioxide. Carbon dioxide emission aggravates the greenhouse effect (Chapter 41) and increases global warming. It is not immediately feasible to prevent large amounts of carbon dioxide entering the atmosphere, but less is produced by burning natural gas than by burning oil or coal; burning coal produces most carbon dioxide for each unit of energy produced. When coal and oil are burnt they also produce sulphur dioxide which causes acid rain. The sulphur dioxide can be extracted from the waste gases so it does not enter the atmosphere or the sulphur can be removed from the fuel before combustion, but these are both costly processes which increase the price of electricity produced using these measures.

b) Nuclear fuels

The energy released in a nuclear reactor (Chapter 59) from uranium, found as an ore in the ground, can be used to produce electricity. Nuclear fuels do not pollute the atmosphere with carbon dioxide or sulphur dioxide but they do generate radioactive waste materials with very long half-lives (Chapter 58); safe ways of storing this waste for perhaps thousands of years must be found. As long as a reactor is operating normally it does not pose a radiation risk, but if an accident occurs, dangerous radioactive material can leak from the reactor and spread over a large area.

Two advantages of all non-renewable fuels are

(i) their high **energy density** (i.e. they are concentrated sources) and the relatively small size of the energy transfer device (e.g. a furnace) which releases their energy, and

(ii) their ready **availability** when energy demand increases suddenly or fluctuates seasonally.

Renewable energy sources

These cannot be exhausted and are generally non-polluting.

a) Solar energy

The energy falling on the Earth from the Sun is mostly in the form of light and in an hour equals the total energy used by the world in a year. Unfortunately its low energy density requires large collecting devices and its availability varies. Its greatest potential use is as an energy source for low-temperature water heating. This uses **solar panels** as the energy transfer devices, Figure 42.1, which convert light into heat energy. They are used increasingly to produce domestic hot water at about 70 °C and to heat swimming pools.

Solar energy can also be used to produce high-temperature heating, up to 3000 °C or so, if a large curved mirror (a **solar furnace**) focuses the Sun's rays on to a small area. The energy can then be used to turn water to steam for driving the turbine of an electric generator in a power station.

Figure 42.1 Solar panels on a house provide hot water

Solar cells, made from semiconducting materials, convert sunlight into electricity directly. Panels of cells connected together supply the electronic equipment in communication and other satellites. They are also used for small-scale power generation in remote areas of developing countries where there is no electricity supply. Recent developments have made large-scale generation more cost-effective and there is now a large solar power plant in California. There are many designs for prototype light vehicles run on solar power, Figure 42.2.

Figure 42.2 Solar-powered car which won the World Solar Challenge race in 1999

b) Wind energy

Giant windmills called **wind turbines** with two or three blades each up to 30 m long drive electrical generators. 'Wind farms' of 20 to 100 turbines spaced about 400 m apart, Figure 2b, p.x, supply about 400 MW (enough electricity for 250 000 homes) in the UK and provide a useful 'top-up' to the National Grid.

They can be noisy and may be considered unsightly so there is some environmental objection to wind farms, especially as the best sites are often in coastal areas of great natural beauty.

c) Wave energy

The rise and fall of sea waves has to be transferred by some kind of wave-energy converter into the rotary motion required to drive a generator. It is a difficult problem and the large-scale production of electricity by this means is unlikely in the near future, but small systems are being developed to supply island communities with power.

d) Tidal and hydroelectric energy

The flow of water from a higher to a lower level from behind a tidal barrage (barrier) or the dam of a hydroelectric scheme is used to drive a water turbine (water wheel) connected to a generator.

One of the largest working tidal schemes is the *La Grande I* project in Canada, Figure 42.3. Feasibility studies have shown that a 10 mile long barrage across the Severn estuary could produce about 7% of today's electrical energy consumption in England and Wales. Such schemes have significant implications for the environment, as they may destroy wildlife habitats of wading birds for example, and also for shipping routes.

In the UK hydroelectric power stations generate about 2% of the electricity supply. Most are located in Scotland and Wales where the average rainfall is higher than in other areas. With good management hydroelectric energy is a reliable energy source, but there are risks connected with the construction of dams, and a variety of problems may result from the impact of a dam on the environment. Land previously used for forestry or farming may have to be flooded.

Figure 42.3 Tidal barrage in Canada

e) Geothermal energy

If cold water is pumped down a shaft into hot rocks below the Earth's surface, it may be forced up another shaft as steam. This can be used to drive a turbine and generate electricity or to heat buildings. The energy that heats the rocks is constantly being released by radioactive elements deep in the Earth as they decay (Chapter 58).

Geothermal power stations are in operation in the USA, New Zealand and Iceland.

f) Biomass (vegetable fuels)

These include cultivated crops (e.g. oil-seed rape), crop residues (e.g. cereal straw), natural vegetation (e.g. gorse), trees (e.g. spruce) grown for their wood, animal dung and sewage. **Biofuels** such as alcohol (ethanol) and methane gas are obtained from them by fermentation using enzymes or by decomposition by bacterial action in the absence of air. Liquid biofuels can replace petrol, Figure 42.4; although they have up to 50% less energy per litre, they are lead- and sulphur-free and so cleaner. **Biogas** is a mix of methane and carbon dioxide with an energy content about two-thirds that of natural gas. In developing countries it is produced from animal and human waste in 'digesters', Figure 42.5, and used for heating and cooking.

Figure 42.4 Filling up with biofuel in Brazil

Figure 42.5 Feeding a biogas digester in rural India

■ *Power stations*

The processes involved in the production of electricity at power stations depend on the energy source being used.

a) Non-renewable sources

These are used in **thermal** power stations to produce heat energy that turns water into steam. The steam drives turbines which in turn drive the generators that produce electrical energy as described in Chapter 55. If fossil fuels are the energy source (usually coal but natural gas is favoured in new stations), the steam is obtained from a boiler. If nuclear fuel is used, i.e. uranium or plutonium, the steam is produced in a heat exchanger as explained in Chapter 59.

The action of a **steam turbine** resembles that of a water wheel but moving steam not moving water causes the motion. Steam enters the turbine and is directed by the **stator** or **diaphragm** (sets of fixed blades) on to the **rotor** (sets of blades on a shaft that can rotate), Figure 42.6. The rotor revolves and drives the electrical generator. The steam expands as it passes through the turbine and the size of the blades increases along the turbine to allow for this.

Figure 42.6 The rotor of a steam turbine

The overall efficiency of thermal power stations is only about 30%. They require cooling towers to condense steam from the turbine to water and this is a waste of energy. A block diagram and an energy-transfer diagram for a thermal power station are given in Figure 42.7.

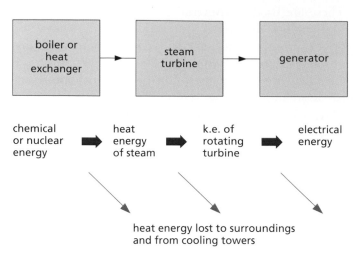

Figure 42.7 Energy transfers in a thermal power station

In gas-fired power stations, natural gas is burnt in a **gas turbine** linked directly to an electricity generator. The hot exhaust gases from the turbine are not released into the atmosphere but used to produce steam in a boiler. The steam is then used to generate more electricity from a steam turbine driving another generator. The efficiency is claimed to be over 50% without any extra fuel consumption. Furthermore, the gas turbines have a near-100% combustion efficiency so very little harmful exhaust gas (i.e. unburnt methane) is produced, and natural gas is almost sulphur-free so the environmental pollution caused is much less than for coal.

b) Renewable sources

In most cases the renewable energy source is used to drive turbines directly, as explained earlier in the cases of hydroelectric, wind, wave, tidal and geothermal schemes.

The block diagram and energy-transfer diagram for a hydroelectric scheme like that in Figure 24.3d (p.97) are shown in Figure 42.8. The efficiency of a large installation can be as high as 85–90% since many of the causes of loss in thermal power stations (e.g. water cooling towers) are absent. In some cases the generating costs are half those of thermal stations.

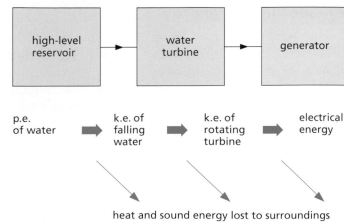

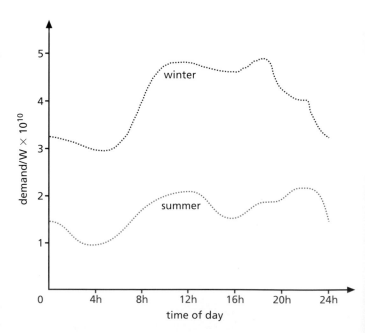

Figure 42.8 Energy transfers in a hydroelectric power station

A feature of some hydroelectric stations is **pumped storage**. Electrical energy cannot be stored on a large scale but must be used as it is generated. The demand varies with the time of day and the season, Figure 42.9, so in a pumped-storage system electricity generated at off-peak periods is used to pump water back up from a low-level reservoir to a higher-level one. It is easier to do this than to reduce the output of the generator. At peak times the p.e. of the water in the high-level reservoir is converted back into electrical energy; three-quarters of the electrical energy that was used to pump the water is generated.

Figure 42.9 Variation in power demand

Economic, environmental and social issues

When considering the large-scale generation of electricity, the economic and environmental costs of using various energy sources have to be weighed against the benefits that electricity brings to society as a 'clean', convenient and fairly 'cheap' energy supply.

Environmental problems such as polluting emissions that arise with different energy sources were outlined when each was discussed previously. Apart from people using less energy, how far pollution can be reduced by, for example, installing desulphurization processes in coal-fired power stations, is often a matter of cost.

Although there are no fuel costs associated with electricity generation from renewable energy sources such as wind power, the energy is so 'dilute' that the capital costs of setting up the generating installation are high. Similarly, although fuel costs for nuclear power stations are relatively low, the costs of building the stations and of dismantling them at the end of their useful life is higher than for gas or coal-fired stations.

It has been estimated that it currently (2001) costs between 1.5p and 2.5p to produce a unit of electricity in a gas or coal-fired power station. The cost for a nuclear power station is around 2.5p per unit. Wind energy costs vary, depending upon location, but are in the range 2.0p to 4.0p per unit. In the most favourable locations it competes with coal and gas generation.

The reliability of a source has also to be considered, as well as how easily production can be started up and shut down as demand for electricity varies. Natural gas power stations have a short start-up time, while coal and then oil power stations take successively longer to start up; nuclear power stations take longest. They are all reliable in that they can produce electricity at any time of day and in any season of the year as long as fuel is available. Hydroelectric power stations are also very reliable and have a very short start-up time which means they can be switched on when the demand for electricity peaks. The electricity output of a tidal power station, although predictable, is not as reliable because it depends on the height of the tide which varies over daily, monthly and seasonal time scales. The wind and the Sun are even less reliable sources of energy since the output of a wind turbine changes with the strength of the wind and that of a solar cell with the intensity of light falling on it; the output may not be able to match the demand for electricity at a particular time.

Renewable sources are still only being used on a small scale globally. The contribution of the main energy sources to the world's total energy consumption at present is given in Table 42.1. (The use of biofuels is not well documented.) The pattern in the UK is similar but France generates nearly three-quarters of its electricity from nuclear plants; for Japan and Taiwan the proportion is one-third, and it is in the developing economies of East Asia where interest in nuclear energy is growing most dramatically. However, the great dependence on fossil fuels worldwide is evident. It is clear the world has an energy problem, Figure 42.10.

Table 42.1 World use of energy sources

Oil	Coal	Gas	Nuclear	Hydroelectric
41%	26%	24%	7%	2%

Consumption varies from one country to another; North America and Europe are responsible for about two-thirds of the world's energy consumption each year. Table 42.2 shows approximate values for the annual consumption per head of population for different areas. These figures include the 'hidden' consumption in the manufacturing and transporting of goods.

Table 42.2 Energy consumption per head per year/$J \times 10^9$

N. America	UK	Japan	S. America	China	Africa
270	135	120	40	25	15

The world average consumption is 61×10^9 J per head per year.

Figure 42.10 An energy supply crisis in California forces a blackout on stock exchange traders

Questions

1 The pie chart in Figure 42.11 shows the percentages of the main energy sources used by a certain country.
 a What percentage is supplied by water power?
 b Which of the sources is/are renewable?
 c What is meant by 'renewable'?
 d Name two other renewable sources.
 e Why, if energy is always conserved, is it important to develop renewable sources?

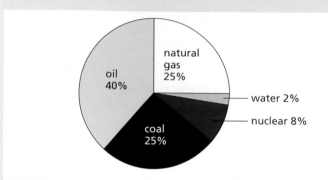

Figure 42.11

2 List *six* properties which you think the ideal energy source should have for generating electricity in a power station.

3 a List *six* social everyday benefits for which electrical energy is responsible.
 b Draw up *two* lists of suggestions for saving energy
  (i) in the home, and
  (ii) globally.

4 A group of houses uses solar panels and windmills as alternative energy sources.

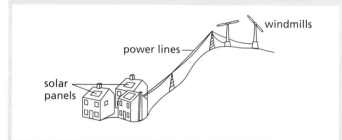

Figure 42.12

 a (i) Write down one **advantage** of using these alternative energy sources.
  (ii) Write down one **disadvantage** of using solar panels.
  (iii) Write down one **disadvantage** of using windmills.
 b Jan works out the efficiency of one of the windmills. The energy of the air hitting the blades of the windmills is 20 000 J each second. The energy transferred to the power lines is 5000 J each second.
  Calculate the efficiency of the windmill. Use the equation below. You *must* show how you work out your answer.

$$\text{energy efficiency} = \frac{\text{useful energy output}}{\text{total energy input}}$$

 c Energy can be transferred by **conduction**, **convection** and **radiation**.
  Figures 42.13a, b show details of a solar panel.

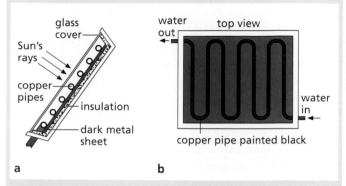

Figure 42.13

Explain why the water coming out is a lot warmer than the water going in. Use your ideas about energy transfer.

(OCR Foundation, June 99)

5 This question is about generating electricity.
 A dam has been built across a river where it meets the sea. The levels of the water go up and down twice a day. Figure 42.14 shows the water levels around the dam at high tide. The dam contains a tidal power station.

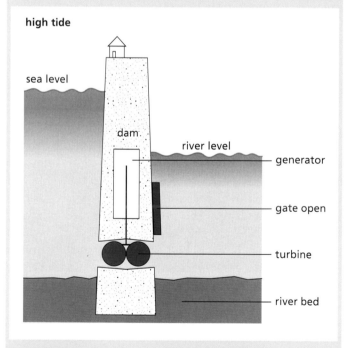

Figure 42.14

 a Describe how the energy of the water can be used to produce electricity.
 b Tidal and hydro-electric power are **renewable** sources of energy. Write down the name of *one* other renewable source of energy.
 c The gate on the dam was closed until low tide. This keeps the water level in the river high as shown in Figure 42.15.

low tide

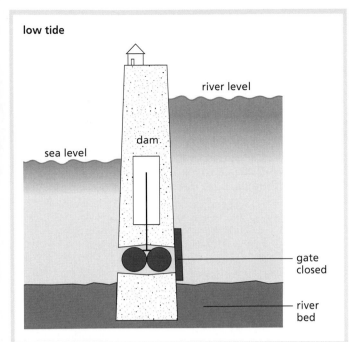

Figure 42.15

Suggest *one* problem this can cause to the environment.

(OCR Foundation, June 99)

6 The map in Figure 42.16 shows the position of towns **A** and **B** on the banks of a large river estuary. **A** is an important fishing and ferry port. The wind usually blows from the west. The major roads and railways are shown.

A power station is to be built in area **X** to generate electricity for the region. The choice is between a nuclear power station and a coal-fired power station.

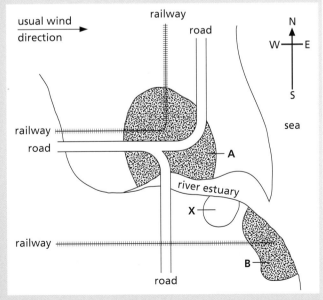

Figure 42.16

a State the advantages and disadvantages of the two methods of generating electrical energy.

b Which method would you choose for this site? Explain the reasons for your choice.

(NEAB Higher, June 98)

■ *Checklist*

After studying this chapter you should be able to

■ distinguish between **renewable** and **non-renewable** energy sources,

■ give some advantages and some disadvantages of using non-renewable fuels,

■ describe the different ways of harnessing **solar**, **wind**, **wave**, **tidal**, **hydroelectric**, **geothermal** and **biomass** energy,

■ describe the energy transfer processes in a thermal and a hydroelectric power station,

■ compare and contrast the advantages and disadvantages of using different energy sources to generate electricity,

■ discuss the environmental and economic issues of electricity production and consumption.

Heat and energy
Additional questions

Thermometers

1 a State the advantages and disadvantages of mercury and alcohol as thermometric liquids.
  b A clinical thermometer should not be sterilized in boiling water. Why?

Expansion of solids and liquids

2 When a metal bar is heated the increase in length is greater if

  1 the bar is long
  2 the temperature rise is large
  3 the bar has a large diameter.

Which statement(s) is (are) correct?

  A 1, 2, 3   **B** 1, 2   **C** 2, 3   **D** 1   **E** 3

3 A bimetallic thermostat for use in an iron is shown in the diagram below.

  1 It operates by the bimetallic strip bending away from the contact.
  2 Metal A has a greater expansivity than metal B.
  3 Screwing in the control knob raises the temperature at which the contacts open.

Which statement(s) is (are) correct?

  A 1, 2, 3   **B** 1, 2   **C** 2, 3   **D** 1   **E** 3

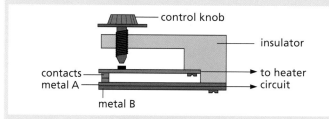

4 Which one of the graphs **A** to **E** most nearly shows how the volume of water changes with temperature between −5 °C and +10 °C?

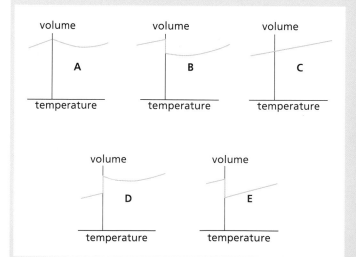

The gas laws

5 Carbon dioxide is used to make fizzy drinks. It is stored at high pressure in a cast iron cylinder. The diagram represents the particles in a cylinder of carbon dioxide.

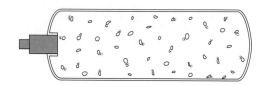

a Describe how the particles of carbon dioxide exert pressure.
b The temperature of the gas in the cylinder is increased.
  (i) Describe the effect this has on the movement of the carbon dioxide particles.
  (ii) Explain how this affects the pressure exerted by the gas.
  (iii) The cylinders are painted black. Explain why the cylinders should *not* be stored outside in direct sunlight.

(London Foundation, June 99)

6 Anna investigates whether a trapped column of air in a thin glass tube acts like a thermometer. The tube is sealed at the bottom and open at the top. The air is trapped by a short length of coloured liquid.

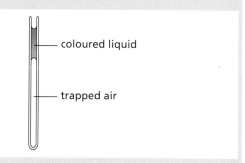

She measures the length of the column of air. She plots these results on a graph (opposite).

	Temperature	Length of air column
at room temperature	20 °C	78 mm
when the tube is in melting ice	0 °C	73 mm
when the tube is in boiling water	100 °C	100 mm

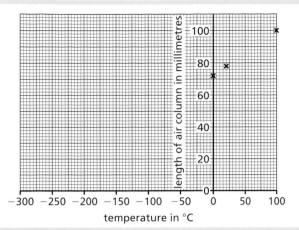

a (i) Use the graph to find the temperature at which the length of the air column is zero.
(ii) What do we call this temperature?
(iii) Explain why the pressure of a gas at this temperature is zero.
b Use your ideas about particles in air to explain why the column of liquid does not fall to the bottom of the tube at room temperature.

(*OCR Higher, June 99*)

Specific heat capacity; latent heat

7 A certain liquid has a specific heat capacity of 4.0 J/(g °C). How much heat must be supplied to raise the temperature of 10 g of the liquid from 20 °C to 50 °C?

8 The diagrams show how the rate of heat loss from a jogger running at a steady pace changes with different air temperatures.

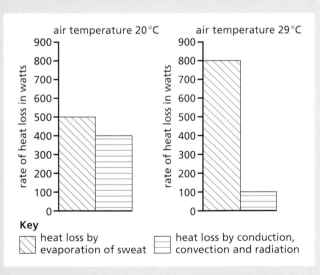

Key
▨ heat loss by evaporation of sweat
▥ heat loss by conduction, convection and radiation

a (i) Use the information from the diagrams to calculate how much energy, in joules, the jogger will lose by evaporation in one hour when the air temperature is 20 °C.
(ii) Use the following equation

energy exchanged = mass × specific latent heat

to calculate the mass of sweat produced by the jogger in one hour. Show clearly how you work out your final answer and give the unit.
(Take the specific latent heat of evaporation of sweat as 2 400 000 J/kg.)

b For the first 10 minutes after the jogger has finished running, her body continues to lose heat at an average rate of 700 watts. Use the following equation

energy exchanged = mass × specific heat capacity × temperature change

to calculate the drop in body temperature which will occur in this time. Show clearly how you work out your final answer and give the unit.
(Take the specific heat capacity of the jogger as 4000 J/kg °C and the mass of the jogger as 50 kg.)
c The diagrams show that as the air temperature goes up the jogger loses more heat by the evaporation of sweat and less by conduction, convection and radiation. Explain why.

(*SEG Higher, Summer 98*)

Conduction and convection; radiation

9 Explain why on a cold day the metal handlebars of a bicycle feel colder than the rubber grips.

10 This question is about keeping a house warm.
A house has been insulated in these two ways.

1 The windows are double glazed.
2 There is shiny aluminium foil on the wall behind the radiators.

a Describe how each of these ways helps to keep the house warm. Use your ideas about conduction, convection and radiation.
(i) Double glazing.

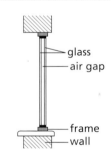

(ii) Putting shiny aluminium foil on the wall behind a radiator.

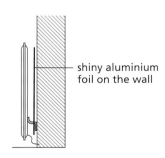

b Write down one *other* type of insulation that can be used in houses.

(*OCR Foundation, June 99*)

■ **189**

11 A heating engineer designs a storage heater, which must contain either concrete or oil. Electric elements are used to heat up the heater at night when electricity costs less.

a The storage heater contains a box measuring 1.0 m long, 0.5 m high and 0.2 m deep.

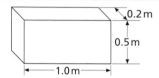

Calculate the volume of this box. Use the equation below. You *must* show how you work out your answer.

volume = length × depth × height

b The engineer works out the mass of concrete which would fill the box. He writes this in the table.

Material	Energy to raise the temperature of 1 kg by 1 deg C	Density	Mass of material to fill box
concrete	3400 J	2200 kg/m³	*220* kg
oil	2000 J	760 kg/m³	. . . kg

Finish the table by writing the mass of the oil.

c The engineer decides to use concrete blocks to store energy. Look at all the information in the table above. Write down *two* reasons which support this decision.

d The box is filled with concrete blocks. Each of the concrete blocks has a mass of 5 kg. One of the blocks is heated from 15 °C to 65 °C. Calculate the energy gained by this block. Use the equation below. You *must* show how you work out your answer.

**energy gained = mass
× specific heat capacity
× temperature rise**

e The hot concrete block is put in contact with a cold concrete block. Both blocks are covered with a good insulator so that no heat is lost. The graph shows how the temperature of each block changes with time.

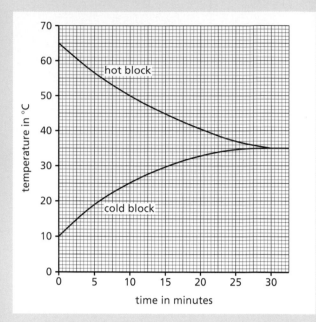

(i) Use the graph to calculate the average rate of cooling of the hot block between 5 and 10 minutes. You *must* show with reference to the graph how you work out your answer.

(ii) How can you tell from the graph that the cold concrete block has a larger mass than the hot concrete block?

f The graph shows that
• initially the hot block cools down and the cold block warms up.
• after 30 minutes the temperatures of both blocks become steady.

A model for explaining how energy transfers between the blocks involves the random transfer of tiny packets of energy. Use your ideas about these energy packets and the rate at which they transfer energy to explain these observations.

(OCR Higher, June 99)

12 Calculate the rate of heat energy loss through the four double-brick air-cavity walls of a building measuring 10 m by 10 m along the ground. The walls are 5 m high and the temperatures outside and inside the building are 3 °C and 18 °C respectively.
(*U*-value of walls = 2.0 W/m² °C)

Electricity and electromagnetic effects

43 *Static electricity*

Figure 43.1 A flash of lightning is nature's most spectacular static electricity effect

Clothes containing nylon often crackle when they are taken off. We say they are 'charged with static electricity'; the crackles are caused by tiny electric sparks which can be seen in the dark. Pens and combs made of certain plastics become charged when rubbed on your sleeve and can then attract scraps of paper.

■ *Positive and negative charges*

When a strip of polythene (white) is rubbed with a cloth it becomes charged. If it is hung up and another rubbed polythene strip is brought near, repulsion occurs, Figure 43.2. Attraction occurs when a rubbed strip of cellulose acetate (clear) approaches.

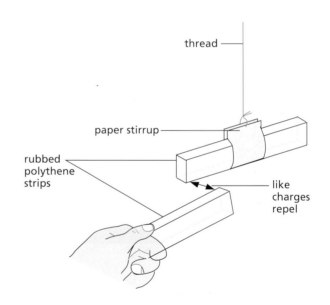

Figure 43.2 Investigating charges

This shows there are two kinds of electric charge. That on cellulose acetate is taken as **positive** (+) and that on polythene is **negative** (−). It also shows that:

Like charges (+ and + or − and −) repel, while unlike charges (+ and −) attract.

The force between electric charges decreases as their separation increases.

Charges, atoms and electrons

There is evidence (Chapter 59) that we can picture an atom as made up of a small central nucleus containing positively charged particles called **protons**, surrounded by an equal number of negatively charged **electrons**. The charges on a proton and an electron are equal and so an atom as a whole is normally electrically neutral, i.e. has no net charge.

Hydrogen is the simplest atom with one proton and one electron, Figure 43.3. A copper atom has 29 protons in the nucleus and 29 surrounding electrons. Every nucleus except hydrogen also contains uncharged particles called **neutrons**.

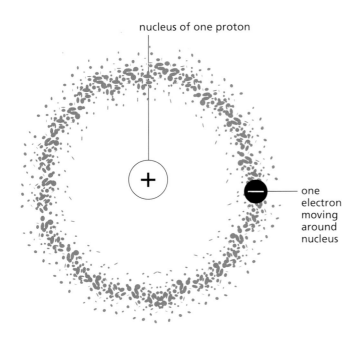

Figure 43.3 Hydrogen atom

The production of charges by rubbing can be explained by supposing that electrons are transferred from one material to the other. For example, when cellulose acetate is rubbed, electrons go from the acetate to the cloth, leaving the acetate short of electrons, i.e. positively charged. The cloth now has more electrons than protons and becomes negatively charged. Note that it is only electrons which move; the protons remain fixed in the nucleus.

How does polythene become charged when rubbed?

Practical work

Gold-leaf electroscope

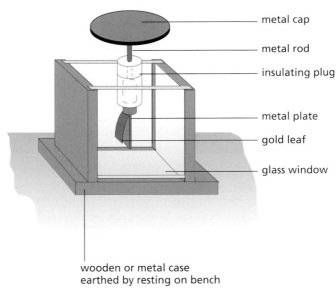

Figure 43.4 Gold-leaf electroscope

A gold-leaf electroscope consists of a metal cap on a metal rod at the foot of which is a metal plate having a leaf of gold foil attached, Figure 43.4. The rod is held by an insulating plastic plug in a case with glass sides to protect the leaf from draughts.

a) Detecting a charge

Bring a charged polythene strip towards the cap: the leaf rises away from the plate. On removing the charged strip the leaf falls again. Repeat with a charged acetate strip.

b) Charging by contact

Draw a charged polythene strip *firmly across the edge* of the cap. The leaf should rise and stay up when the strip is removed. If it does not, repeat the process but press harder. The electroscope has now become negatively charged by contact with the polythene strip, from which electrons have been transferred.

c) Insulators and conductors

Touch the cap of the charged electroscope with different things, e.g. a piece of paper, a wire, your finger, a comb, a cotton handkerchief, a piece of wood, a glass rod, a plastic pen, rubber tubing.

When the leaf falls, charge is passing to or from the ground through you and the material touching the cap. If the fall is rapid the material is a **good conductor**, if slow, it is a poor conductor and if the leaf does not alter the material is a **good insulator**. Record your results.

■ Electrons, insulators and conductors

In an insulator all electrons are bound firmly to their atoms; in a conductor some electrons can move freely from atom to atom. An insulator can be charged by rubbing because the charge produced cannot move from where the rubbing occurs, i.e. the electric charge is static. A conductor will become charged only if it is held with an insulating handle; otherwise electrons are transferred between the conductor and the ground via the person's body.

Good insulators are plastics such as polythene, cellulose acetate, Perspex, nylon. All metals and carbon are good conductors. In between are materials that are both poor conductors and (because they conduct to some extent) poor insulators. Examples are wood, paper, cotton, the human body, the Earth. Water conducts and if it were not present in materials like wood and on the surface of, for example, glass, these would be good insulators. Dry air insulates well.

■ Electrostatic induction

This effect may be shown by bringing a negatively charged polythene strip near to an insulated metal sphere X which is touching a similar sphere Y, Figure 43.5a. Electrons in the spheres are repelled to the far side of Y.

If X and Y are separated, with the charged strip still in position, X is left with a positive charge (deficient of electrons) and Y with a negative charge (excess of electrons), Figure 43.5b. The signs of the charges can be tested by removing the charged strip (c) and taking X up to the cap of a positively charged electroscope. Electrons will be drawn towards X, making the leaf more positive so that it rises. If Y is taken towards the cap of a *negatively* charged electroscope the leaf again rises; can you explain why, in terms of electron motion?

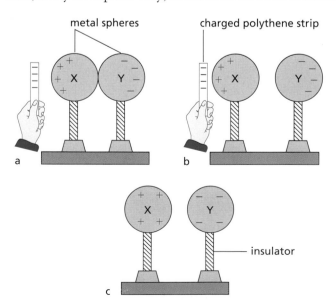

Figure 43.5 Electrostatic induction

■ Attraction between uncharged and charged objects

The attraction of an uncharged object by a charged object near it is due to electrostatic induction.

In Figure 43.6a a small piece of aluminium foil is attracted to a negatively charged polythene rod held just above it. The charge on the rod pushes free electrons to the bottom of the foil (aluminium is a conductor), leaving the top of the foil short of electrons, i.e. with a net positive charge, and the bottom negatively charged. The top of the foil is nearer the rod than the bottom. Hence the force of attraction between the negative charge on the rod and the positive charge on the top of the foil is greater than the force of repulsion between the negative charge on the rod and the negative charge on the bottom of the foil. The foil is therefore pulled to the rod.

A small scrap of paper, although an insulator, is also attracted by a charged rod. There are no free electrons in the paper but the charged rod pulls the electrons of the atoms in the paper slightly closer (by electrostatic induction) and so distorts the atoms. In the case of a negatively charged polythene rod, the paper behaves as if it had a positively charged top and a negative charge at the bottom.

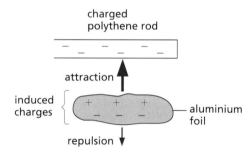

Figure 43.6a An uncharged object is attracted to a charged one

Figure 43.6b A slow stream of water being bent by electrostatic attraction

In Figure 43.6b a slow, uncharged stream of water is attracted by a charged polythene rod, due to similar distortion of the water molecules.

■ Dangers of static electricity

a) Lightning

A tall building is protected by a lightning conductor consisting of a thick copper strip on the outside of the building connecting metal spikes at the top to a metal plate in the ground, Figure 43.7.

Thunderclouds carry charges; a negatively charged one passing overhead repels electrons from the spikes to the Earth. The points of the spikes are left with a large positive charge (charge concentrates on sharp points) which removes electrons from nearby air molecules, so charging them positively and causing them to be repelled from the spike. This effect, called **action at points**, results in an 'electric wind' of positive air molecules streaming upwards to cancel some of the charge on the cloud. If a flash occurs it is now less violent and the conductor gives it an easy path to ground.

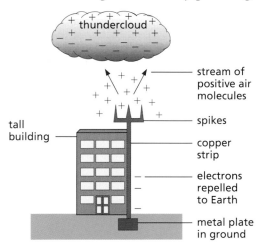

Figure 43.7 Lightning conductor

b) Refuelling

Sparks from static electricity can be dangerous when flammable vapour is present. For this reason, the tanks in an oil-tanker may be cleaned in an atmosphere of nitrogen – otherwise oxygen in the air could promote a fire.

An aircraft in flight may become charged by 'rubbing' the air. Its tyres are made of conducting rubber which lets the charge pass harmlessly to ground on landing, otherwise an explosion could be 'sparked off' when the aircraft refuels. What precautions are taken at petrol pumps when a car is refuelled?

c) Operating theatres

Dust and germs are attracted by charged objects and so it is essential to ensure that equipment and medical personnel are well 'earthed' allowing electrons to flow to and from the ground, e.g. by conducting rubber.

d) Computers

Computers require similar 'anti-static' conditions as they are vulnerable to electrostatic damage.

■ Uses of static electricity

a) Flue-ash precipitation

An electrostatic precipitator removes dust and ash that goes up the chimneys of coal-burning power stations. It consists of a charged fine wire mesh which gives a similar charge to the rising particles of ash. They are then attracted to plates with an opposite charge. These are tapped from time to time to remove the ash which falls to the bottom of the chimney, from where it is removed.

b) Photocopiers

These contain a charged drum and when the paper to be copied is laid on the glass plate, the light reflected from the white parts of the paper causes the charge to disappear from the corresponding parts of the drum opposite. The charge pattern remaining on the drum corresponds to the dark-coloured printing on the original. Special **toner** powder is then dusted over the drum and sticks to those parts which are still charged. When a sheet of paper passes over the drum, the particles of toner are attracted to it and fused into place by a short burst of heat.

c) Inkjet printers

In an inkjet printer tiny drops of ink are forced out of a fine nozzle, charged electrostatically and then passed between two oppositely charged plates; a negatively charged drop will be attracted towards the positive plate causing it to be deflected as shown in Figure 43.8. The amount of deflection and hence the position at which the ink strikes the page is determined by the charge on the drop and the p.d. between the plates; both of these are controlled by a computer. About 100 precisely located drops are needed to make up an individual letter but very fast printing speeds can be achieved.

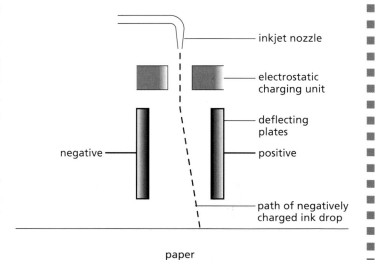

Figure 43.8 Inkjet printer

▪ *Van de Graaff generator*

This produces a continuous supply of charge on a large metal dome when a rubber belt is driven by an electric motor or by hand, Figure 43.9.

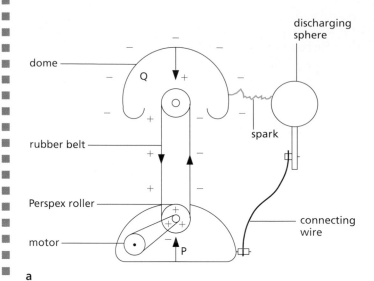

a

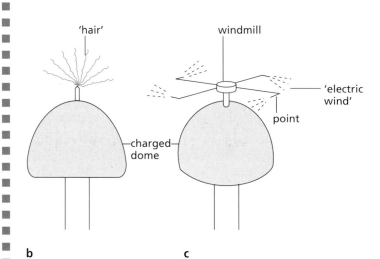

b   **c**

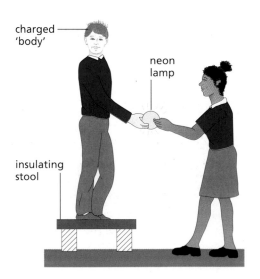

d

Figure 43.9

a) Demonstrations

In Figure 43.9a sparks jump between the dome and the discharging sphere. Electrons flow round a complete path (circuit) from the dome. Can you trace it? In b why does the 'hair' stand on end? In c the 'windmill' revolves due to the reaction that arises from the 'electric wind' caused by the **action at points** effect, explained above for the lightning conductor. In d the 'body' on the insulating stool first gets charged by touching the dome and then lights a neon lamp.

The dome can be discharged harmlessly by bringing your elbow close to it.

b) Action

Initially a positive charge is produced on the motor-driven Perspex roller due to it rubbing the belt. This induces a negative charge on the 'comb' of metal points P, Figure 43.9a, which are sprayed off by 'action at points' onto the outside of the belt and carried upwards. A positive charge is then induced in the comb of metal points Q and negative charge is repelled to the dome.

Questions

1 Two identical conducting balls, suspended on nylon threads, come to rest with the threads making equal angles with the vertical, Figure 43.10. This shows that:

A the balls are equally and oppositely charged
B the balls are oppositely charged but not necessarily equally charged
C one ball is charged and the other is uncharged
D the balls both carry the same type of charge
E one is charged and the other may or may not be charged.

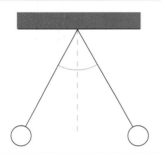

Figure 43.10

2 Explain in terms of electron movement what happens when a polythene rod becomes charged negatively by being rubbed with a cloth.

3 a One method of painting a car uses electrostatics. A paint spray produces paint droplets, all of which are given a positive charge. The car body is given a negative charge, Figure 43.11.

positively charged paint droplets   negatively charged car body

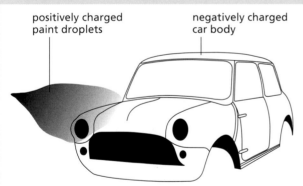

Figure 43.11

(i) Explain why it is important to give all of the paint droplets a positive charge.
(ii) Explain why it is important to give the car body a negative charge.

b Figure 43.12 shows a light aircraft being refuelled after a flight.

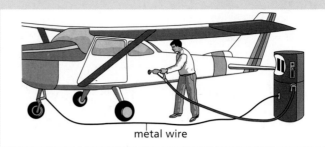

metal wire

Figure 43.12

Explain why it is important that, before refuelling starts, the aircraft is first connected to the fuel pump by a metal wire.

(SEG Foundation, Summer 98)

■ *Checklist*

After studying this chapter you should be able to

■ describe how positive and negative charges are produced by rubbing,

■ recall that like charges repel and unlike charges attract,

■ explain the charging of objects in terms of the motion of negatively charged electrons,

■ describe the gold-leaf electroscope, and explain how it can be used to compare electrical conductivities of different materials,

■ explain the differences between insulators and conductors,

■ describe how a conductor can be charged by induction,

■ explain how a charged object can attract uncharged objects,

■ give examples of the dangers and the uses of static electricity,

■ describe the action of a van de Graaff generator and explain effects observed with its use.

44 Electric current

Effects of a current
The ampere and the coulomb
Circuit diagrams
Series and parallel circuits

Direct and alternating current
Practical work
Measuring current.

An electric current consists of moving electric charges. In Figure 44.1 when the van de Graaff is working, the table tennis ball shuttles rapidly to and fro between the plates and the meter records a small current. As the ball touches each plate it becomes charged and is repelled to the other plate. In this way charge is carried across the gap. This also shows that 'static' charges cause a deflection on a meter just as current electricity produced by a battery does.

In a metal, each atom has one or more loosely held electrons that are free to move. When a van de Graaff or a battery is connected across the ends of such a conductor, the free electrons drift slowly along it in the direction from the negative to the positive terminal of a battery. There is then a current of negative charge.

■ *Effects of a current*

An electric current has three effects that reveal its existence and which can be shown with the circuit of Figure 44.2.

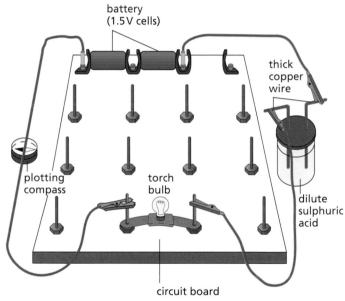

Figure 44.2 Investigating the effects of a current

a) Heating and lighting

The bulb lights due to a small wire in it (the filament) being made white hot by the current.

b) Magnetic

The plotting compass is deflected when it is placed in the magnetic field produced round any wire carrying a current.

c) Chemical

Bubbles of gas are given off at the wires in the acid because of the chemical action of the current.

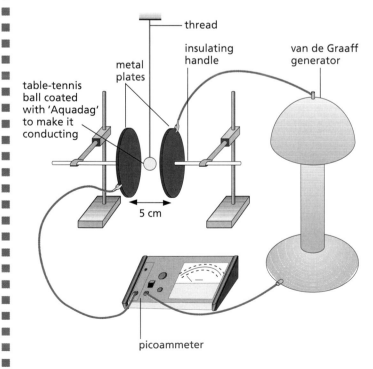

Figure 44.1 Demonstrating that an electric current consists of moving charges

■ *The ampere and the coulomb*

The unit of current is the **ampere** (A) which is defined using the magnetic effect. One milliampere (mA) is one-thousandth of an ampere. Current is measured by an **ammeter**.

The unit of charge, the **coulomb** (C), is defined in terms of the ampere.

> One coulomb is the charge passing any point in a circuit when a steady current of 1 ampere flows for 1 second. That is, $1\,C = 1\,As$.

A charge of 3 C would pass each point in 1 s if the current was 3 A. In 2 s, $3\,A \times 2\,s = 6\,As = 6\,C$ would pass. In general, if a steady current I (amperes) flows for time t (seconds) the charge Q (coulombs) passing any point is given by

$$Q = I \times t$$

This is a useful expression connecting charge and current.

■ *Circuit diagrams*

Current must have a complete path (a circuit) of conductors if it is to flow. Wires of copper are used to connect batteries, lamps, etc. in a circuit since copper is a good electrical conductor. If the wires are covered with insulation, e.g. plastic, the ends are bared for connecting up.

The signs or symbols used for various parts of an electric circuit are shown in Figure 44.3.

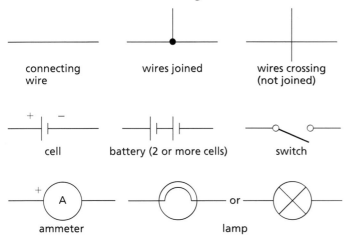

Figure 44.3 Circuit symbols

Before the electron was discovered scientists agreed to think of current as positive charges moving round a circuit in the direction from positive to negative of a battery. This agreement still stands. Arrows on circuit diagrams show the direction of what we call the **conventional current**, i.e. the direction in which **positive** charges would flow.

Practical work

Measuring current

(a) Connect the circuit of Figure 44.4a (on a circuit board if possible) ensuring that the + of the cell (the metal stud) goes to the + of the ammeter (marked red). Note the current.

(b) Connect the circuit of Figure 44.4b. The cells are **in series** (+ of one to − of the other), as are the lamps. Record the current. Measure the current at B, C and D by disconnecting the circuit at each point in turn and inserting the ammeter. What do you find?

(c) Connect the circuit of Figure 44.4c. The lamps are **in parallel**. Read the ammeter. Also measure the currents at P, Q and R. What is your conclusion?

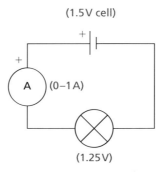

a

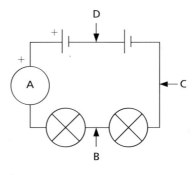

b

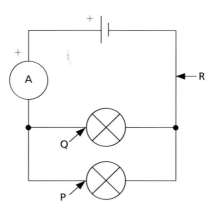

c

Figure 44.4

■ *Series and parallel circuits*

a) Series

In a series circuit, Figure 44.4b, p.199, the different parts follow one after the other and there is just one path for the current to follow. You should have found in the previous experiment that the reading on the ammeter when in the position shown (e.g. 0.2 A) is also obtained at B, C and D. That is, current is not used up.

> The current is the same at all points in a series circuit.

b) Parallel

In a parallel circuit, Figure 44.4c, the lamps are side by side and alternative paths are provided for the current which splits. Some goes through one lamp and the rest through the other. For example, if the ammeter reading was 0.4 A in the position shown, then if the lamps are identical the reading at P would be 0.2 A, as it would be at Q, giving a total of 0.4 A. Whether the current splits equally or not depends on the lamps (as we will see later); it might divide so that 0.3 A goes one way and 0.1 A by the other branch.

> The sum of the currents in the branches of a parallel circuit equals the current entering or leaving the parallel section.

■ *Direct and alternating current*

a) Difference

In a **direct current (d.c.)** the electrons flow in one direction only. Graphs for steady and varying d.c. are shown in Figure 44.5. In an **alternating current (a.c.)** the direction of flow reverses regularly, Figure 44.6. The circuit sign for a.c. is ⟶ ⌇ •⟶

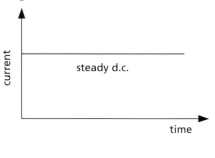

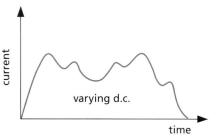

Figure 44.5 Direct current

The pointer of an ammeter for measuring d.c. is deflected one way by d.c.; a.c. makes it move to and fro about the zero if the changes are slow enough, otherwise there is no deflection.

Batteries give d.c.; generators can produce either d.c. or a.c.

b) Frequency of a.c.

The number of complete alternations or cycles in 1 second is the **frequency** of the a.c. The unit of frequency is the **hertz** (Hz). The frequency of the a.c. in Figure 44.6 is 2 Hz, which means there are two cycles per second, or one cycle lasts $\frac{1}{2} = 0.5$ s. The mains supply in the UK is a.c. of frequency 50 Hz; each cycle lasts $\frac{1}{50}$th of a second. This regularity was used in the tickertape timer (Chapter 28) and is relied upon in mains-operated clocks.

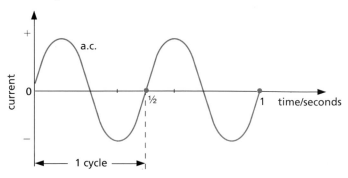

Figure 44.6 Alternating current

Questions

1 If the current through a floodlamp is 5 A, what charge passes in
  a 1 s,
  b 10 s,
  c 5 minutes?

2 What is the current in a circuit if the charge passing each point is
  a 10 C in 2 s,
  b 20 C in 40 s,
  c 240 C in 2 minutes?

3 Here are some circuit symbols.

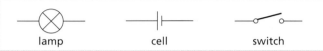

lamp   cell   switch

Figure 44.7

  a (i) Draw a circuit diagram showing two lamps, one cell and a switch connected in series.
   (ii) How can you change the brightness of the lamps?
  b (i) Draw a circuit diagram showing two lamps in parallel, together with a cell and a switch that works both lamps.
   (ii) An ammeter is used to measure the current in the cell. Mark an (A) on your diagram to show where the ammeter should be placed.
                         (*London Foundation, June 99*)

4 Study the circuits in Figure 44.8. The switch S is open (there is a break in the circuit at this point). In which circuit would lamps Q and R light but not lamp P?

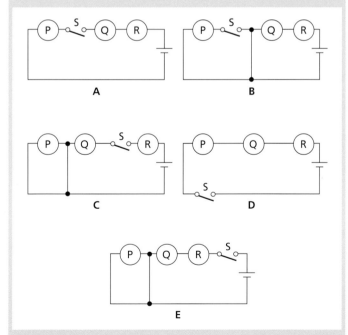

Figure 44.8

5 Using the circuit in Figure 44.9, which of the following statements is correct?

  A When S_1 and S_2 are closed A and B are lit.
  B With S_1 open and S_2 closed A is lit and B is not lit.
  C With S_2 open and S_1 closed A and B are lit.

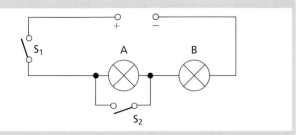

Figure 44.9

6 If the lamps are both the same in Figure 44.10 and if A_1 reads 0.50 A, what do A_2, A_3, A_4 and A_5 read?

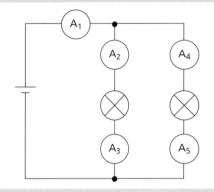

Figure 44.10

■ *Checklist*

After studying this chapter you should be able to

■ describe a demonstration which shows that an electric current is a flow of charge,

■ recall that an electric current in a metal is a flow of negative electrons from the negative to the positive terminal of the battery round a circuit,

■ state the three effects of an electric current,

■ state the unit of electric current and recall that current is measured by an ammeter,

■ define the unit of charge in terms of the unit of current,

■ recall the relation $Q = It$ and use it to solve problems,

■ use circuit symbols for wires, cells, switches, ammeters and lamps,

■ draw and connect simple series and parallel circuits, observing correct polarities for meters,

■ recall that the current in a series circuit is the same everywhere in the circuit,

■ recall that the sum of the currents in the branches of a parallel circuit equals the current entering or leaving the parallel section,

■ distinguish between electron flow and conventional current,

■ distinguish between direct current and alternating current,

■ recall that frequency of a.c. is the number of cycles per second.

45 Potential difference

Energy transfers and p.d.
Model of a circuit
The volt
Cells and batteries

Voltages round a circuit
Practical work
Measuring voltage.

A battery transforms chemical energy to electrical energy. Because of the chemical action going on inside it, it builds up a surplus of electrons at one of its terminals (the negative) and creates a shortage at the other (the positive). It is then able to maintain a **flow of electrons**, i.e. an **electric current**, in any circuit connected across its terminals so long as the chemical action lasts.

The battery is said to have a **potential difference** (**p.d.** for short) at its terminals. Potential difference is measured in **volts** (V) and the term **voltage** is sometimes used instead of p.d. The p.d. of a car battery is 12 V and of the domestic mains supply in the UK 230 V.

■ *Energy transfers and p.d.*

In an electric circuit electrical energy is supplied from a source such as a battery and is transferred to other forms of energy by devices in the circuit. A lamp produces heat and light.

If the circuits of Figure 45.1 are connected up, it will be found from the ammeter readings that the current is about the same (0.4 A) in each lamp. However, the mains lamp with a p.d. of 230 V applied to it gives much more light and heat than the car lamp with 12 V across it. In terms of energy, the mains lamp transfers a great deal more electrical energy in a second than the car lamp.

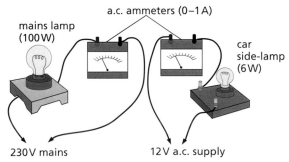

Figure 45.1 Investigating the effect of p.d. on energy transfer

Evidently the p.d. across a device affects the rate at which it transfers electrical energy. This gives us a way of defining the unit of p.d. – the volt.

■ *Model of a circuit*

It may help you to understand the definition of the volt, i.e. what a volt is, if you *imagine* that the current in a circuit is formed by 'drops' of electricity, each having a charge of 1 coulomb and carrying equal-sized 'bundles' of electrical energy. In Figure 45.2, Mr Coulomb represents one such 'drop'. As a 'drop' moves around the circuit it gives up all its energy which is changed to other forms of energy. **Note that electrical energy, not charge or current, is 'used up'.**

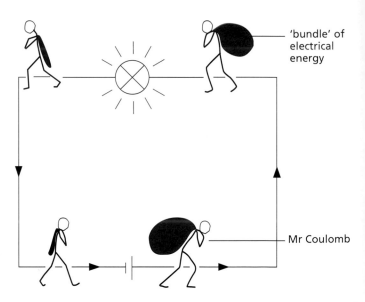

Figure 45.2 Model of a circuit

In our imaginary representation, Mr Coulomb travels round the circuit and unloads energy as he goes, most of it in the lamp. We think of him receiving a fresh 'bundle' every time he passes through the battery, which suggests he must be travelling very fast. In fact, as we found earlier (Chapter 44), the electrons drift along quite slowly. As soon as the circuit is complete, energy is delivered at once to the lamp, not by electrons directly from the battery but from electrons that were in the connecting wires. The model is helpful but not an exact representation.

■ *The volt*

The demonstrations of Figure 45.1 show that the greater the voltage at the terminals of a supply, the larger is the 'bundle' of electrical energy given to each coulomb and the greater is the rate at which light and heat are produced in a lamp.

> The p.d. between two points in a circuit is 1 volt if 1 joule of electrical energy is transferred to other forms of energy when 1 coulomb passes from one point to the other.

That is, 1 volt = 1 joule per coulomb ($1\,V = 1\,J/C$). If 2J are given up by each coulomb, the p.d. is 2V. If 6J are transferred when 2C pass, the p.d. is $6\,J/2\,C = 3\,V$.

In general if W (joules) is the energy transferred (i.e. the work done) when charge Q (coulombs) passes between two points, the p.d. V (volts) between the points is given by

$$V = W/Q \quad \text{or} \quad W = Q \times V$$

If Q is in the form of a steady current I (amperes) flowing for time t (seconds) then $Q = I \times t$ (Chapter 44) and

$$W = I \times t \times V$$

■ *Cells and batteries*

A 'battery' consists of two or more **electric cells** (see Chapter 48). Greater voltages are obtained when cells are joined in series, i.e. + of one to − of next. In Figure 45.3a the two 1.5V cells give a voltage of 3V at the terminals A, B. Every coulomb in a circuit connected to this battery will have 3J of electrical energy.

The cells in Figure 45.3b are in opposition and the voltage at X, Y is zero.

If two 1.5V cells are connected in parallel, Figure 45.3c, the voltage at terminals P, Q is still 1.5V but the arrangement behaves like a larger cell and will last longer.

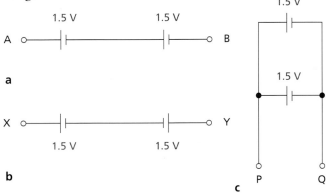

Figure 45.3

Practical work

Measuring voltage

A **voltmeter** is an instrument for measuring voltage or p.d. It looks like an ammeter but has a scale marked in volts. Whereas an ammeter is inserted **in series** in a circuit to measure the current, a voltmeter is connected across that part of the circuit where the voltage is required, i.e. **in parallel**. (We will see later that a voltmeter should have a high resistance and an ammeter a low resistance.)

To prevent damage the + terminal (marked red) must be connected to the point nearest the + of the battery.

(a) Connect the circuit of Figure 45.4a. The voltmeter gives the voltage across the lamp. Read it.

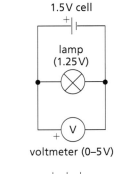

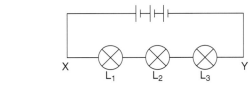

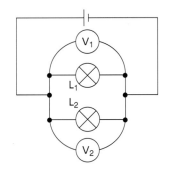

Figure 45.4

(b) Connect the circuit of Figure 45.4b. Measure:

(i) the voltage V between X and Y,
(ii) the voltage V_1 across lamp L_1,
(iii) the voltage V_2 across lamp L_2,
(iv) the voltage V_3 across lamp L_3.

How does the value of V compare with $V_1 + V_2 + V_3$?

(c) Connect the circuit of Figure 45.4c, so that two lamps L_1 and L_2 are in parallel across one 1.5V cell. Measure the voltages, V_1 and V_2, across each lamp in turn. How do V_1 and V_2 compare?

■ *Voltages round a circuit*

a) Series

In the previous experiment you should have found in the circuit of Figure 45.4b that

$$V = V_1 + V_2 + V_3$$

For example, if $V_1 = 1.4\,V$, $V_2 = 1.5\,V$ and $V_3 = 1.6\,V$, then V should be $(1.4 + 1.5 + 1.6) = 4.5\,V$.

> The voltage at the terminals of a battery equals the sum of the voltages across the devices in the external circuit from one battery terminal to the other.

b) Parallel

In the circuit of Figure 45.4c

$$V_1 = V_2$$

> The voltages across devices in parallel in a circuit are equal.

Questions

1 The p.d. across the lamp in Figure 45.5 is 12V. How many joules of electrical energy are changed into light and heat when
   a a charge of 1C passes through it,
   b a charge of 5C passes through it,
   c a current of 2A flows through it for 10s?

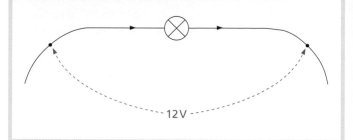

Figure 45.5

2 Three 2V cells are connected in series and used as the supply for a circuit.
   a What is the p.d. at the terminals of the supply?
   b How many joules of electrical energy does 1C gain on passing through (i) one cell, (ii) all three cells?

3 The symbol for a 1.5V cell is ⊣⊢ . Which of the arrangements in Figure 45.6 would produce a battery with a p.d. of 6V?

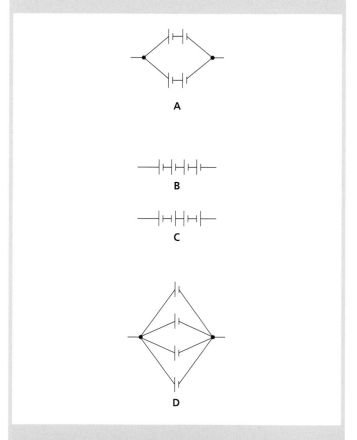

Figure 45.6

4 The lamps and the cells in all the circuits of Figure 45.7 are the same. If the lamp in **a** has its full, normal brightness, what can you say about the brightness of the lamps in **b**, **c**, **d**, **e** and **f**?

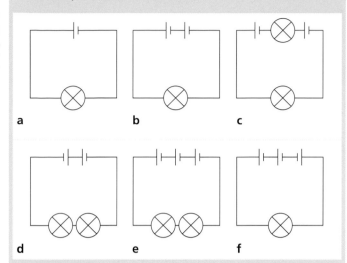

Figure 45.7

5 Three voltmeters (V), (V₁), (V₂) are connected as in Figure 45.8.
   a If (V) reads 18V and (V₁) reads 12 V, what does (V₂) read?
   b If the ammeter (A) reads 0.5A, how much electrical energy is changed to heat and light in L_1 in one minute?
   c Copy Figure 45.8 and mark with a + the positive terminals of the ammeter and voltmeters for correct connection.

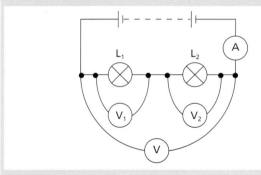

Figure 45.8

6 Three voltmeters are connected as in Figure 45.9. What are the voltmeter readings x, y and z in the table below (which were obtained with three different batteries)?

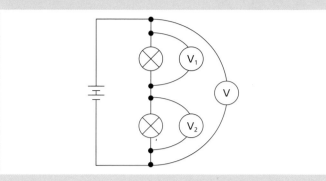

Figure 45.9

V/V	V_1/V	V_2/V
x	12	6
6	4	y
12	z	4

Checklist

After studying this chapter you should be able to

- describe simple experiments to show the transfer of electrical energy to other forms (e.g. in a lamp),
- recall the definition of the unit of p.d. and that p.d. (also called 'voltage') is measured by a voltmeter,
- demonstrate that the sum of the voltages across any number of components in series equals the voltage across all of those components,
- demonstrate that the voltages across any number of components in parallel are the same,
- work out the voltages of cells connected in series and parallel.

46 Resistance

Electrons move more easily through some conductors than others when a p.d. is applied. The opposition of a conductor to current is called its **resistance**. A good conductor has a low resistance and a poor conductor has a high resistance. The resistance of a wire of a certain material

(i) increases as its length increases,
(ii) increases as its cross-section area decreases,
(iii) depends on the material.

A long thin wire has more resistance than a short thick one of the same material. Silver is the best conductor, but copper, the next best, is cheaper and is used for connecting wire and domestic electric cables.

■ The ohm

If the current through a conductor is I when the voltage across it is V, Figure 46.1a, its resistance R is defined by

$$R = \frac{V}{I}$$

This is a reasonable way to measure resistance since the smaller I is for a given V, the greater is R. If V is in volts and I in amperes, R is in **ohms** (Ω: pronounced omega). For example, if $I = 2\,\text{A}$ when $V = 12\,\text{V}$, then $R = 12\,\text{V}/2\,\text{A} = 6\,\Omega$.

The ohm is the resistance of a conductor in which the current is 1 ampere when a voltage of 1 volt is applied across it.

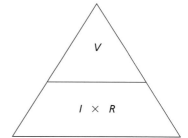

a

b

Figure 46.1

Alternatively, if R and I are known, V can be found from

$$V = IR$$

Also, knowing V and R, I can be calculated from

$$I = \frac{V}{R}$$

The triangle in Figure 46.1b is an aid to remembering the three equations. It is used like the 'density triangle' in Chapter 18.

■ *Resistors*

Conductors intended to have resistance are called **resistors** (symbol ─▭─) and are made either from wires of special alloys or from carbon. Those in radio and television sets have values from a few ohms up to millions of ohms, Figure 46.2a.

Variable resistors are used in electronics (and are then called **potentiometers**) as volume and other controls, Figure 46.2b. Larger current versions are useful in laboratory experiments and consist of a coil of constantan wire (an alloy of 60% copper, 40% nickel) wound on a tube with a sliding contact on a metal bar above the tube, Figure 46.3.

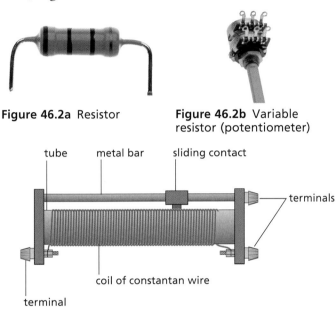

Figure 46.2a Resistor   **Figure 46.2b** Variable resistor (potentiometer)

Figure 46.3 Large variable resistor

There are two ways of using such a variable resistor. It may be used as a **rheostat** for changing the current in a circuit; only one end connection and the sliding contact are then required. In Figure 46.4a moving the sliding contact to the left reduces the resistance and increases the current. It can also act as a **potential divider** for changing the p.d. applied to a device, all three connections being used. In Figure 46.4b any fraction from the total p.d. of the battery to zero can be 'tapped off' by moving the sliding contact down.

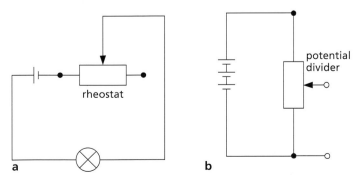

Figure 46.4 A variable resistor can be used as a rheostat or as a potential divider

Practical work

Measuring resistance

The resistance R of a conductor can be found by measuring the current I through it when a p.d. V is applied across it and then using $R = V/I$. This is called the **ammeter–voltmeter method**.

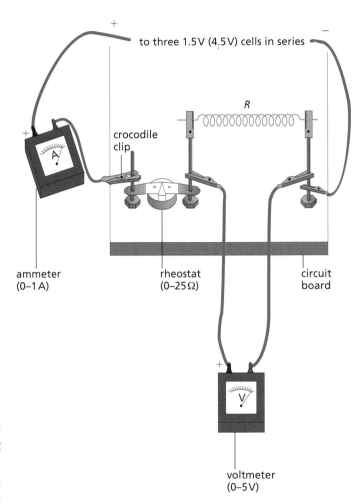

Figure 46.5

Set up the circuit of Figure 46.5 in which the unknown resistance R is 1 metre of SWG 34 constantan wire. Altering the rheostat changes both the p.d. V and the current I. Record in a table, with three columns, five values of I (e.g. 0.10, 0.15, 0.20, 0.25 and 0.30 A) and the corresponding values of V. Work out R for each pair of readings.

Repeat the experiment, but instead of the wire use (i) a torch bulb (e.g. 2.5 V, 0.3 A), (ii) a semiconductor diode (e.g. 1 N4001) connected first one way then the other way round, (iii) a thermistor (e.g TH 7).

■ *I–V graphs: Ohm's law*

The results of the previous experiment allow graphs of *I* against *V* to be plotted for different conductors.

a) Metallic conductors

Metals and some alloys give *I–V* graphs which are a straight line through the origin, Figure 46.6a, so long as their temperature is constant. *I* is directly proportional to *V*, i.e. $I \propto V$. Doubling *V* doubles *I*, etc. Such conductors obey **Ohm's law**, stated as follows.

> The current through a metallic conductor is directly proportional to the voltage across its ends if the temperature and other conditions are constant.

They are called **ohmic** or **linear** conductors and since $I \propto V$, it follows that V/I = a constant (obtained from the slope of the *I–V* graph). The resistance of an ohmic conductor therefore does not change when the voltage does.

b) Semiconductor diodes

The typical *I–V* graph in Figure 46.6b shows that current passes when the voltage is applied in one direction but is almost zero when it acts in the opposite direction. A diode has a small resistance when connected one way round but a very large resistance when the voltage is reversed. It conducts in one direction only and is a **non-ohmic** conductor. This makes it useful as a **rectifier** for changing alternating current (a.c.) to direct current (d.c.).

c) Filament lamp

For a filament lamp, e.g. a torch bulb, the *I–V* graph bends over as *V* and *I* increase, Figure 46.6c. That is, the resistance (V/I) increases as *I* increases and makes the filament hotter.

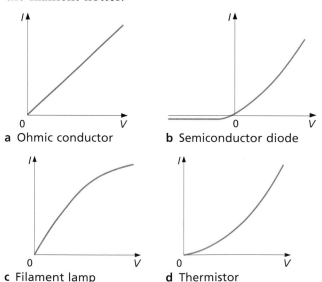

a Ohmic conductor **b** Semiconductor diode
c Filament lamp **d** Thermistor

Figure 46.6 *I–V* graphs

d) Variation of resistance with temperature

In general, an increase of temperature increases the resistance of metals, as for the filament lamp in Figure 46.6c, but decreases the resistance of semiconductors. The resistance of most **thermistors** (p. 283) decreases if their temperature rises, i.e. their *I–V* graph bends up, Figure 46.6d.

If a resistor and a thermistor are connected as a potential divider, Figure 46.7, the voltage across the resistor increases as the temperature of the thermistor increases; the circuit can be used to monitor temperature, for example in a car radiator.

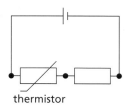

thermistor

Figure 46.7 Potential divider circuit for monitoring temperature

e) Variation of resistance with light intensity

The resistance of some semiconducting materials decreases when the intensity of light falling on them increases. This property is made use of in light dependent resistors, LDRs (Chapter 60). The *I–V* graph for an LDR is similar to that shown in Figure 46.6d for a thermistor.

■ *Resistors in series*

The resistors in Figure 46.8 are in series. The **same current** *I* flows through each and the total voltage *V* across all three equals the separate voltages across them, i.e.

$$V = V_1 + V_2 + V_3$$

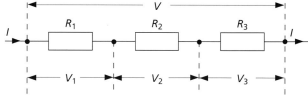

Figure 46.8 Resistors in series

But $V_1 = IR_1$, $V_2 = IR_2$ and $V_3 = IR_3$. Also, if *R* is the combined resistance, $V = IR$ and so

$$IR = IR_1 + IR_2 + IR_3$$

Dividing both sides by *I*,

$$R = R_1 + R_2 + R_3$$

Resistors in parallel

The resistors in Figure 46.9 are in parallel. The **voltage V between the ends of each is the same** and the total current I equals the sum of the currents in the separate branches, i.e.

$$I = I_1 + I_2 + I_3$$

But $I_1 = V/R_1$, $I_2 = V/R_2$ and $I_3 = V/R_3$. Also, if R is the combined resistance, $V = IR$ and so

$$\frac{V}{R} = \frac{V}{R_1} + \frac{V}{R_2} + \frac{V}{R_3}$$

Dividing both sides by V,

$$\frac{1}{R} = \frac{1}{R_1} + \frac{1}{R_2} + \frac{1}{R_3}$$

For the simpler case of *two* resistors in parallel

$$\frac{1}{R} = \frac{1}{R_1} + \frac{1}{R_2} = \frac{R_2}{R_1 R_2} + \frac{R_1}{R_1 R_2}$$

$$\therefore \qquad \frac{1}{R} = \frac{R_2 + R_1}{R_1 R_2}$$

Inverting both sides,

$$R = \frac{R_1 R_2}{R_1 + R_2} = \frac{\text{product of resistances}}{\text{sum of resistances}}$$

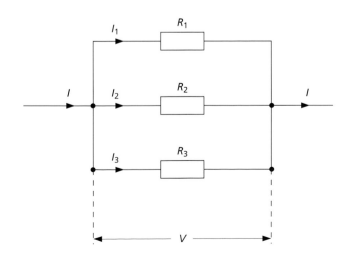

Figure 46.9 Resistors in parallel

Worked example

A p.d. of 24 V from a battery is applied to the network of resistors in Figure 46.10a.

a What is the combined resistance of the $6\,\Omega$ and $12\,\Omega$ resistors in parallel?
b What is the current in the $8\,\Omega$ resistor?
c What is the voltage across the parallel network?
d What is the current in the $6\,\Omega$ resistor?

a Let $R_1 = $ resistance of $6\,\Omega$ and $12\,\Omega$ in parallel

$$\therefore \qquad \frac{1}{R_1} = \frac{1}{6} + \frac{1}{12} = \frac{2}{12} + \frac{1}{12} = \frac{3}{12}$$

$$\therefore \qquad R_1 = \frac{12}{3} = 4\,\Omega$$

b Let $R = $ *total* resistance of circuit $= 4 + 8 = 12\,\Omega$. The equivalent circuit is shown in Figure 46.10b and if I is the current in it then since $V = 24\,V$

$$I = \frac{V}{R} = \frac{24\,V}{12\,\Omega} = 2\,A$$

$$\therefore \qquad \text{current in } 8\,\Omega \text{ resistor} = 2\,A$$

c Let $V_1 = $ voltage across parallel network

$$\therefore \qquad V_1 = I \times R_1 = 2\,A \times 4\,\Omega = 8\,V$$

d Let $I_1 = $ current in $6\,\Omega$ resistor, then since $V_1 = 8\,V$

$$I_1 = \frac{V_1}{6\,\Omega} = \frac{8\,V}{6\,\Omega} = \frac{4}{3}\,A$$

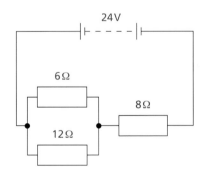

a

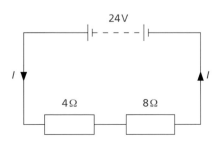

b

Figure 46.10

■ *Resistor colour code*

Resistors have colour coded bands as shown in Figure 46.11. In the orientation shown the first two bands on the left give digits 2 and 7; the third band gives the number of noughts (3) and the fourth band gives the resistor's 'tolerance' (or accuracy, here $\pm10\%$). So the resistor has a value of $27\,000\,\Omega$ ($\pm10\%$).

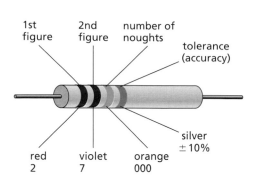

resistor value = $27\,000\,\Omega(\pm10\%)$
= $27\,k\Omega(\pm10\%)$

Figure	Colour
0	black
1	brown
2	red
3	orange
4	yellow
5	green
6	blue
7	violet
8	grey
9	white

Tolerance	
$\pm5\%$	gold
$\pm10\%$	silver
$\pm20\%$	no band

Figure 46.11 Colour code for resistors

■ *Resistivity*

Experiments show that the resistance R of a wire of a given material is

(i) directly proportional to its length l, i.e. $R \propto l$,
(ii) inversely proportional to its cross-section area A, i.e. $R \propto l/A$ (doubling A halves R).

Combining these two statements, we get

$$R \propto l/A \quad \text{or} \quad R = \rho l/A$$

where ρ is a constant, called the **resistivity** of the material. If we put $l=1\,m$ and $A=1\,m^2$, then $\rho=R$.

> The resistivity of a material is numerically equal to the resistance of a 1 m length of it of cross-section area $1\,m^2$.

The unit of ρ is the **ohm-metre** ($\Omega\,m$) as can be seen by rearranging the equation to give $\rho=AR/l$ and inserting units for A, R and l. Knowing ρ for a material, the resistance of any sample of it can be calculated. The resistivities of metals increase at higher temperatures; for most other materials they decrease.

■ *Worked example*

Calculate the resistance of a copper wire 1.0 km long and 0.50 mm diameter if the resistivity of copper is $1.7 \times 10^{-8}\,\Omega\,m$.

Converting all units to metres, we get

$$\text{length } l = 1.0\,km = 1000\,m = 10^3\,m$$
$$\text{diameter } d = 0.50\,mm = 0.50 \times 10^{-3}\,m$$

If r is the radius of the wire, the cross-section area $A = \pi r^2 = \pi(d/2)^2 = (\pi/4)d^2$

$$\therefore \quad A = \frac{\pi}{4}(0.50 \times 10^{-3})^2\,m^2 \approx 0.20 \times 10^{-6}\,m^2$$

$$R = \frac{\rho l}{A} = \frac{(1.7 \times 10^{-8}\,\Omega\,m) \times (10^3\,m)}{0.20 \times 10^{-6}\,m^2} = 85\,\Omega$$

■ *Potential divider*

In the circuit shown in Figure 46.12, two resistors R_1 and R_2 are in series with a supply of voltage V. The current in the circuit is

$$I = \frac{\text{supply voltage}}{\text{total resistance}} = \frac{V}{(R_1 + R_2)}$$

So the voltage across R_1 is

$$V_1 = I \times R_1 = \frac{V \times R_1}{(R_1 + R_2)}$$

and the voltage across R_2 is

$$V_2 = I \times R_2 = \frac{V \times R_2}{(R_1 + R_2)}$$

Also the ratio of the voltages across each resistor is

$$\frac{V_1}{V_2} = \frac{R_1}{R_2}$$

Returning to Figure 46.7, can you now explain why the voltage across the resistor increases when the resistance of the thermistor decreases?

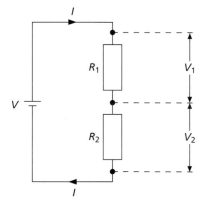

Figure 46.12 Potential divider circuit

Questions

1 What is the resistance of a lamp when a voltage of 12 V across it causes a current of 4 A?

2 Calculate the p.d. across a 10 Ω resistor carrying a current of 2 A.

3 The p.d. across a 3 Ω resistor is 6 V. What is the current flowing (in ampere)?

A $\frac{1}{2}$ **B** 1   **C** 2   **D** 6   **E** 8

4 The resistors R_1, R_2, R_3 and R_4 in Figure 46.13 are all equal in value. What would you expect the voltmeters A, B and C to read, assuming that the connecting wires in the circuit have negligible resistance?

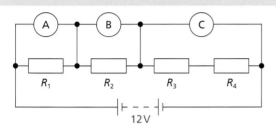

Figure 46.13

5 Calculate the effective resistance between A and B in Figure 46.14.

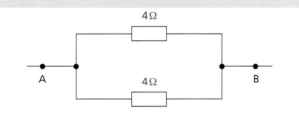

Figure 46.14

6 What is the effective resistance in Figure 46.15 between
 a A and B,
 b C and D?

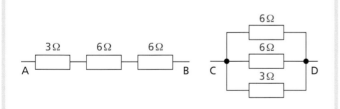

Figure 46.15

7 Figure 46.16 shows three resistors. Their combined resistance in ohms is

A $1\frac{5}{7}$ **B** 14   **C** $1\frac{1}{5}$ **D** $7\frac{1}{2}$ **E** $6\frac{2}{3}$

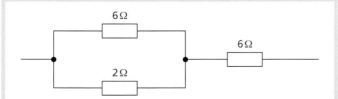

Figure 46.16

8 A student investigates how the current flowing through a filament lamp changes with the voltage across it. She is given a filament lamp and connecting wire. She decides to use a 15 V power supply, a variable resistor, an ammeter, a voltmeter and a switch.
 a Complete the circuit diagram, started in Figure 46.17, to show how she should set up the circuit.

15 V

Figure 46.17

 b The student obtains the following results.

Voltage (V)	0	3.0	5.0	7.0	9.0	11.0
Current (A)	0	1.0	1.4	1.7	1.9	2.1

 (i) Plot a graph of current against voltage.
 (ii) Use your graph to find the current when the voltage is 10 V.
 (iii) Use your answer to (ii) to calculate the resistance of the lamp when the voltage is 10 V.
 c (i) What happens to the resistance of the lamp as the current through it increases?
 (ii) Explain your answer.

(AQA (NEAB) Higher, June 99)

47 Capacitors

Capacitance
Types of capacitor

Charging and discharging a capacitor
Effect of capacitors in d.c. and a.c. circuits

A capacitor stores electric charge and is useful in many electronic circuits. In its simplest form it consists of two parallel metal plates separated by an insulator, called the **dielectric**, Figure 47.1.

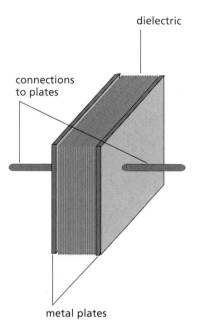

a Parallel-plate capacitor

b Symbol for a capacitor

Figure 47.1

■ Capacitance

The more charge a capacitor can store, the greater is its **capacitance** (*C*). The capacitance is large when the plates have a large area and are close together. It is measured in **farads** (F) but smaller units such as the **microfarad** (μF) are more convenient.

$$1\,\mu F = 1 \text{ millionth of a farad} = 10^{-6}\,F$$

■ Types of capacitor

Practical capacitors, with values ranging from about $0.01\,\mu F$ to $100\,000\,\mu F$, often consist of two long strips of metal foil separated by long strips of dielectric, rolled up like a 'Swiss roll' as in Figure 47.2. The arrangement allows plates of large area to be close together in a small volume. Plastics (e.g. polyesters) are commonly used as the dielectric, with films of metal being deposited on the plastic to act as the plates, Figure 47.3.

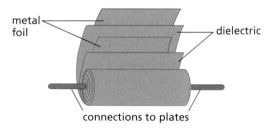

Figure 47.2 Construction of a practical capacitor

Figure 47.3 Polyester capacitor

The **electrolytic** type, Figure 47.4a, has a very thin layer of aluminium oxide as the dielectric between two strips of aluminium foil, giving large capacitances. It is polarized, i.e. it has positive and negative terminals, Figure 47.4b, and these *must* be connected to the + and − respectively of the voltage supply.

a Electrolytic capacitor

b Symbol for an electrolytic capacitor showing polarity

Figure 47.4

Charging and discharging a capacitor

a) Charging

A capacitor can be charged by connecting a battery across it. In Figure 47.5a, the + of the battery attracts electrons (since they have a negative charge) from plate X and the − of the battery repels electrons to plate Y. A positive charge builds up on plate X (since it loses electrons) and an equal negative charge builds up on Y (since it gains electrons).

During the charging, there is a *brief* flow of electrons round the circuit from X to Y (but not through the dielectric). A momentary current would be detected by a sensitive ammeter. The voltage builds up between X and Y and opposes the battery voltage. Charging stops when these two voltages are equal; the electron flow, i.e. the charging current, is then zero.

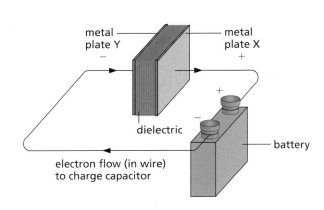

a

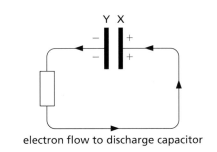

electron flow to discharge capacitor

b

Figure 47.5 Charging and discharging a capacitor

b) Discharging

When a conductor is connected across a charged capacitor, there is a brief flow of electrons from the negatively charged plate to the positively charged one, i.e. from Y to X in Figure 47.5b. The charge stored by the capacitor falls to zero, as does the voltage across it.

c) Demonstration

The circuit in Figure 47.6 uses a two-way switch which charges C in position 1 and discharges it in position 2. The larger the values of R and C the longer it takes for the capacitor to charge or discharge. The direction of the deflection of the centre-zero milliammeter reverses

for each process. The corresponding changes of capacitor charge (measured by the voltage across it) with time are shown by the graphs in Figures 47.7a, b. These can be plotted directly if the voltmeter is replaced by a datalogger and computer.

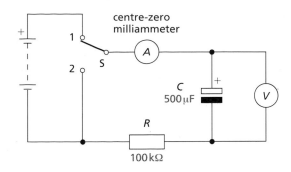

Figure 47.6 Demonstration circuit

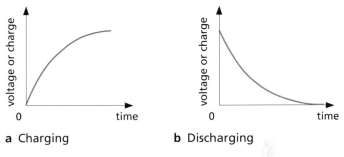

a Charging   b Discharging

Figure 47.7 Graphs

Effect of capacitors in d.c. and a.c. circuits

a) d.c.

In Figure 47.8a the supply is d.c. but the lamp does not light, i.e. a capacitor blocks d.c.

b) a.c.

In Figure 47.8b the supply is a.c. and the lamp lights, suggesting that a capacitor passes a.c. In fact, no current actually passes *through* the capacitor since its plates are separated by an insulator. But as the a.c. reverses direction, the capacitor charges and discharges causing electrons to flow to and fro rapidly in the wires joining the plates. Thus effectively a.c. flows round the circuit, lighting the lamp.

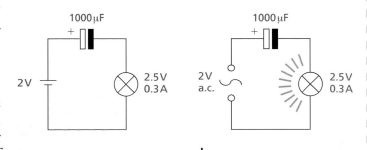

a   b

Figure 47.8 A capacitor blocks d.c. and passes a.c.

Questions

1 **a** Describe the basic construction of a capacitor.
   b What does a capacitor do?
   c State two ways of increasing the capacitance of a capacitor.
   d Name a unit of capacitance.

2 **a** When a capacitor is being charged, is the value of the charging current maximum or zero
   (i) at the start and
   (ii) at the end of charging?
   b Repeat **a** for a capacitor discharging.

3 How does a capacitor behave in a circuit with
   a a d.c. supply,
   b an a.c. supply?

■ *Checklist*

After studying this chapter you should be able to

■ state what a capacitor does and what its capacitance depends on,
■ state the unit of capacitance,
■ describe the construction of practical capacitors,
■ describe in terms of electron motion how a capacitor can be charged and discharged, and sketch graphs of the capacitor voltage with time for charging and discharging through a resistor,
■ recall that a capacitor blocks d.c. but passes a.c. and explain why.

48 Electrolysis and cells

■ Conduction in liquids

The apparatus of Figure 48.1 can be used to find which liquids conduct electricity. There is no effect with distilled water; it is not a conductor. However, if a little dilute sulphuric acid or common salt is added to the water the bulb lights up, showing that current is flowing through the solution. Bubbles of gas appear at the copper strips and are evidence of chemical action. Tap water usually contains dissolved salts and allows a small current to flow.

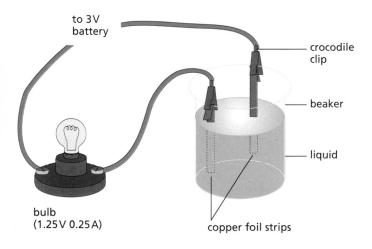

Figure 48.1 Demonstrating electrolysis

The production of chemical action in a liquid by an electric current is called **electrolysis**. The liquid is called an **electrolyte**. Solutions in water of acids, bases and salts are electrolytes. Mercury conducts electricity but no chemical change occurs and so it is not an electrolyte.

The two conductors (wires or plates) where current enters and leaves the liquid are called **electrodes**. The one joined to the positive terminal of the battery at which the current (conventional) enters is the **anode**; the other is the **cathode**.

■ Electrolysis of copper sulphate solution with copper electrodes

The apparatus and circuit of Figure 48.1 can be used with a strong copper sulphate solution (10 g to 100 cm^3 of water) in the beaker. No bubbles appear at the electrodes but if they are removed after a few minutes the cathode will be seen to be covered with a fresh layer of copper while the anode is dull. It can be shown that the mass of copper lost by the anode (if it is pure) equals that gained by the cathode. Also, the concentration of the copper sulphate solution is unchanged. It seems that copper is transferred from the anode to the cathode.

■ Ionic theory

Current is considered to be carried in an electrolyte by **ions**. An ion is an atom or group of atoms which has either a positive charge due to losing one or more electrons or a negative charge due to gaining one or more electrons, Figure 48.2a.

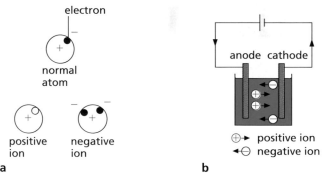

Figure 48.2 In an electrolyte the charge carriers are positive and negative ions

According to the ionic theory, when some chemical compounds are dissolved in water or are melted, they form ions. In electrolysis positive ions are attracted to the negative cathode and negative ions to the positive anode, Figure 48.2b.

Copper sulphate solution contains copper ions and sulphate ions and since copper is deposited on the cathode during electrolysis, the copper ions must be positively charged.

■ 215

■ Electroplating

Articles of one metal are often electrically plated with another to improve their appearance, or to prevent rusting if they are made of iron or steel. Cutlery may be made of nickel, plated with silver. Steel is often chromium plated, for example in bicycle parts. The object to be plated is effectively made the cathode, with a salt of the plating metal as electrolyte, e.g. a silver salt for silver plating.

Figure 48.3 Industrial electroplating

■ Electric cells

In an electric cell an electrolyte reacts chemically with two electrodes, making one have a positive electric charge (the anode) and the other a negative charge (the cathode). When the two electrodes (or terminals) are connected in a conducting circuit, a current flows and chemical energy is transferred to electrical energy. This is the reverse of what happens in electrolysis.

A 'battery' is strictly two or more cells together but the term is commonly used to mean a single cell.

216 ■

■ Primary cells

A primary cell is one which is discarded when the chemicals are used up.

a) Simple cell

This has a voltage of about 1.0 V but stops working after a short time due to 'polarization', i.e. the collection of hydrogen gas bubbles on the copper plate. The cell is depolarized by adding potassium dichromate which oxidizes the hydrogen to water. A second defect is 'local action'. This is due to impurities in the zinc and results in the zinc being used up even when current is not supplied. The simple cell is not used except for demonstration.

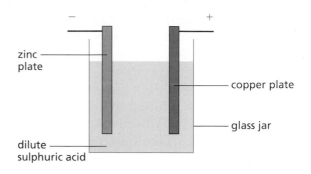

Figure 48.4 Simple cell

b) Zinc–carbon (Leclanché or dry) cell,

This has a zinc cathode, a manganese(IV) oxide anode (which also acts as a slow depolarizer) and the electrolyte is a solution of ammonium chloride. The carbon rod is in contact with the anode (but is not involved in the chemical reaction) and is called the 'current collector'. The voltage is 1.5 V. This is the most popular cell for low current (e.g. 0.3 A) or occasional use, e.g. in torches.

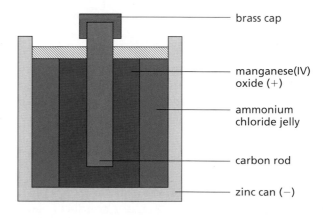

Figure 48.5 Zinc–carbon cell

c) Other primary cells

Alkaline manganese batteries are more expensive than zinc–carbon ones but are longer lasting; compact button cells, for use in watches and calculators, are made from silver oxide or lithium manganese.

Secondary cells

Secondary cells can be recharged by passing a current through them in the opposite direction to that in which they supply one. They are much more expensive than primary cells.

In the **lead–acid cell** or 'accumulator', Figure 48.6, the anode is lead(IV) oxide (brown) and the cathode lead (grey). The electrolyte is dilute sulphuric acid. The voltage is steady at 2.0 V and quite large continuous currents can be supplied. A 12 V car battery consists of six lead–acid cells in series.

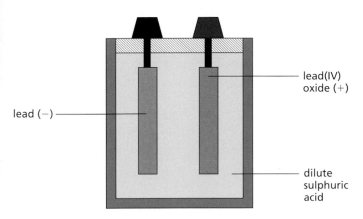

Figure 48.6 Lead–acid cell

Nickel–cadmium (NiCad), and nickel–metal hydride (NiMH) batteries are more compact types of re-chargeable cell, used in portable computers and mobile phones.

Figure 48.7 shows some different types of cell (or 'battery') including compact rechargeable cells.

Figure 48.7 Compact cells

Questions

1 a Name the current-carrying particles in an electrolyte.
  b In the electrolysis of copper sulphate solution what would be the effect of increasing
    (i) the current,
    (ii) the time for which it passes? (**Hint.** Remember that the current is the charge passing per unit time.)

2 a What is the voltage of
    (i) a simple cell,
    (ii) a zinc–carbon cell,
    (iii) a lead–acid cell?
  b What materials are used for the anode and cathode and the electrolyte of each of the cells in **a**?

3 a What is the difference between a primary and a secondary cell?
  b Suggest advantages and disadvantages of each.

▪ Checklist

After studying this chapter you should be able to

▪ recall that **electrolysis** is the production of chemical action in a liquid by an electric current, that the liquid is called an **electrolyte**, and that the positive electrode is the **anode** and the negative electrode the **cathode**,

▪ describe a demonstration of the effects of electrolysis with copper electrodes in copper sulphate solution,

▪ explain electrolysis in terms of ion movement,

▪ recall that in an electric cell a chemical reaction produces positively and negatively charged electrodes,

▪ draw and describe the construction of a simple cell, a zinc–carbon cell and a lead–acid cell,

▪ distinguish between a **primary** cell and a **secondary** cell.

49 Electric power

Power in electric circuits
Electric lighting
Electric heating

Joulemeter
Practical work
Measuring electric power.

Power in electric circuits

In many circuits it is important to know the rate at which electrical energy is transferred into other forms of energy. Earlier (Chapter 24) we said that **energy transfers were measured by the work done** and power was defined by the equation

$$\text{power} = \frac{\text{work done}}{\text{time taken}} = \frac{\text{energy transfer}}{\text{time taken}}$$

In symbols
$$P = \frac{W}{t} \tag{1}$$

where if W is in joules (J) and t in seconds (s) then P is in J/s or watts (W).

From the definition of p.d. (Chapter 45) we saw that if W is the electrical energy transferred when a steady current I (in amperes) passes for time t (in seconds) through a device (e.g. a lamp) with a p.d. V (in volts) across it, Figure 49.1, then

$$W = ItV \tag{2}$$

Substituting for W in (1) we get

$$P = \frac{W}{t} = \frac{ItV}{t}$$

or

$$P = IV$$

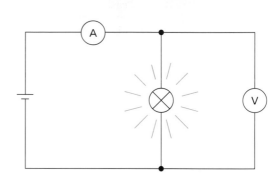

Figure 49.1

Therefore to calculate the power P of an electrical appliance we multiply the current I through it by the p.d. V across it. For example if a lamp on a 240 V supply has a current of 0.25 A through it, its power is 240 V × 0.25 A = 60 W. The lamp is transferring 60 J of electrical energy into heat and light each second. Larger units of power are the **kilowatt** (kW) and the **megawatt** (MW) where

$$1\,\text{kW} = 1000\,\text{W} \quad \text{and} \quad 1\,\text{MW} = 1\,000\,000\,\text{W}$$

In units

$$\text{watts} = \text{amperes} \times \text{volts} \tag{3}$$

It follows from (3) that since

$$\text{volts} = \frac{\text{watts}}{\text{amperes}} \tag{4}$$

the volt can be defined as a **watt per ampere** and p.d. calculated from (4).

If all the energy is transferred to heat in a resistor of resistance R, then $V = IR$ and the rate of production of heat is given by

$$P = V \times I = IR \times I = I^2R$$

That is, if the current is doubled, four times as much heat is produced per second. Also, $P = V^2/R$.

Practical work

Measuring electric power

a) Lamp

Connect the circuit of Figure 49.2. Note the ammeter and voltmeter readings and work out the electric power supplied to the bulb in watts.

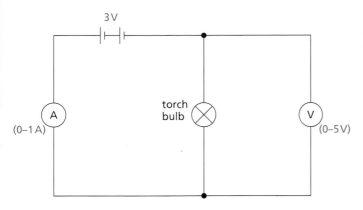

Figure 49.2

b) Motor

Replace the bulb in Figure 49.2 by a small electric motor. Attach a known mass m (in kg) to the axle of the motor with a length of thin string and find the time t (in s) required to raise the mass through a known height h (in m) at a steady speed. Then the power output P_o (in W) of the motor is given by

$$P_o = \frac{\text{work done in raising mass}}{\text{time taken}} = \frac{mgh}{t}$$

If the ammeter and voltmeter readings I and V are noted while the mass is being raised, the power input P_i (in W) can be found from

$$P_i = IV$$

The efficiency of the motor is given by

$$\text{efficiency} = \frac{P_o}{P_i} \times 100\%$$

Also investigate the effect of a greater mass on (i) the speed, (ii) the power output and (iii) the efficiency of the motor at its rated p.d.

■ *Electric lighting*

a) Filament lamps

The filament is a small coiled coil of tungsten wire, Figure 49.3, which becomes white hot when current flows through it. The higher the temperature of the filament the greater is the proportion of electrical energy transferred to light and for this reason it is made of tungsten, a metal with a high melting point (3400 °C).

Most lamps are gas-filled and contain nitrogen and argon, not air. This reduces evaporation of the tungsten which would otherwise condense on the bulb and blacken it. The coil is coiled compactly so that it is cooled less by convection currents in the gas.

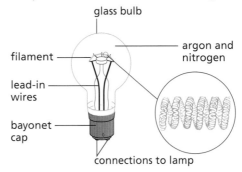

Figure 49.3 A filament lamp

b) Fluorescent strips

A filament lamp transfers only 10% of the electrical energy supplied to light; the other 90% becomes heat. Fluorescent strip lamps, Figure 49.4a, are five times as efficient and may last 3000 hours compared with the 1000-hour life of filament lamps. They cost more to install but running costs are less.

When a fluorescent strip lamp is switched on, the mercury vapour emits invisible ultraviolet radiation which makes the powder on the inside of the tube fluoresce (glow), i.e. visible light is emitted. Different powders give different colours.

c) Compact fluorescent lamps

These energy-saving fluorescent lamps, Figure 49.4b, fit straight into normal light sockets, either bayonet or screw-in. They last up to eight times longer (typically 8000 hours) and use about five times less energy than filament lamps for the *same* light output. For example, a 20 W compact fluorescent is equivalent to a 100 W filament lamp.

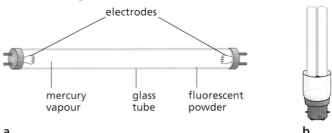

Figure 49.4 Fluorescent lamps

■ *Electric heating*

a) Heating elements

In domestic appliances such as electric fires, cookers, kettles and irons the 'elements', Figure 49.5, are made from Nichrome wire. This is an alloy of nickel and chromium which does not oxidize (and so become brittle) when the current makes it red hot.

The elements in **radiant** electric fires are at red heat (about 900 °C) and the radiation they emit is directed into the room by polished reflectors. In **convector** types the element is below red heat (about 450 °C) and is designed to warm air which is drawn through the heater by natural or forced convection. In **storage** heaters the elements heat fire-clay bricks during the night using 'off-peak' electricity. On the following day these cool down, giving off the stored heat to warm the room.

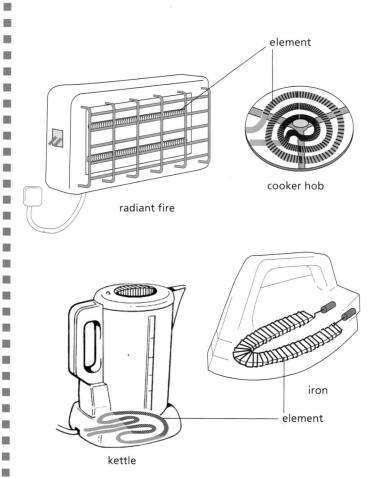

Figure 49.5 Heating elements

b) Three-heat switch

This is sometimes used to control heating appliances. It has three settings and uses two identical elements. On 'high', the elements are in parallel across the supply voltage, Figure 49.6a; on 'medium', current only passes through one, Figure 49.6b; on 'low', they are in series, Figure 49.6c.

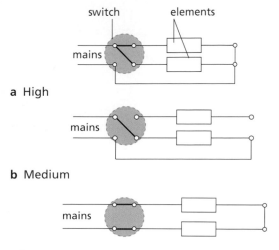

a High

b Medium

c Low

Figure 49.6 Three-heat switch

c) Fuses

A fuse (see also Chapter 50) is a short length of wire of material with a low melting point, often tinned copper, which melts and breaks the circuit when the current through it exceeds a certain value. Two reasons for excessive currents are 'short circuits' due to worn insulation on connecting wires, and overloaded circuits. Without a fuse the wiring would become hot in these cases and could cause a fire. **A fuse should ensure that the current-carrying capacity of the wiring is not exceeded.** In general the thicker a cable is, the more current it can carry, but each size has a limit.

Two types of fuse are shown in Figure 49.7. **Always switch off before replacing a fuse,** and always replace with one of the same value as recommended by the manufacturer of the appliance.

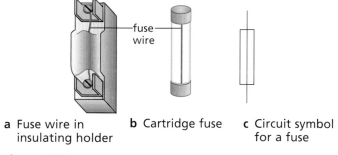

a Fuse wire in insulating holder

b Cartridge fuse

c Circuit symbol for a fuse

Figure 49.7

■ *Joulemeter*

Instead of using an ammeter and a voltmeter to measure the electrical energy transferred by an appliance, a **joulemeter** can be used to obtain it directly in joules. The circuit connections are shown in Figure 49.8. A household electricity meter, Figure 50.5, is a joulemeter.

Figure 49.8 Connections to a joulemeter

Questions

1 How much electrical energy in **joules** does a 100 watt lamp transfer in
   a 1 second,
   b 5 seconds,
   c 1 minute?

2 a What is the power of a lamp rated at 12 V 2 A?
   b How many joules of electrical energy are transferred per second by a 6 V 0.5 A lamp?

3 The largest number of 100 W bulbs connected in parallel which can safely be run from a 230 V supply with a 5 A fuse is

   A 2   **B** 5   **C** 11   **D** 12   **E** 20

4 What is the maximum power in kilowatts of the appliance(s) that can be connected safely to a 13 A 230 V mains socket?

5 Sue connects the circuit shown in Figure 49.9.

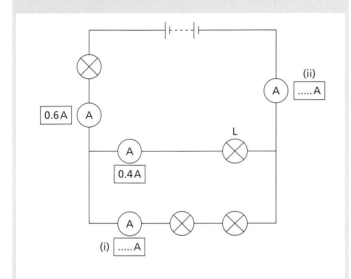

Figure 49.9

   a She measures the current at four places in the circuit. Two of her readings are shown on the diagram. What are the readings on the other *two* ammeters?
   b Sue uses a voltmeter to measure the voltage across lamp L.
      (i) Copy the diagram and show where the voltmeter should be connected. Use the correct circuit symbol.
      (ii) The reading on the voltmeter is 10 V. Calculate the power being transferred by lamp L. You *must* show how you work out your answer.
      (OCR Foundation, June 99)

6 Figure 49.10 shows a circuit for emergency lighting.

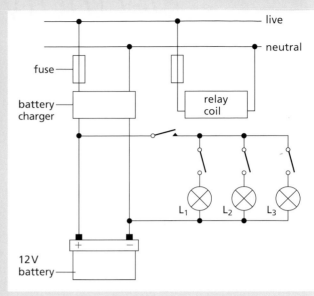

Figure 49.10

If the mains supply fails the relay switch closes and the lamps operate from the 12 V battery.
   a Write down *one* reason for connecting the lamps in parallel.
   b When switched on, the current in each lamp is 5 A.
      (i) Calculate the current in the battery when all three lamps are switched on.
      (ii) Calculate the power of each lamp.
   c Lamps in emergency lighting circuits are connected to the battery using thick wires. Mains lamps of the same power are connected using thinner wires. Explain why.
      (London Higher, June 99)

■ *Checklist*

After studying this chapter you should be able to

■ recall the relations $W = ItV$ and $P = IV$ and use them to solve simple problems on energy transfers,

■ describe experiments to measure electric power,

■ describe electric lamps, heating elements and fuses,

■ recall that a joulemeter measures electrical energy.

50 Electricity in the home

■ House circuits

Electricity usually comes to our homes by an underground cable containing two wires, the **live** (L) and the **neutral** (N). The neutral is earthed at the local substation and so there is no p.d. between it and earth. The supply is a.c. (Chapter 44) and the live wire is alternately positive and negative. Study the typical house circuit shown in Figure 50.1.

a) Circuits in parallel

Every circuit is connected in parallel with the supply, i.e. across the live and neutral, and receives the full mains p.d. of 230 V.

b) Switches and fuses

These are always in the live wire. If they were in the neutral, light switches and power sockets would be 'live' when switches were 'off' or fuses 'blown'. A fatal shock could then be obtained by, for example, touching the element of an electric fire when it was switched off.

c) Staircase circuit

The light is controlled from two places by the two two-way switches.

d) Ring main circuit

The live and neutral wires each run in two complete rings round the house, and the power sockets, each rated at 13 A, are tapped off from them. Thinner wires can be used since the current to each socket flows by two paths, i.e. in the whole ring. The ring has a 30 A fuse and if it has, say, ten sockets all can be used so long as the total current does not exceed 30 A, otherwise the wires overheat. A house may have several ring circuits, each serving a different area.

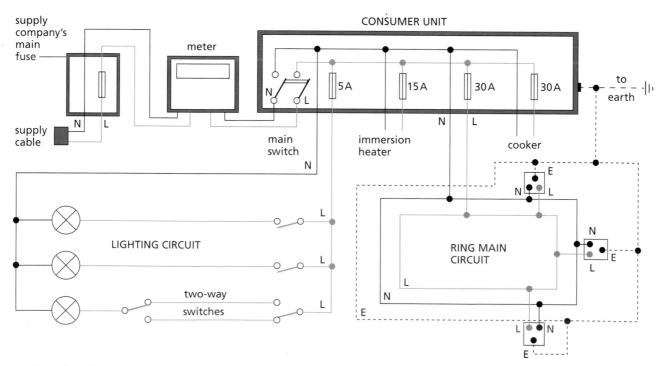

Figure 50.1 Electric circuits in a house

e) Fused plug

Only one type of plug is used in a ring main circuit. It is wired as in Figure 50.2a – note the colours of the wire coverings. It has its own cartridge fuse, 3 A (red) for appliances with powers up to 720 W, or 13 A (brown) for those between 720 W and 3 kW. Typical power ratings for various appliances are shown in Table 50.1, p. 224.

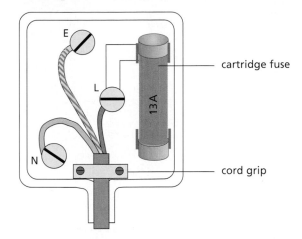

a Wiring of a plug

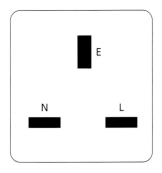

b Socket

Figure 50.2

f) Earthing and safety

A ring main has a third wire which goes to the top sockets on all power points, Figure 50.1, and is earthed by being connected either to a **metal** water pipe entering the house or to an earth connection on the supply cable. This third wire is a safety precaution to prevent electric shock should an appliance develop a fault.

The earth pin on a three-pin plug is connected to the metal case of the appliance which is thus joined to earth by a path of almost zero resistance. If then, for example, the element of an electric fire breaks or sags and touches the case, a large current flows to earth and 'blows' the fuse. Otherwise the case would become 'live' and anyone touching it would receive a shock which might be fatal, especially if they were 'earthed' by, say, standing in a damp environment, e.g. on a wet concrete floor.

g) Circuit breakers

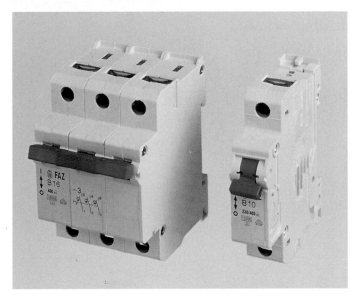

Figure 50.3 Circuit breakers

Circuit breakers, Figure 50.3, are now used in consumer units instead of fuses. They contain an electromagnet (Chapter 52) which, when the current exceeds the rated value of the circuit breaker, becomes strong enough to separate a pair of contacts and breaks the circuit. They operate much faster than fuses and have the advantage that they can be reset by pressing a button.

The **residual current circuit breaker** (RCCB), also called a **residual current device** (RCD), is an adapted circuit breaker which is used when the resistance of the earth path between the consumer and the sub-station is not small enough for a fault-current to blow the fuse or trip the circuit breaker. It works by detecting any difference between the currents in the live and neutral wires; when these become unequal due to an earth fault (i.e. some of the current returns to the sub-station via the case of the appliance and earth) it breaks the circuit before there is any danger. They have high sensitivity and a quick response.

An RCD should be plugged into a socket supplying power to a portable appliance such as an electric lawnmower or hedge trimmer. In these cases the risk of electrocution is greater because the user is generally making a good earth connection through the feet.

h) Double insulation

Appliances such as vacuum cleaners, hairdryers and food mixers are usually double-insulated. Connection to the supply is by a two-core insulated cable, with no earth wire, and the appliance is enclosed in an insulating plastic case. Any metal attachments which the user might touch are fitted into this case so that they do not make a direct connection with the internal electrical parts, e.g. a motor. There is then no risk of a shock should a fault develop.

Practical work

House circuits

On a circuit board connect up and investigate

a the staircase circuit of Figure 50.4a,
b the ring main circuit of Figure 50.4b.

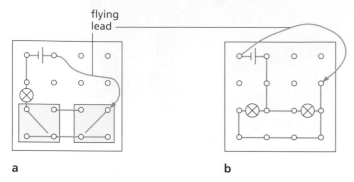

Figure 50.4

■ *Paying for electricity*

Electricity supply companies charge for the **electrical energy** they supply. A joule is a very small amount of energy and a larger unit, the **kilowatt-hour** (kWh) is used.

> A kilowatt-hour is the electrical energy used by a 1 kW appliance in 1 hour.

$$1\,\text{kWh} = 1000\,\text{J/s} \times 3600\,\text{s}$$
$$= 3\,600\,000\,\text{J} = 3.6\,\text{MJ}$$

A 3 kW electric fire working for 2 hours uses 6 kWh of electrical energy – usually called 6 'units'. Electricity meters, which are joulemeters (Chapter 49), are marked in kWh: the latest have digital readouts like the one in Figure 50.5. At present a 'unit' costs about 8p.

Typical powers of some appliances are given in Table 50.1.

Table 50.1 Power of some appliances

video recorder	20 W	iron	1 kW
hi-fi	40 W	fire	1, 2, 3 kW
light bulbs	60, 100 W	kettle	2 kW
television	100 W	immersion heater	3 kW
fridge	150 W	cooker	8 kW

Note that the current required by an 8 kW cooker is given by

$$I = \frac{P}{V} = \frac{8000\,\text{W}}{230\,\text{V}} = 35\,\text{A}$$

This is too large a current to draw from the ring main and so a separate circuit must be used.

Figure 50.5 Electricity meter with digital display

■ *Safety with electricity*

1 Switch off the electrical supply before starting repairs.
2 Use plugs that have a rubber or plastic case, the correct fuse, an earth pin and a cord grip.
3 Do not overload circuits by using too many adaptors.
4 Do not have long cables trailing across a room, or under a carpet that is walked over regularly.
5 Do not let appliances or cables come into contact with water, e.g. holding a hairdryer with wet hands in a bathroom can be dangerous.
6 Do not connect appliances that use large amounts of power (e.g. an electric fire) to a lighting circuit.

■ *Electric shock*

Electric current passing through the heart can be fatal. The following action should be taken in cases of electric shock.

1 **Switch off the supply** if the shocked person is still touching live equipment.
2 **Send for qualified medical assistance.**
3 **If breathing has stopped apply the 'kiss of life'.**
4 **If the heart has stopped try to restart it** by striking the chest smartly three times over the heart.

Questions

1 The circuits of Figure 50.6a, b show 'short circuits' between the live (L) and neutral (N) wires. In both, the fuse has blown but whereas circuit a is now safe, b is still dangerous even though the lamp is out which suggests the circuit is safe. Explain.

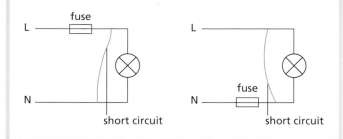

Figure 50.6

2 What steps should be taken before replacing a blown fuse in a plug?

3 What size fuse (3 A or 13 A) should be used in a plug connected to
  a a 150 W television,
  b a 900 W iron,
  c a 2 kW kettle,
  if the supply is 230 V?

4 What is the cost of heating a tank of water with a 3000 W immersion heater for 80 minutes if electricity costs 10p per kWh?

5 This question is about electricity in the home.
  a Figure 50.7 shows the inside of a fused 13 A plug.

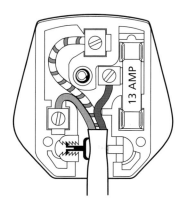

Figure 50.7

  (i) Write down the **names** of the three wires in the plug.
  (ii) What **colour** wire should you connect to the fuse in the plug?

  blue or **brown** or **green and yellow**

  (iii) The fuse helps to prevent fire caused by electrical faults. Explain how it does this.

b Mr Thomas uses a kettle to boil water. It is a 2 kW kettle. It takes 6 minutes to boil the water.
  (i) Calculate the energy transferred in **kilowatt-hours**. Use the equation below. You *must* show how you work out your answer.

  energy in kWh = power in kW × time in hours

  (ii) Electricity costs 10p per kWh. How much does it cost Mr Thomas to boil the water? You *must* show how you work out your answer.
c Mrs Thomas uses a 2 kW electric lawn mower. The lawn mower is marked with this symbol.

This means it is double-insulated. Explain what is meant by the term **double-insulated**.
(OCR Foundation, Summer 99)

6 a The table below shows the current in three different electrical appliances when connected to the 240 V mains a.c. supply.

Appliance	Current in A
kettle	8.5
lamp	0.4
toaster	4.8

  (i) Which appliance has the greatest electrical resistance? How does the data show this?
  (ii) The lamp is connected to the main supply using thin, twin-cored cable, consisting of live and neutral connections. State *two* reasons why this cable should not be used for connecting the kettle to the mains supply.
b (i) Calculate the power rating of the kettle when it is operated from the 240 V a.c. mains supply.
  (ii) A holiday-maker takes the kettle abroad where the mains supply is 120 V. What is the current in the kettle when it is operated from the 120 V supply? You can assume that the resistance of the kettle does not change.
  (iii) The kettle is filled with water. Explain how the time it takes to boil the kettle changes when it is operated from the 120 V supply.
(London Higher, June 98)

7 a Figure 50.8 shows an electric fence, designed to keep horses in a field.

Figure 50.8

When a horse touches the wire the horse receives a mild electric shock. Explain how.

b Figure 50.9 shows how a person could receive an electric shock from a faulty electrical appliance. Using a residual circuit breaker (RCB) can help to protect the person against receiving a serious shock.

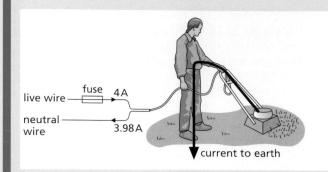

Figure 50.9

(i) Compare the action of an RCB to that of a fuse.
(ii) The graph in Figure 50.10 illustrates how the severity of an electric shock depends upon both the size of the current and the time for which the current flows through the body.

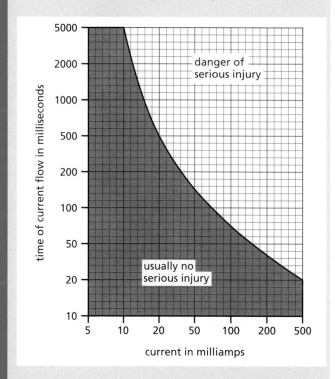

Figure 50.10

Within how long must the RCB cut off the current if the person using the lawnmower is to be in no danger of serious injury?

(AQA (SEG) Higher, Summer 99)

■ *Checklist*

After studying this chapter you should be able to

■ describe with the aid of diagrams a house wiring system and explain the functions and positions of switches, fuses, circuit breakers and earth,

■ wire a mains plug and recall the international insulation colour code,

■ perform calculations of the cost of electrical energy in joules and kilowatt-hours,

■ recall safety precautions for domestic wiring and appliances,

■ recall the procedure for dealing with someone who has received an electric shock.

Electricity
Additional questions

Static electricity; electric current; potential difference; resistance

1 In the process of electrostatic induction

   A a conductor is rubbed with an insulator
   B a charge is produced by friction
   C negative and positive charges are separated
   D a positive charge induces a positive charge
   E electrons are 'sprayed' into an object.

2 a A strip of plastic is rubbed with a duster. The plastic is then brought near to some small pieces of paper. State and explain what happens.
   b Use words from this list to complete the sentences.

attract   gains   loses   negative

positive   repel

   (i) When a balloon is rubbed with a duster, the balloon gains electrons and the duster ...... electrons.
   (ii) The balloon gains a ...... charge and the duster is left with a ...... charge.
   (iii) When the balloon and the duster are held close to each other, they will ...... each other.
   c The diagram shows a circuit which can be used to cover a spoon with a layer of silver.

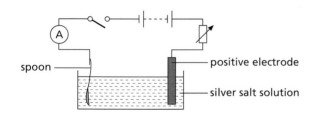

When the switch is closed, a current flows around the circuit. The current is a flow of charged particles.
(i) Explain, as fully as you can, how the current travels through the wires.
(ii) Explain, as fully as you can, how the current travels through the silver salt solution.
(NEAB Foundation, June 98)

3 If two resistors of 25 Ω and 15 Ω are joined together in series and then placed in parallel with a 40 Ω resistor, the effective resistance in ohms of the combination is

   A 0.1   **B** 10   **C** 20   **D** 40   **E** 400

4 V_1, V_2, V_3 are the p.ds across the 1 Ω, 2 Ω, and 3 Ω resistors in the following diagram and the current is 5 A. Which one of the columns **A** to **E** shows the correct values of V_1, V_2 and V_3 measured in volts?

	A	B	C	D	E
V_1	1.0	5	0.2	5.0	4.0
V_2	2.0	10	0.4	2.5	3.0
V_3	3.0	15	0.6	1.6	2.0

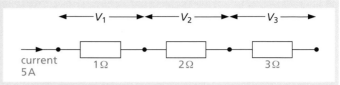

5 a The graph below illustrates how the p.d. across the ends of a conductor is related to the current flowing through it.
   (i) What law may be deduced from the graph?
   (ii) What is the resistance of the conductor?
   b Draw diagrams to show how six 2 V lamps could be lit to normal brightness when using a
   (i) 2 V supply,
   (ii) 6 V supply,
   (iii) 12 V supply.

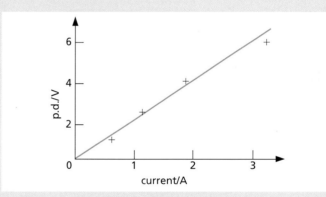

6 When a 4 Ω resistor is connected across the terminals of a 12 V battery, the number of coulombs passing through the resistor per second is

   A 0.3   **B** 3   **C** 4   **D** 12   **E** 48

7 The diagram below shows a 2 V cell connected to an arrangement of resistors, part in series and part in parallel.
   a What is the total resistance of the two 4 Ω resistors in parallel?
   b What is the current flowing in the 1 Ω resistor?
   c What is the current in one of the 4 Ω resistors?

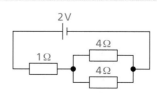

8 A 4 Ω coil and a 2 Ω coil are connected in parallel. What is their combined resistance? A total current of 3 A passes through the coils. What current flows through the 2 Ω coil?

9 Sam is investigating how the resistance of a lamp changes as she alters the current through it. She uses this circuit.

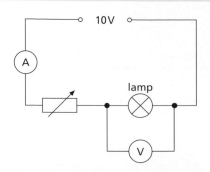

a She adjusts the setting of the variable resistor. Explain how this affects the current.

b She records the values of the voltage across the lamp as the current changes. She plots this graph.

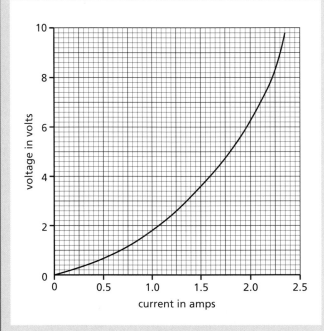

(i) Use the graph to find the value of the current when the voltage is 4.0 V.

(ii) Calculate the resistance of the lamp when the voltage is 4.0 V. You *must* show how you work out your answer.

c How can you tell from the graph that the resistance of the lamp increases between 4.0 V and 8.0 V?

(OCR Higher, June 99)

10 What is the resistance of a wire of length 300 m and cross-section area 1.0 mm² made of material of resistivity $1.0 \times 10^{-7}\,\Omega\,m$?

11 Calculate the resistance of an aluminium cable of length 10 km and diameter 2.0 mm if the resistivity of aluminium is $2.7 \times 10^{-8}\,\Omega\,m$.

Electric power; electricity in the home

12 a Below is a list of wattages of various appliances. State which is most likely to be the correct one for each of the appliances named.

60 W   250 W   850 W   2 kW   3.5 kW

(i) kettle
(ii) table lamp
(iii) iron

b What current will be taken by a 920 W appliance if the supply voltage is 230 V?

13 a The diagram shows an incorrectly wired 13 amp plug.

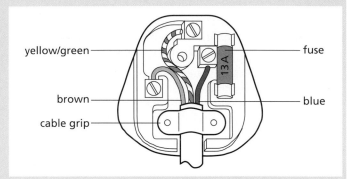

(i) What is wrong with the way this plug has been wired?

(ii) Why do plugs have a fuse?

b The diagram shows an immersion heater which can be used to boil water in a mug.

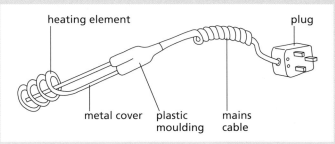

(i) Which part of the immersion heater should be connected to the earth pin of the plug?

(ii) Complete the sentence by choosing the correct words from the box. Each word may be used once or not at all.

chemical	electrical	heat	light

When the immersion heater is switched on ...... energy is transferred to ...... energy.

(iii) When the immersion heater is used correctly, the heating element is at the bottom of the mug of water. Describe how the water at the top of the mug becomes hot.

(SEG Foundation, Summer 99)

14 The information plate on a hairdrier is shown.

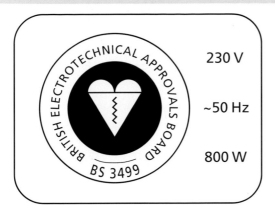

230 V

~50 Hz

800 W

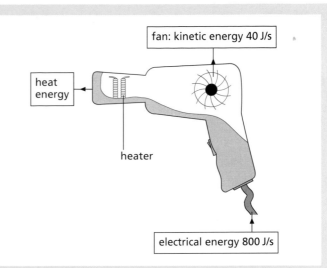

fan: kinetic energy 40 J/s

heat energy

heater

electrical energy 800 J/s

a What is the power rating of the hairdrier?
b (i) Write down the equation which links current, power and voltage.
(ii) Calculate the current in amperes, when the hairdrier is being used. Show clearly how you work out your answer.
(iii) Which *one* of the following fuses, 3 A, 5 A or 13 A, should you use with this hairdrier?
c The hairdrier transfers electrical energy to heat energy and kinetic energy.

Use the following equation to calculate the efficiency of the hairdrier in transferring electrical energy into heat energy.

$$\text{efficiency} = \frac{\text{useful energy output}}{\text{total energy input}}$$

d One kilowatt-hour of electricity costs 6p. Use the following equation to calculate how much it will cost to use the hairdrier for 10 minutes.

$$\text{cost of electricity} = \text{energy transferred} \times \text{price per unit}$$

(SEG Foundation, Summer 98)

51 Magnetic fields

■ Properties of magnets

a) Magnetic materials

Magnets only attract strongly certain materials such as iron, steel, nickel, cobalt, which are called 'ferro-magnetics'.

b) Magnetic poles

The poles are the places in a magnet to which magnetic materials are attracted, e.g. iron filings. They are near the ends of a bar magnet and occur in pairs of equal strength.

c) North and south poles

If a magnet is supported so that it can swing in a horizontal plane it comes to rest with one pole, the north-seeking or N pole, always pointing roughly towards the Earth's north pole. A magnet can therefore be used as a compass.

d) Law of magnetic poles

If the N pole of a magnet is brought near the N pole of another magnet repulsion occurs. Two S (south-seeking) poles also repel. By contrast, N and S poles always attract. The law of magnetic poles summarizes these facts and states:

Like poles repel, unlike poles attract.

The force between magnetic poles decreases as their separation increases.

■ Magnetization of iron and steel

Chains of small iron nails and steel paper clips can be hung from a magnet, Figure 51.1. Each nail or clip magnetizes the one below it and the unlike poles so formed attract.

If the iron chain is removed by pulling the top nail away from the magnet, the chain collapses, showing that **magnetism induced in iron is temporary**. When the same is done with the steel chain, it does not collapse; **magnetism induced in steel is permanent**.

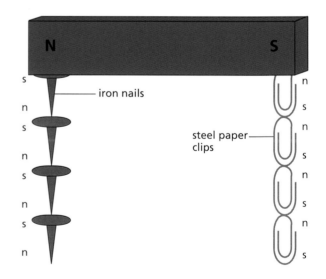

Figure 51.1 Investigating the magnetization of iron and steel

Magnetic materials like iron which magnetize easily but do not keep their magnetism are said to be **soft**. Those like steel which are harder to magnetize than iron but stay magnetized are **hard**. Both types have their uses; very hard ones are used to make permanent magnets.

■ *Magnetic fields*

The space surrounding a magnet where it produces a magnetic force is called a **magnetic field**. The force around a bar magnet can be detected and shown to vary in direction using the apparatus in Figure 51.2. If the floating magnet is released near the N pole of the bar magnet, it is repelled to the S pole and moves along a curved path known as a **line of force** or a **field line**. It moves in the opposite direction if its south pole is uppermost.

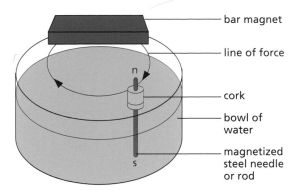

Figure 51.2 Detecting magnetic force

It is useful to consider that a magnetic field has a direction and to represent the field by lines of force. It has been decided that **the direction of the field at any point should be the direction of the force on a N pole**. To show the direction, arrows are put on the lines of force and point away from a N pole towards a S pole.

Practical work

Plotting lines of force

a) Plotting compass method

A plotting compass is a small pivoted magnet in a glass case with non-magnetic metal walls, Figure 51.3a.

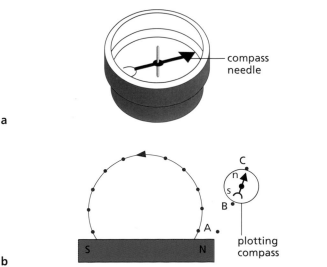

b

Figure 51.3

Lay a bar magnet on a sheet of paper. Place the plotting compass at a point such as A, Figure 51.3b, near one pole of the magnet. Mark the position of the poles n, s of the compass by pencil dots B, A. Move the compass so that pole s is exactly over B, mark the new position of n by dot C.

Continue this process until the S pole of the bar magnet is reached. Join the dots to give one line of force and show its direction by putting an arrow on it. Plot other lines by starting at different points round the magnet.

A typical field pattern is shown in Figure 51.4.

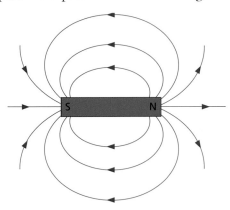

Figure 51.4 Magnetic field lines around a bar magnet

The combined field due to two neighbouring magnets can also be plotted to give patterns like those in Figures 51.5a, b. In a, where two like poles are facing each other, the point X is called a **neutral point**. At X the field due to one magnet cancels out that due to the other and there are no lines of force.

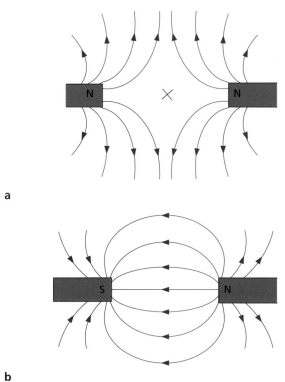

a

b

Figure 51.5 Field lines due to two neighbouring magnets

b) Iron filings method

Place a sheet of paper *on top of* a bar magnet and sprinkle iron filings *thinly and evenly* on to the paper from a 'pepper pot'.

Tap the paper gently with a pencil and the filings should form patterns of the lines of force. Each filing turns in the direction of the field when the paper is tapped.

This method is quick but no use for weak fields. Figures 51.6a, b show typical patterns with two magnets. Why are they different? What combination of poles would give the observed patterns?

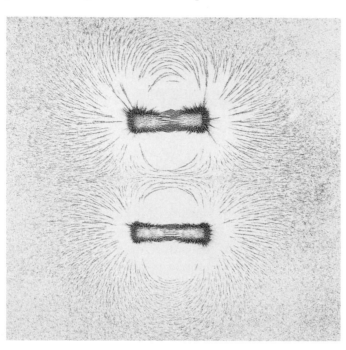

a

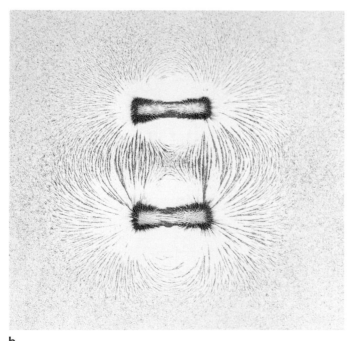

b

Figure 51.6 Field lines round two bar magnets shown by iron filings

Earth's magnetic field

If lines of force are plotted on a sheet of paper with no magnets near, a set of parallel straight lines is obtained. They run roughly from S to N geographically, Figure 51.7, and represent a small part of the Earth's magnetic field in a horizontal plane.

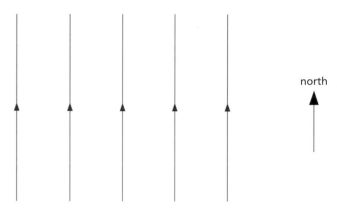

north

Figure 51.7 Lines of force due to the Earth's field

At most places on the Earth's surface a magnetic compass points slightly east or west of true north, i.e. the Earth's geographical and magnetic north poles do not coincide. The angle between magnetic north and true north is called the **declination**, Figure 51.8. In London at present (2000) it is 6° W of N and is decreasing. By about the year 2140 it should be 0°.

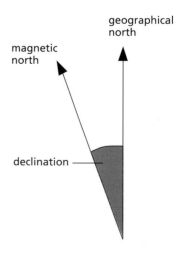

geographical
north

magnetic
north

declination

Figure 51.8 The Earth's geographical and magnetic poles do not coincide

Questions

1 A magnet attracts

A plastics   **B** any metal   **C** iron and steel
D aluminium   **E** carbon.

2 Copy Figure 51.9 which shows a plotting compass and a magnet. Label the N pole of the magnet and draw the field line on which the compass lies.

Figure 51.9

3 The three diagrams in Figure 51.10 show the lines of force (field lines) between the poles of two magnets. Identify the poles A, B, C, D, E, F.

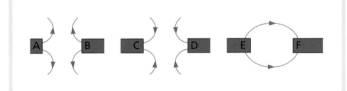

Figure 51.10

4 a Figure 51.11 shows the magnetic field between two magnets.

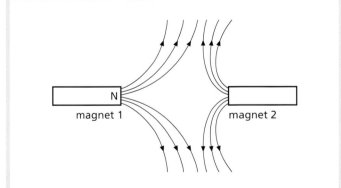

Figure 51.11

(i) Copy the diagram and label the other poles of the magnets.
(ii) Which is the weaker magnet?
b A child has accidentally swallowed some small metal objects. These objects have stuck in the child's throat. A doctor uses the tool in Figure 51.12 to remove them.

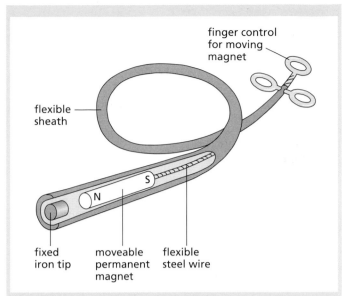

Figure 51.12

(i) What happens to the fixed iron tip when the permanent magnet is moved towards it?
(ii) Why is it an advantage to make the tip out of iron?
(iii) The tool is pushed down the child's throat and guided to the metal object. Explain why the sheath needs to be flexible.
(iv) The doctor tries to use the tool to remove a steel paper clip and an aluminium washer from the child's throat. Which object can the doctor remove? Explain your answer.

(London Foundation, June 98)

■ *Checklist*

After studying this chapter you should be able to

- ■ state the properties of magnets,
- ■ explain what is meant by **soft** and **hard** magnetic materials,
- ■ recall that a magnetic field is the region round a magnet where a magnetic force is exerted and is represented by lines of force whose direction at any point is the direction of the force on a N pole,
- ■ map magnetic fields (by the plotting compass and iron filings methods) round (a) one magnet, (b) two magnets,
- ■ recall that at a neutral point the field due to one magnet cancels that due to another,
- ■ define **declination**.

■ *Oersted's discovery*

In 1819 Oersted accidentally discovered the magnetic effect of an electric current. His experiment can be repeated by holding a wire over and parallel to a compass needle which is pointing N and S, Figure 52.1. The needle moves when the current is switched on. Reversing the current causes the needle to move in the opposite direction.

Evidently around a wire carrying a current there is a magnetic field. As with the field due to a permanent magnet, we represent the field due to a current by field lines or lines of force. Arrows on the lines show the direction of the field, i.e. the direction in which a N pole points.

Different field patterns are given by differently shaped conductors.

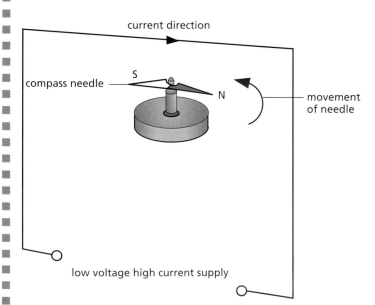

Figure 52.1 An electric current produces a magnetic effect

■ *Field due to a straight wire*

If a straight vertical wire passes through the centre of a piece of card held horizontally and a current is passed through the wire, Figure 52.2, iron filings sprinkled on the card set in concentric circles when the card is tapped.

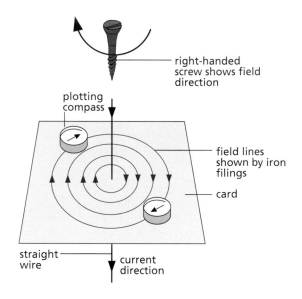

Figure 52.2 Field due to a straight wire

Plotting compasses placed on the card set along the field lines and show the direction of the field at different points. When the current direction is reversed, the compasses point in the opposite direction showing that the direction of the field reverses when the current reverses.

If the current direction is known, the direction of the field can be predicted by the **right-hand screw rule**:

If a right-handed screw moves forwards in the direction of the current (conventional), the direction of rotation of the screw gives the direction of the field.

Field due to a circular coil

The field pattern is shown in Figure 52.3. At the centre of the coil the field lines are straight and at right angles to the plane of the coil. The right-hand screw rule again gives the direction of the field at any point.

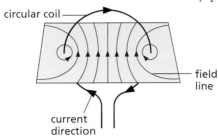

Figure 52.3 Field due to a circular coil

Field due to a solenoid

A solenoid is a long cylindrical coil. It produces a field similar to that of a bar magnet; in Figure 52.4a, end A behaves like a N pole and end B like a S pole. The polarity can be found as before by applying the right-hand screw rule to a short length of one turn of the solenoid. Alternatively the **right-hand grip rule** can be used. This states that if the fingers of the right hand grip the solenoid in the direction of the current (conventional), the thumb points to the N pole, Figure 52.4b. Figure 52.4c shows how to link the end-on view of the current direction in the solenoid to the polarity.

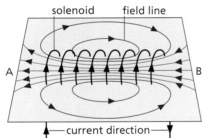

a Field due to a solenoid

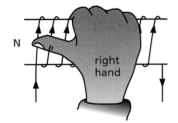

b The right-hand grip rule

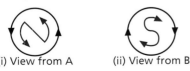

(i) View from A   (ii) View from B

c End-on views
Figure 52.4

The field inside a solenoid can be made very strong if it has a large number of turns or a large current. Permanent magnets can be made by allowing molten ferromagnetic metal to solidify in such fields.

Practical work

Simple electromagnet

An electromagnet is a coil of wire wound on a soft iron core. A 5 cm iron nail and 3 m of PVC-covered copper wire (SWG 26) are needed.

(a) Leave about 25 cm at one end of the wire (for connecting to the circuit) and then wind about 50 cm as a single layer on the nail. **Keep the turns close together and always wind in the same direction.** Connect the circuit of Figure 52.5, setting the rheostat at its maximum resistance.

Find the number of paper clips the electromagnet can support when known currents between 0.2 A and 2.0 A pass through it. Record the results in a table. How does the 'strength' of the electromagnet depend on the current?

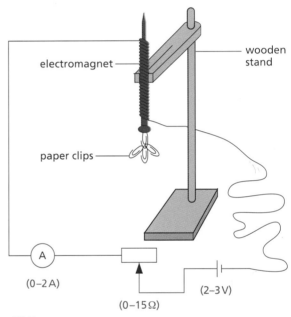

Figure 52.5

(b) Add another two layers of wire to the nail, winding in the *same direction* as the first layer. Repeat the experiment. What can you say about the 'strength' of an electromagnet and the number of turns of wire?

(c) Place the electromagnet on the bench and under a sheet of paper. Sprinkle iron filings on the paper, tap it gently and observe the field pattern. How does it compare with that given by a bar magnet?

(d) Use the right-hand screw (or grip) rule to predict which end of the electromagnet is a N pole. Check with a plotting compass.

■ Electromagnets

The magnetism of an electromagnet is *temporary* and can be switched on and off, unlike that of a permanent magnet. It has a core of soft iron which is magnetized only when current flows in the surrounding coil.

The strength of an electromagnet increases if

(i) the current in the coil increases,
(ii) the number of turns on the coil increases,
(iii) the poles are closer together.

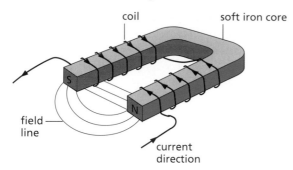

Figure 52.6 C-core or horseshoe electromagnet

In C-core (or horseshoe) electromagnets condition (iii) is achieved, Figure 52.6. Note that the coil is wound in *opposite* directions on each limb of the core.

As well as being used as a crane to lift iron objects, scrap iron, etc., Figure 52.7, an electromagnet is an essential part of many electrical devices.

Figure 52.7 Electromagnet being used to lift scrap metal

■ Electric bell

When the circuit in Figure 52.8 is completed, current flows in the coils of the electromagnet which becomes magnetized and attracts the soft iron bar (the armature). The hammer hits the gong but the circuit is now broken at the point C of the contact screw.

The electromagnet loses its magnetism and no longer attracts the armature. The springy metal strip is then able to pull the armature back, remaking contact at C and so completing the circuit again. This cycle is repeated so long as the bell push is depressed and continuous ringing occurs.

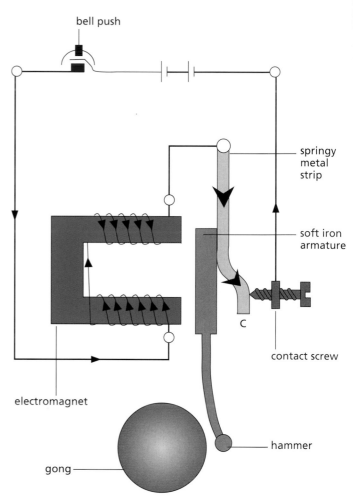

Figure 52.8 Electric bell

■ *Relay and circuit breaker*

a) Relay

A relay is a switch based on the principle of an electromagnet. It is useful if we want one circuit to control another, especially if the current and power are larger in the second circuit (see question 2, p.255). Figure 52.9 shows a typical relay. When a current is in the coil from the circuit connected to AB, the soft iron core is magnetized and attracts the L-shaped iron armature. This rocks on its pivot and closes the contacts at C in the circuit connected to DE. The relay is then 'energized' or 'on'.

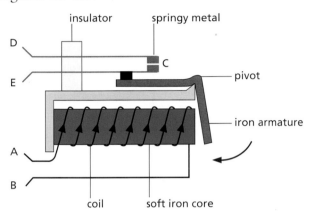

Figure 52.9 Relay

The current needed to operate a relay is called the **pull-on** current, and the **drop-off** current is the smaller current in the coil when the relay just stops working. If the coil resistance R of a relay is $185\,\Omega$ and its operating p.d. V is $12\,\text{V}$, the pull-on current $I = V/R = 12/185 = 0.065\,\text{A} = 65\,\text{mA}$. The symbols for relays with normally open and normally closed contacts are given in Figure 52.10a, b.

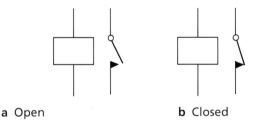

a Open   **b** Closed

Figure 52.10 Symbols for a relay

Some examples of the use of relays in circuits appear in Chapter 60.

b) Circuit breaker

A circuit breaker (p.223) acts in a similar way to a normally closed relay; when the current in the electromagnet exceeds a critical value the contact points are separated and the circuit is broken. In the design shown in question 4 overleaf, when the iron bolt is attracted far enough towards the electromagnet, the plunger is released and the push switch opens, breaking contact to the rest of the circuit.

■ *Telephone*

A telephone contains a microphone at the speaking end and a receiver at the listening end.

a) Carbon microphone

When someone speaks into a carbon microphone, Figure 52.11, sound waves cause the diaphragm to move backwards and forwards. This varies the pressure on the carbon granules between the movable carbon dome which is attached to the diaphragm and the fixed carbon cup at the back. When the pressure increases, the granules are squeezed closer together and their electrical resistance decreases. A decrease of pressure has the opposite effect. The current passing through the microphone varies in a similar way to the sound wave variations.

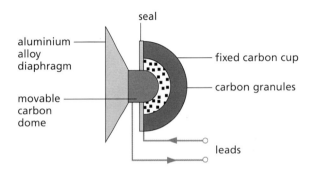

Figure 52.11 Carbon microphone

b) Receiver

The coils are wound in opposite directions on the *two* S poles of a magnet as in Figure 52.12. If the current goes round one in a clockwise direction, it goes round the other anticlockwise, so making one S pole stronger and the other weaker. This causes the iron armature to rock on its pivot towards the stronger S pole. When the current reverses, the armature rocks the other way due to the S pole which was the stronger before becoming the weaker. These armature movements are passed on to the diaphragm, making it vibrate and produce sound of the same frequency as the alternating current in the coil (received from the microphone).

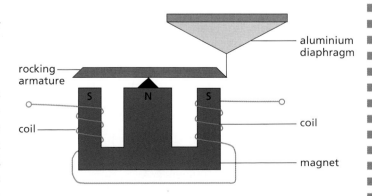

Figure 52.12 Telephone receiver

Questions

1 The vertical wire in Figure 52.13 is at right angles to the card. In what direction will a plotting compass at A point when
a there is no current in the wire,
b current flows upwards?

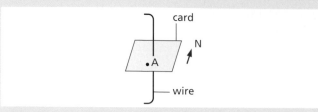

Figure 52.13

2 Figure 52.14 shows a solenoid wound on a core of soft iron. Will the end A be a N pole or S pole when the current flows in the direction shown?

Figure 52.14

3 Figure 52.15 shows an electromagnet being used to lift some weights.

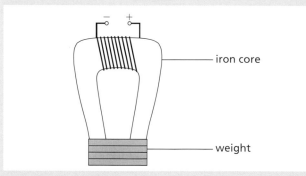

Figure 52.15

The graph in Figure 52.16 shows how the weight lifted by the electromagnet depends on the current in the coil.

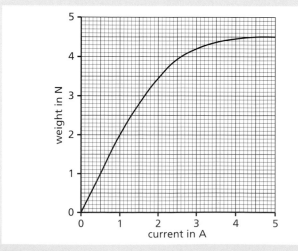

Figure 52.16

a (i) What is the heaviest weight that the electromagnet can lift?
(ii) Suggest how the electromagnet can be changed so that it will lift a heavier weight.
b The electromagnet is used to pick up iron cans filled with fruit juice. Each can weighs 2 N.
(i) What current is needed for the electromagnet to lift one can?
(ii) When this current is doubled, is the electromagnet able to lift two cans at the same time? Use data from the graph to give the reason for your answer.

(London Foundation, June 99)

4 A fault in an electrical circuit can cause too great a current to flow. Some circuits are switched off by a circuit breaker.

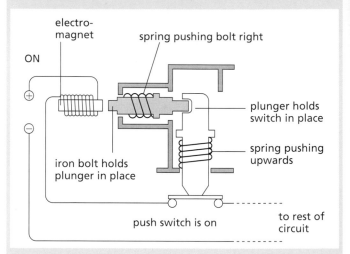

Figure 52.17

One type of circuit breaker is shown in Figure 52.17. A normal current is flowing. Explain, in detail, what happens when a current which is bigger than normal flows.

(AQA (NEAB) Foundation, June 99)

■ *Checklist*

After studying this chapter you should be able to

- ■ describe and draw sketches of the magnetic fields round current-carrying, straight and circular conductors and solenoids,
- ■ recall the right-hand screw and grip rules for relating current direction and magnetic field direction,
- ■ make a simple electromagnet,
- ■ describe uses of electromagnets,
- ■ explain the action of an electric bell, a relay, a circuit breaker, and the microphone and receiver of a telephone.

53 Electric motors

Electric motors form the heart of a whole host of electrical devices ranging from domestic appliances such as vacuum cleaners and washing machines to electric trains and lifts. In a car the windscreen wipers are usually driven by one and the engine is started by another.

■ *The motor effect*

A wire carrying a current in a magnetic field experiences a force. If the wire can move it does so.

a) Demonstration

In Figure 53.1 the flexible wire is loosely supported in the strong magnetic field of a C-shaped magnet (permanent or electro). When the switch is pressed, current flows in the wire which jumps upwards as shown. If either the direction of the current or the direction of the field is reversed, the wire moves downwards. **The force increases if the strength of the field increases and if the current increases.**

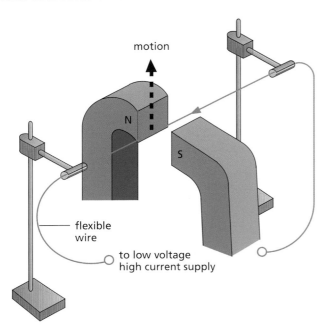

Figure 53.1 A wire carrying a current in a magnetic field experiences a force

b) Explanation

Figure 53.2a is a side view of the magnetic field lines due to the wire and the magnet. Those due to the wire are circles and we will assume their direction is as shown. The dotted lines represent the field lines of the magnet and their direction is to the right.

The resultant field obtained by combining both fields is shown in Figure 53.2b. There are more lines below than above the wire since both fields act in the same direction below but in opposition above. If we *suppose* the lines are like stretched elastic, those below will try to straighten out and in so doing will exert an upward force on the wire.

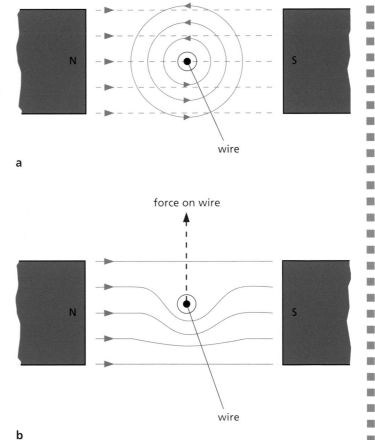

Figure 53.2

■ 239

■ *Fleming's left-hand rule*

The direction of the force or thrust on the wire can be found by this rule which is also called the 'motor rule', Figure 53.3.

> Hold the thumb and first two fingers of the left hand at right angles to each other with the **F**irst finger pointing in the direction of the **F**ield and the se**C**ond finger in the direction of the **C**urrent, then the **Th**umb points in the direction of the **Th**rust.

If the wire is not at right angles to the field, the force is smaller and is zero if the wire is parallel to the field.

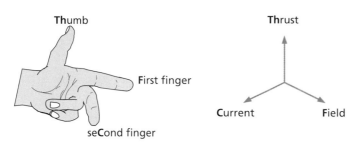

Figure 53.3 Fleming's left-hand (motor) rule

■ *Simple d.c. electric motor*

A simple motor to work from direct current (d.c.) consists of a rectangular coil of wire mounted on an axle which can rotate between the poles of a C-shaped magnet, Figure 53.4. Each end of the coil is connected to half of a split ring of copper, called the **commutator**, which rotates with the coil. Two carbon blocks, the **brushes**, are pressed lightly against the commutator by the springs. The brushes are connected to an electrical supply.

If Fleming's left–hand rule is applied to the coil in the position shown, we find that side **ab** experiences an upward force and side **cd** a downward force. (No forces act on **ad** and **bc** since they are parallel to the field.) These two forces form a **couple** which rotates the coil in a clockwise direction until it is vertical.

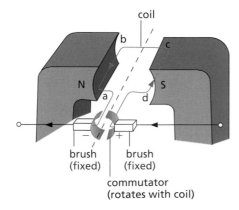

Figure 53.4 Simple d.c. motor

The brushes are then in line with the gaps in the commutator and the current stops. However, because of its inertia, the coil overshoots the vertical and the commutator halves change contact from one brush to the other. This reverses the current through the coil and so also the directions of the forces on its sides. Side **ab** is on the right now, acted on by a downward force, while **cd** is on the left with an upward force. The coil thus carries on rotating clockwise.

■ *Practical motors*

Practical motors have:

(a) a coil of many turns wound on a soft iron cylinder or core which rotates with the coil. This makes it more powerful. The coil and core together are called the **armature**.

(b) several coils each in a slot in the core and each having a pair of commutator segments. This gives increased power and smoother running. The motor of an electric drill is shown in Figure 53.5.

(c) an electromagnet (usually) to produce the field in which the armature rotates.

Most electric motors used in industry are **induction motors**. They work off a.c. (alternating current) on a different principle to the d.c. motor.

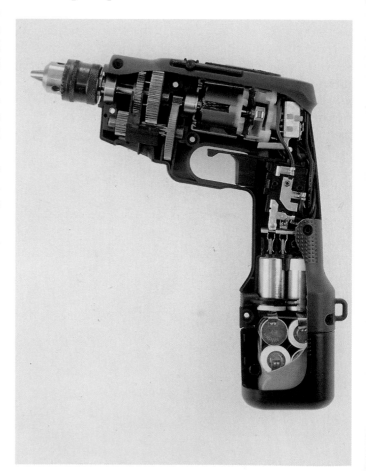

Figure 53.5 Motor inside an electric drill

Practical work

A model motor

The motor is shown in Figure 53.6 and is made from a kit of parts.

1 Wrap Sellotape round one end of the metal tube which passes through the wooden block.
2 Cut two rings off a piece of narrow rubber tubing; slip them onto the Sellotaped end of the metal tube.
3 Remove the insulation from one end of a $1\frac{1}{2}$ metre length of SWG 26 PVC-covered copper wire and fix it under both rubber rings so that it is held tight against the Sellotape. This forms one end of the coil.
4 Wind 10 turns of the wire in the slot in the wooden block and finish off the second end of the coil by removing the PVC and fixing this too under the rings but on the *opposite* side of the tube from the first end. The bare ends act as the **commutator**.
5 Push the axle through the metal tube of the wooden base so that the block spins freely.
6 Arrange two $\frac{1}{2}$ metre lengths of wire to act as **brushes** and leads to the supply, as shown. Adjust the brushes so that they are vertical and each touches one bare end of the coil when the plane of the coil is horizontal. **The motor will not work if this is not so.**
7 Slide the base into the magnet with *opposite poles facing*. Connect to a 3 V battery (or other low voltage d.c. supply) and a slight push of the coil should set it spinning at high speed.

Moving-coil loudspeaker

Varying currents from a radio, disc player, etc. pass through a short cylindrical coil whose turns are at right angles to the magnetic field of a magnet with a central pole and a surrounding ring pole, Figure 53.7a.

A force acts on the coil which, according to Fleming's left-hand rule, makes it move in and out. A paper cone attached to the coil moves with it and sets up sound waves in the surrounding air, Figure 53.7b.

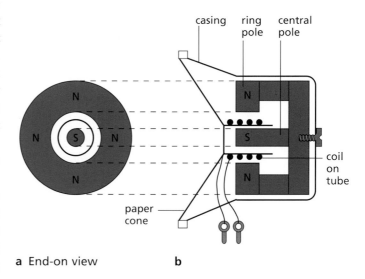

a End-on view **b**

Figure 53.7 Moving-coil loudspeaker

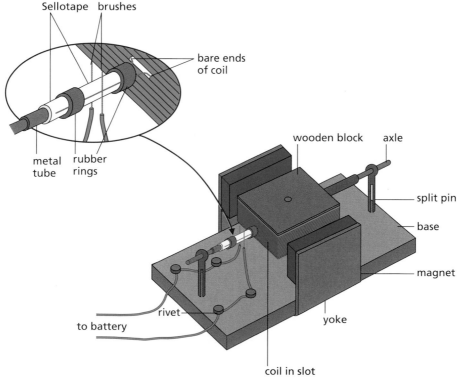

Figure 53.6 A model motor

Questions

1 A current flows in a wire running between the N and S poles of a magnet lying horizontally as shown in Figure 53.8. The force on the wire due to the magnet is directed

A from N to S
B from S to N
C opposite to the current direction
D in the direction of the current
E vertically upwards.

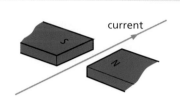

current

Figure 53.8

2 In the simple electric motor of Figure 53.9, the coil rotates anticlockwise as seen by the eye from the position X when current flows in the coil. Is the current flowing clockwise or anticlockwise around the coil when viewed from above?

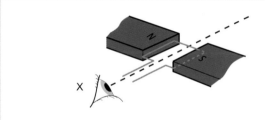

X

Figure 53.9

3 Explain how a moving-coil loudspeaker produces sound. What factors affect
a the loudness,
b the pitch of the sound?

4 An electric motor is a device which transfers

A mechanical energy to electrical energy
B heat energy to electrical energy
C electrical energy to heat only
D heat to mechanical energy
E electrical energy to mechanical energy and heat.

■ *Checklist*

After studying this chapter you should be able to

■ describe a demonstration to show that a force acts on a current-carrying conductor in a magnetic field and recall that it increases with the strength of the field and the size of the current,

■ draw the resultant field pattern for a current-carrying conductor which is at right angles to a uniform magnetic field,

■ explain why a rectangular current-carrying coil experiences a couple in a uniform magnetic field,

■ draw a diagram of a simple d.c. electric motor and explain how it works,

■ describe a practical d.c. motor,

■ draw a diagram of a moving-coil loudspeaker and explain how it works.

54 Electric meters

Moving-coil galvanometer
Ammeters and shunts

Voltmeters and multipliers
Multimeters

■ Moving-coil galvanometer

A galvanometer detects small currents or small p.ds, often of the order of milliamperes (mA) or millivolts (mV).

In the moving-coil **pointer-type** meter, a coil is pivoted on jewelled bearings between the poles of a permanent magnet, Figure 54.1a. Current enters and leaves the coil by hair springs above and below it. When current flows, a couple acts on the coil (as in an electric motor), causing it to rotate until stopped by the springs. The greater the current, the greater the deflection which is shown by a pointer attached to the coil.

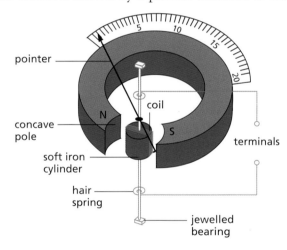

pointer

concave pole

soft iron cylinder

hair spring

coil

N   S

terminals

jewelled bearing

a

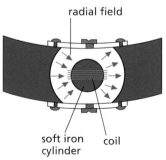

radial field

soft iron cylinder   coil

b View from above

Figure 54.1 Moving-coil pointer-type galvanometer

The soft iron cylinder at the centre of the coil is fixed and along with the concave poles of the magnet it produces a **radial** field, Figure 54.1b, i.e. the field lines are directed to the centre of the cylinder. The scale on the meter is then even or linear, i.e. all divisions are the same size.

The sensitivity of a galvanometer is increased by having

(i) more turns on the coil,
(ii) a stronger magnet,
(iii) weaker hair springs or a wire suspension,
(iv) as a pointer, a long beam of light reflected from a mirror on the coil.

The last two are used in **light-beam** meters which have a full-scale deflection of a few microamperes (μA). $(1 \, \mu A = 10^{-6} \, A)$

■ Ammeters and shunts

An ammeter is a galvanometer having a known low resistance, called a **shunt**, in parallel with it to take most of the current, Figure 54.2. An ammeter is placed in *series* in a circuit and must have a *low resistance* otherwise it changes the current to be measured.

galvanometer

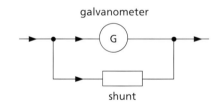

shunt

Figure 54.2 An ammeter

■ Voltmeters and multipliers

A voltmeter is a galvanometer having a known high resistance, called a **multiplier**, in series with it, Figure 54.3. A voltmeter is placed in *parallel* with the part of the circuit across which the p.d. is to be measured and must have a *high resistance* – otherwise the total resistance of the whole circuit is reduced so changing the current and the p.d. required.

galvanometer

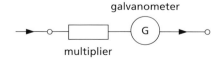

multiplier

Figure 54.3 A voltmeter

■ *Multimeters*

These can have analogue or digital displays (see Figures 60.32a and b, p. 290) and can be used to measure a.c. or d.c. currents or voltages and also resistance. The required function is first selected, say a.c. current, and then a suitable range chosen. For example if a current of a few milliamps is expected, the 10 mA range might be selected and the value of the current (in mA) read from the display; if the reading is off-scale, the sensitivity should be reduced by changing to the higher, perhaps 100 mA, range.

For the measurement of resistance, the resistance function is chosen and the appropriate range selected. The terminals are first short-circuited to check the zero of resistance, then the unknown resistance is disconnected from any circuit and reconnected across the terminals of the meter in place of the short circuit.

Analogue multimeters are adaptions of moving-coil galvanometers. Digital multimeters are constructed from integrated circuits (see p. 281). On the voltage setting they have a very high input resistance (10 MΩ), i.e. they affect most circuits very little and so give very accurate readings.

Questions

1 What does a galvanometer do?
2 Why should the resistance of
   a an ammeter be very small,
   b a voltmeter be very large?

■ *Checklist*

After studying this chapter you should be able to

■ draw a diagram of a simple moving-coil galvanometer and explain how it works,
■ explain how a moving-coil galvanometer can be modified for use as (a) an ammeter and (b) a voltmeter,
■ explain why (a) an ammeter should have a very low resistance and (b) a voltmeter should have a very high resistance.

55 Generators

An electric current creates a magnetic field. The reverse effect of producing electricity from magnetism was discovered in 1831 by Faraday and is called **electromagnetic induction**. It led to the construction of generators for producing electrical energy in power stations.

■ *Electromagnetic induction*

Two ways of investigating the effect follow.

a) Straight wire and U-shaped magnet

First the wire is held at rest between the poles of the magnet. It is then moved in each of the six directions shown in Figure 55.1 and the meter observed. Only *when it is moving upwards* (direction 1) or *downwards* (direction 2) is there a deflection on the meter, indicating an induced current in the wire. The deflection is in opposite directions in each case and only lasts while the wire is in motion.

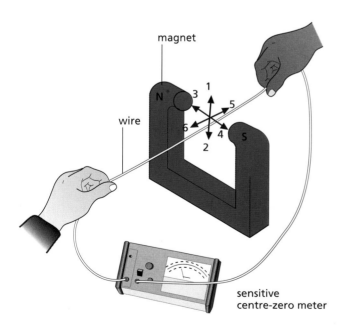

Figure 55.1 A current is induced in the wire when it moves up or down between the magnet poles

b) Bar magnet and coil

The magnet is pushed into the coil one pole first, Figure 55.2, then held still inside it. It is next withdrawn. The meter shows that current is induced in the coil in one direction as the magnet *moves in* and in the opposite direction as it is *removed*. There is no deflection when the magnet is at rest. The results are the same if the coil is moved instead of the magnet, i.e. only **relative motion** is needed.

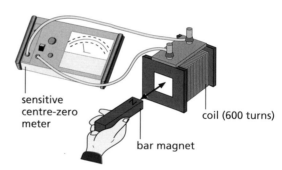

Figure 55.2 A current is induced in the coil when the magnet is moved in or out

■ *Faraday's law*

To 'explain' electromagnetic induction Faraday suggested that a voltage is induced in a conductor whenever it 'cuts' magnetic field lines, i.e. moves *across* them, but not when it moves along them or is at rest. If the conductor forms part of a complete circuit, an induced current is also produced.

Faraday found, and it can be shown with apparatus like that in Figure 55.2, that the induced p.d. or voltage increases with increases of

(i) the speed of motion of the magnet or coil,
(ii) the number of turns on the coil,
(iii) the strength of the magnet.

These facts led him to state a law:

> The size of the induced p.d. is directly proportional to the rate at which the conductor cuts magnetic field lines.

■ Lenz's law

The direction of the induced current can be found by a law due to the Russian scientist, Lenz.

> The direction of the induced current is such as to oppose the change causing it.

In Figure 55.3a the magnet approaches the coil, north pole first. According to Lenz's law the induced current should flow in a direction which makes the coil behave like a magnet with its top a north pole. The downward motion of the magnet will then be opposed since like poles repel.

When the magnet is withdrawn, the top of the coil should become a south pole, Figure 55.3b, and attract the north pole of the magnet, so hindering its removal. The induced current is thus in the opposite direction to that when the magnet approaches.

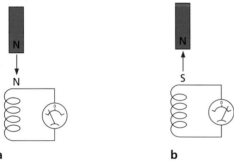

a   **b**

Figure 55.3 The induced current opposes the motion of the magnet

Lenz's law is an example of the principle of conservation of energy. If the currents caused opposite poles to those that they do, electrical energy would be created from nothing. As it is, mechanical energy is provided, by whoever moves the magnet, to overcome the forces that arise.

For a straight wire moving at right angles to a magnetic field a more useful form of Lenz's law is **Fleming's right-hand rule** (the 'dynamo rule'), Figure 55.4.

> Hold the thumb and first two fingers of the right hand at right angles to each other with the **F**irst finger pointing in the direction of the **F**ield and the thu**M**b in the direction of **M**otion of the wire, then the se**C**ond finger points in the direction of the induced **C**urrent.

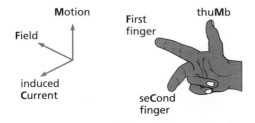

Figure 55.4 Fleming's right-hand (dynamo) rule

■ Simple a.c. generator (alternator)

The simplest alternating current (a.c.) generator consists of a rectangular coil between the poles of a C-shaped magnet, Figure 55.5a. The ends of the coil are joined to two **slip rings** on the axle and against which carbon **brushes** press.

When the coil is rotated it cuts the field lines and a voltage is induced in it. Figure 55.5b shows how the voltage varies over one complete rotation.

As the coil moves through the vertical position with **ab** uppermost, **ab** and **cd** are moving along the lines (**bc** and **da** do so always) and no cutting occurs. The induced voltage is zero.

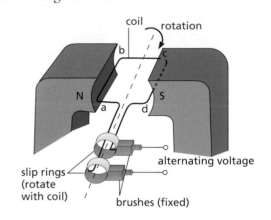

a

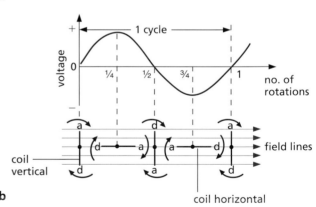

b

Figure 55.5 A simple a.c. generator and its output

During the first quarter rotation the p.d. increases to a maximum when the coil is horizontal. Sides **ab** and **dc** are then cutting the lines at the greatest rate.

In the second quarter rotation the p.d. decreases again and is zero when the coil is vertical with **dc** uppermost. After this, the direction of the p.d. reverses because, during the next half rotation, the motion of **ab** is directed upwards and **dc** downwards.

An alternating voltage is generated which acts first in one direction and then the other; it would cause a.c. to flow in a circuit connected to the brushes. The **frequency** of an a.c. is the number of complete cycles it makes each second and is measured in **hertz** (Hz), i.e. 1 cycle per second = 1 Hz. If the coil rotates twice per second, the a.c. has frequency 2 Hz. The mains supply is a.c. of frequency 50 Hz.

■ *Simple d.c. generator (dynamo)*

An a.c. generator becomes a direct current (d.c.) one if the slip rings are replaced by a **commutator** (like that in a d.c. motor), Figure 55.6a.

The brushes are arranged so that as the coil goes through the vertical, changeover of contact occurs from one half of the split ring of the commutator to the other. In this position the voltage induced in the coil reverses and so one brush is always positive and the other negative.

The voltage at the brushes is shown in Figure 55.6b; although varying in value, it never changes direction and would produce a direct current (d.c.) in an external circuit.

In construction the simple d.c. dynamo is the same as the simple d.c. motor and one can be used as the other. When an electric motor is working it acts as a dynamo and creates a voltage which opposes the applied voltage. The current in the coil is therefore much less once the motor is running.

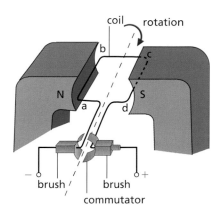

a

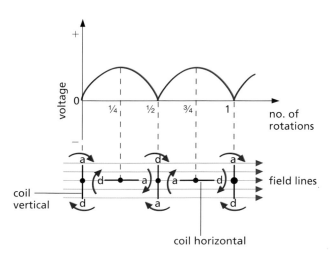

b

Figure 55.6 A simple d.c. generator and its output

■ *Practical generators*

In actual generators several coils are wound in evenly spaced slots in a soft iron cylinder and electromagnets usually replace permanent magnets.

a) Power stations

In power station alternators the electromagnets rotate (the **rotor**, Figure 55.7a) while the coils and their iron core are at rest (the **stator**, Figure 55.7b). The large p.ds and currents (e.g. 25 kV at several thousand amps) induced in the stator are led away through stationary cables, otherwise they would quickly destroy the slip rings by sparking. Instead the relatively small d.c. required by the rotor is fed via the slip rings from a small dynamo (the **exciter**) which is driven by the same turbine as the rotor.

a Rotor (electromagnets)

b Stator (induction coils)

Figure 55.7 The rotor and stator of a power station alternator

The output is three-phase a.c., obtained by having three sets of stator coils and three rotor coils at 120° to one another. This results in a steadier power supply.

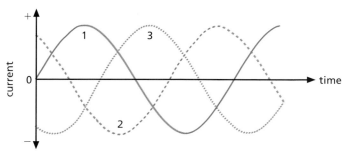

Figure 55.8 Three-phase output of a power station alternator

In a thermal power station (Chapter 42), the turbine is rotated by high-pressure steam obtained by heating water in a coal- or oil-fired boiler or in a nuclear reactor (or by hot gas in a gas-fired power station). A block diagram of a thermal power station is shown in Figure 55.9. The energy transfer diagram was given in Figure 42.7, p. 184.

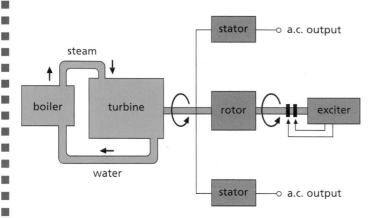

Figure 55.9 Block diagram of a thermal power station

b) Cars

Most cars are now fitted with alternators because they give a greater output than dynamos at low engine speeds.

c) Bicycles

The rotor of a bicycle dynamo is a permanent magnet and the voltage is induced in the coil which is at rest, Figure 55.10.

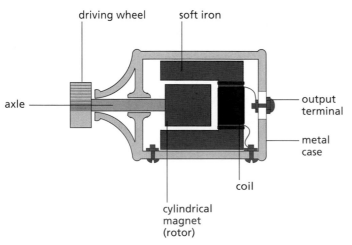

Figure 55.10 Bicycle dynamo

Applications of electromagnetic induction

a) Moving-coil microphone

The moving-coil loudspeaker shown in Figure 53.7 can be operated in reverse mode as a microphone. When sound is incident on the paper cone it vibrates causing the attached coil to move in and out between the poles of the magnet. A varying electric current, representative of the sound, is then induced in the coil by electromagnetic induction.

b) Magnetic recording

Magnetic tapes or disks are used to record information in sound systems and computers (see p. 305). In the recording head shown in Figure 55.11, the tape becomes magnetized when it passes over the gap in the pole piece of the electromagnet and retains a magnetic record of the electrical signal applied to the coil from a microphone or computer. In playback mode, the varying magnetization on the moving tape or disk induces a corresponding electrical signal in the coil as a result of electromagnetic induction.

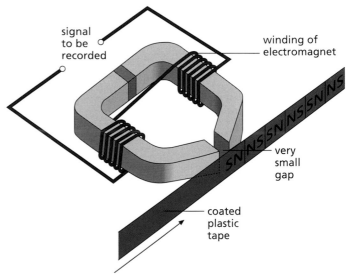

Figure 55.11 Magnetic recording or playback head

Questions

1 a Figure 55.12 shows a magnet being moved into a coil of wire. The reading on the meter is shown in the diagram.

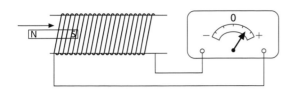

Figure 55.12

Draw the meter reading which you would expect to get in each of the following cases.
(i) The magnet is at rest inside the coil, Figure 55.13.

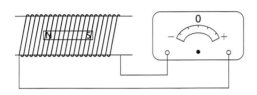

Figure 55.13

(ii) The magnet is moved out of the coil, Figure 55.14.

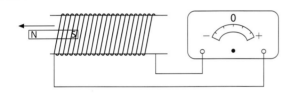

Figure 55.14

b Figure 55.15 shows a bicycle dynamo and part of the wheel.

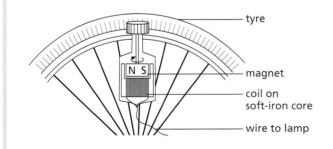

Figure 55.15

Explain, as fully as you can, why a current flows through the bicycle lamp when the wheel of the bicycle turns.

(NEAB Foundation, June 98)

2 A simple generator is shown in Figure 55.16.
a What are A and B called and what is their purpose?
b What changes can be made to increase the p.d. generated?

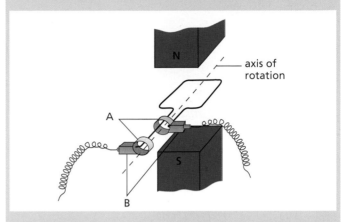

Figure 55.16

■ *Checklist*

After studying this chapter you should be able to

- describe experiments to show electromagnetic induction,
- recall Faraday's explanation of electromagnetic induction,
- predict the direction of the induced current using Lenz's law or Fleming's right-hand rule,
- draw a diagram of a simple a.c. generator and sketch a graph of its output,
- draw a diagram of a simple d.c. generator and sketch a graph of its output,
- describe practical generators,
- recall some applications of electromagnetic induction.

56 Transformers

■ *Mutual induction*

When the current in a coil is switched on or off or changed, a voltage is induced in a neighbouring coil. The effect, called **mutual induction**, is an example of electromagnetic induction and can be shown with the arrangement of Figure 56.1. Coil A is the **primary** and coil B the **secondary**.

Switching on the current in the primary sets up a magnetic field and as its field lines 'grow' outwards from the primary they 'cut' the secondary. A p.d. is induced in the secondary until the current in the primary reaches its steady value. When the current is switched off in the primary, the magnetic field dies away and we can imagine the field lines cutting the secondary as they collapse, again inducing a p.d. in it. Changing the primary current by *quickly* altering the rheostat has the same effect.

The induced p.d. is increased by having a soft iron rod in the coils or, better still, by using coils wound on a complete iron ring. More field lines then cut the secondary due to the magnetization of the iron.

Practical work

Mutual induction with a.c.

An alternating current is changing all the time and if it flows in a primary coil, an alternating voltage and current are induced in a secondary coil.

Connect the circuit of Figure 56.2. The 1 V high current power unit supplies a.c. to the primary and the lamp detects the secondary current.

Find the effect on the brightness of the lamp of

(i) pulling the C-cores apart slightly,
(ii) increasing the secondary turns to 15,
(iii) decreasing the secondary turns to 5.

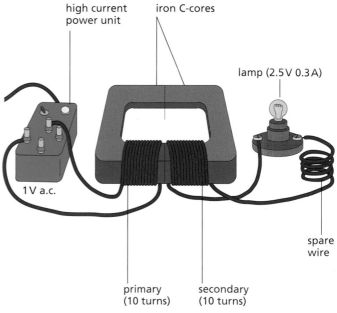

Figure 56.2

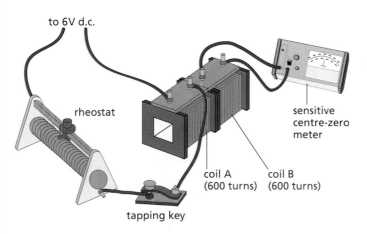

Figure 56.1 A changing current in a primary coil (A) induces a current in a secondary coil (B)

■ *Transformer equation*

A **transformer** transforms (changes) an *alternating* voltage from one value to another of greater or smaller value. It has a primary coil and a secondary coil wound on a complete soft iron core, either one on top of the other, Figure 56.3a, or on separate limbs of the core, Figure 56.3b.

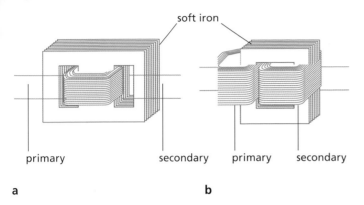

soft iron

primary   secondary   primary   secondary

a   b

Figure 56.3 Primary and secondary coils of a transformer

An alternating voltage applied to the primary induces an alternating voltage in the secondary. The value of the secondary voltage can be shown, for a transformer in which all the field lines cut the secondary, to be given by

$$\frac{\text{secondary voltage}}{\text{primary voltage}} = \frac{\text{secondary turns}}{\text{primary turns}}$$

In symbols

$$\frac{V_s}{V_p} = \frac{N_s}{N_p}$$

A 'step-up' transformer has more turns on the secondary than the primary and V_s is greater than V_p, Figure 56.4a. For example, if the secondary has twice as many turns as the primary, V_s is about twice V_p. In a 'step-down' transformer there are fewer turns on the secondary than the primary and V_s is less than V_p, Figure 56.4b.

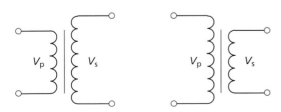

a Step-up: $V_s > V_p$ b Step-down: $V_p > V_s$

Figure 56.4 Symbols for a transformer

■ *Energy losses in a transformer*

If the p.d. is stepped up in a transformer the current is stepped down in proportion. This must be so if we assume that all the electrical energy given to the primary appears in the secondary, i.e. that energy is conserved and the transformer is 100% efficient or 'ideal' (many approach this). Then

$$\text{power in primary} = \text{power in secondary}$$
$$V_p \times I_p = V_s \times I_s$$

where I_p and I_s are the primary and secondary currents respectively.

$$\therefore \qquad \frac{I_s}{I_p} = \frac{V_p}{V_s}$$

So, for the ideal transformer, if the p.d. is doubled the current is halved. In practice, it is more than halved due to small energy losses in the transformer arising from three causes.

a) Resistance of windings

The windings of copper wire have some resistance and heat is produced by the current in them. Large transformers like those in Figure 56.5 have to be oil-cooled to prevent overheating.

Figure 56.5 Step-up transformers at a power station

b) Eddy currents

The iron core is in the changing magnetic field of the primary, and currents, called eddy currents, are induced in it which cause heating. These are reduced by using a **laminated** core made of sheets, insulated from each other to have a high resistance.

c) Leakage of field lines

All the field lines produced by the primary may not cut the secondary, especially if the core has an air gap or is badly designed.

■ *Worked example*

A transformer steps down the mains supply from 230 V to 10 V to operate an answering machine.

a What is the turns ratio of the transformer windings?

b How many turns are on the primary if the secondary has 100 turns?

c What is the current in the primary if the transformer is 100% efficient and the current in the answering machine is 2 A?

a Primary voltage = V_p = 230 V
Secondary voltage = V_s = 10 V
Turns ratio = $N_s/N_p = V_s/V_p$ = 10 V/230 V
= 1/23

b Secondary turns = N_s = 100

From **a**, $$\frac{N_s}{N_p} = \frac{1}{23}$$

$\therefore$ $N_p = 23\,N_s = 23 \times 100$
= 2300 turns

c Efficiency = 100%

$\therefore$ power in primary = power in secondary
$$V_p \times I_p = V_s \times I_s$$

$\therefore$ $$I_p = \frac{V_s \times I_s}{V_p} = \frac{10\,V \times 2\,A}{230\,V}$$

= 2/23 A = 0.09 A

Note In this ideal transformer the current is stepped up in the same ratio as the voltage is stepped down.

■ *Demonstration*

The stepping-up of current can be demonstrated with the 100:1 step-down transformer of Figure 56.6 in which the nail melts spectacularly due to the very large secondary current.

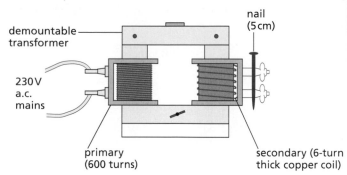

Figure 56.6 Demonstrating the stepping-up of current in a step-down transformer

■ *Transmission of electrical power*

a) Grid system

The Grid is a network of cables, mostly supported on pylons, which connect over 100 power stations throughout Britain to consumers. In the largest modern stations, electricity is generated at 25 000 V (25 kilovolts = 25 kV) and stepped up at once in a transformer to 275 or 400 kV to be sent over long distances on the Supergrid. Later, the p.d. is reduced by substation transformers for distribution to local users, Figure 56.7.

At the National Control Centre engineers direct the flow and re-route it when breakdown occurs. This makes the supply more reliable, and cuts costs by enabling smaller, less efficient stations to be shut down at off-peak periods.

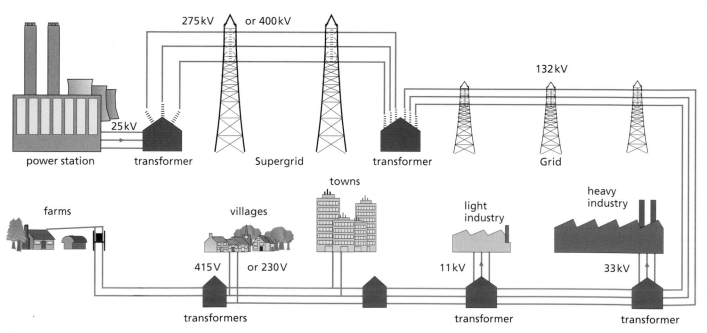

Figure 56.7 The National Grid transmission system

b) Use of high alternating p.ds

If 400 000 W of electrical power has to be sent through cables it can be done as 400 000 V at 1 A or 400 V at 1000 A (since watts = amperes × volts). But the amount of electrical energy transferred to unwanted heat (due to resistance of the cables) is proportional to the **square of the current** and so the power loss (I^2R) is less if transmission occurs at high voltage and low current. On the other hand, high p.ds need good insulation. The efficiency with which transformers step alternating p.ds up and down accounts for the use of a.c. rather than d.c. in power transmission.

The advantages of 'high' alternating voltage power transmission may be shown using the apparatus in Figure 56.8, **so long as the power line is well insulated** (using insulated eureka wire). The eureka resistance wires represent long transmission cables. Without the transformer at each end the lamp at the 'village' end glows dimly due to the power loss in the wires.

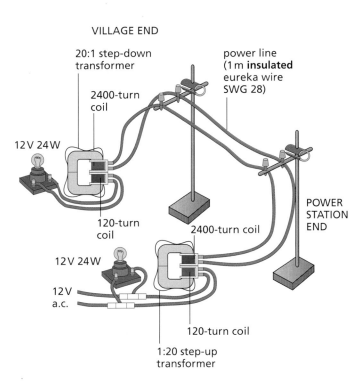

Figure 56.8 Model of a transmission system

When the transformers are connected as shown, the 12 V a.c. supply voltage is stepped up to about 240 V (since $N_p/N_s = 1/20$) at the 'power station' end, then stepped down at the 'village' end to 12 V (since $N_s/N_p = 20/1$). The current in the wires is then much less than without the transformers and so the lamp at the 'village' end is fully lit due to the smaller power loss during transmission.

■ *Applications of eddy currents*

Eddy currents are the currents induced in a piece of metal when it cuts magnetic field lines. They can be quite large due to the low resistance of the metal. They have their uses as well as their disadvantages.

a) Car speedometer

The action depends on the eddy currents induced in a thick aluminium disc when a permanent magnet, near it but *not touching it*, is rotated by a cable driven from the gearbox of the car, Figure 56.9. The eddy currents in the disc make it rotate in an attempt to reduce the relative motion between it and the magnet (see Chapter 55). The extent to which the disc can turn however is controlled by a spring. The faster the magnet rotates the more the disc turns before it is stopped by the spring. A pointer fixed to the disc moves over a scale marked in mph (or km/h) and gives the speed of the car.

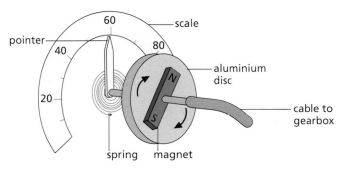

Figure 56.9 Car speedometer

b) Metal detector

The metal detector shown in Figure 56.10 consists of a large primary coil (A) through which an a.c. current is passed and a smaller secondary coil (B). When the detector is swept over a buried metal object (such as a nail, coin or pipe) the fluctuating magnetic field lines associated with the alternating current in coil A 'cut' the hidden metal and induce eddy currents in it. The changing magnetic field lines associated with these eddy currents cut the secondary coil B in turn and induce a current which can be used to operate an alarm. The coils are set at right angles to each other so that their magnetic fields do not interact.

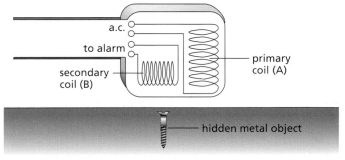

Figure 56.10 Metal detector

Questions

1 Two coils of wire, A and B, are placed near one another, Figure 56.11. Coil A is connected to a switch and battery. Coil B is connected to a centre-reading moving-coil galvanometer.

a If the switch connected to coil A were closed for a few seconds and then opened, the galvanometer connected to coil B would be affected. Explain and describe, step by step, what would actually happen.

b What changes would you expect if a bundle of soft iron wires were placed through the centre of the coils? Give a reason for your answer.

c What would happen if more turns of wire were wound on the coil B?

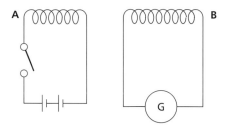

Figure 56.11

2 The main function of a step-down transformer is to

A decrease current
B decrease voltage
C change a.c. to d.c.
D change d.c. to a.c.
E decrease the resistance of a circuit.

3 a Calculate the number of turns on the secondary of a step-down transformer, which would enable a 12 V bulb to be used with a 230 V a.c. mains power, if there are 460 turns on the primary.

b What current will flow in the secondary when the primary current is 0.10 A? Assume there are no energy losses.

4 a Figure 56.12 represents a simple transformer used to light a 12 V lamp. When the power supply is switched on the lamp is very dim.

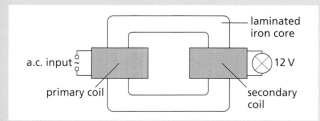

Figure 56.12

(i) Give *one* way to increase the voltage at the lamp without changing the power supply.
(ii) What is meant by the iron core being *laminated*?

b Electrical energy is distributed around the country by a network of high voltage cables, Figure 56.13.

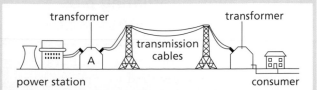

Figure 56.13

(i) For the system to work the power is generated and distributed using alternating current rather than direct current. Why?
(ii) Transformers are an essential part of the distribution system. Explain why.
(iii) The transmission cables are suspended high above the ground. Why?

c The power station generates 100 MW of power at a voltage of 25 kV. Transformer A, which links the power station to the transmission cables, has 44 000 turns in its 275 kV secondary coil.
(i) Write down the equation which links the number of turns in each transformer coil to the voltage across each transformer coil.
(ii) Calculate the number of turns in the primary coil of transformer A. Show clearly how you work out your answer.

d Figure 56.14 shows how the cost of transmitting the electricity along the cables depends upon the thickness of the cable.

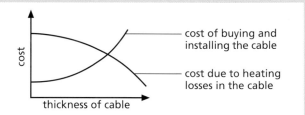

Figure 56.14

(i) Why does the cost due to the heating losses go down as the cable is made thicker?
(ii) By what process is most heat energy lost from the cables?

(AQA (SEG) Higher, Summer 99)

■ *Checklist*

After studying this chapter you should be able to

■ explain the principle of the transformer,
■ recall the transformer equation $V_s/V_p = N_s/N_p$ and use it to solve problems,
■ recall that for an ideal transformer $V_p \times I_p = V_s \times I_s$ and use the relation to solve problems,
■ recall the causes of energy losses in practical transformers,
■ explain why high voltage a.c. is used for transmitting electrical power,
■ explain how eddy currents arise and how they are used in a car speedometer and in a metal detector.

Electromagnetic effects
Additional questions

Electromagnets

1 a Two bar magnets are held close together.

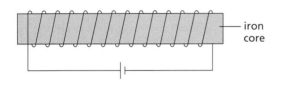

When released the magnets *push away from* each other.
(i) Copy the diagram above and use the letter N to label the north pole of each magnet.
(ii) Write down *one* word which describes what happens to the magnets as they are released.
b The diagram shows an electromagnet. Copy the diagram and draw the shape of the magnetic field of the electromagnet.

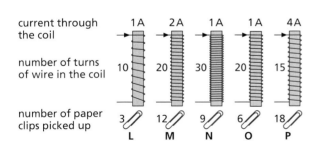

c The diagram shows five electromagnets, **L, M, N, O** and **P**. Each electromagnet is able to pick up a different number of paper clips.

(i) Which *three* electromagnets should you compare if you want to find out how the strength of an electromagnet depends upon the number of turns of wire in the coil?
(ii) What is the connection between the strength of an electromagnet and the number of turns of wire in the coil?

(AQA (SEG) Foundation, Summer 99)

2 Part of the electrical system of a car is shown in the following diagram.
  a Why are connections made to the car body?
  b There are *two* circuits in parallel with the battery. What are they?
  c Why is wire A thicker than wire B?
  d Why is a relay used?

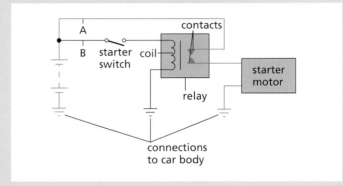

Electric motors, generators and transformers

3 Describe the deflections observed on the sensitive, centre-zero galvanometer G when the copper rod XY is connected to its terminals and is made to vibrate up and down (as shown by the arrows), between the poles of a U-shaped magnet so that it is at right angles to the magnetic field.

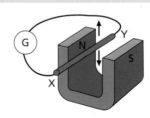

Explain what is happening.

4 A transformer has 1000 turns on the primary coil. The voltage applied to the primary coil is 230 V a.c. How many turns are on the secondary coil if the output voltage is 46 V a.c.?

A 20 **B** 200 **C** 2000 **D** 4000 **E** 8000

5 a Draw a labelled diagram of the essential components of a simple motor. Explain how continuous rotation is produced and show how the direction of rotation is related to the direction of the current.
  b State what would happen to the direction of rotation of the motor you have described if
  (i) the current were reversed,
  (ii) the magnetic field were reversed,
  (iii) both current and field were reversed simultaneously.

6 This question is about electromagnetism.

a Michael is investigating how a short length of copper wire can be made to move in a magnetic field. He uses this apparatus.

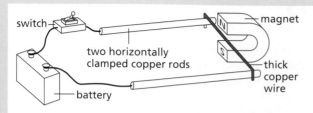

He places the magnet so that the wire is midway between the poles. He writes down these observations.

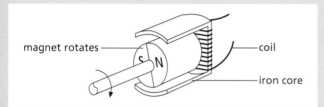

> We can only make the copper wire move along the rods if:
> 1) the switch is closed and
> 2) the poles of the magnet are above and below the wire, not on each side of it.

Explain these *two* observations. Use your ideas about electromagnetism.

b The diagram shows a model generator.

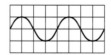

magnet rotates — coil
— iron core

(i) What happens in the coil of wire when the magnet rotates?
(ii) Why does this happen?

c The ends of the coil are connected to a cathode ray oscilloscope (CRO). The diagram shows the trace on the screen as the magnet rotates.

Copy the diagram and draw new traces for each of the following changes. (Assume the settings of the oscilloscope remain the same).
(i) The magnet rotates at the same speed but in the opposite direction.
(ii) The magnet rotates at the same speed, in the same direction as the original, but the number of turns of the coil is doubled.
(iii) The magnet rotates at twice the speed, in the same direction, with the original number of turns of the coil.

d Explain why **iron** is used as the core in the model generator.

e The output from a power station generator is connected to a step-up transformer. The transformer is connected to transmission lines.

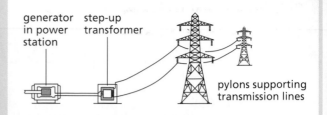

generator step-up
in power transformer
station

pylons supporting
transmission lines

Explain why a step-up transformer is needed. Use your ideas about power losses in transmission.

(OCR Higher, Summer 99)

7 The power output of a generator which supplies power to a consumer along cables of total resistance $16\,\Omega$ is 30 kW at 6 kV. What is

a the current in the cables,
b the power loss in the cables,
c the drop of voltage between the ends of the cables?

8 a When the magnet shown in the diagram below is moving towards the coil, the meter gives a reading to the right.

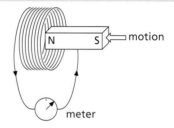

N S ← motion

meter

(i) What is the name of the effect being produced by the moving magnet?
(ii) State what happens to the reading shown on the meter when the magnet is moving *away* from the coil.
(iii) The original experiment is repeated. This time, the magnet is moved towards the coil at a greater speed. State *two* changes you would notice in the reading on the meter.

b The coil is replaced by one which is much longer than the original. The magnet can fit completely inside this coil.
(i) On axes like those below, sketch a graph to show how the reading on the meter will change as the magnet is pushed into, through and then out of the coil at the other end.

meter reading → time

(ii) Describe and explain the main features of the trace you have drawn.

(London Higher, June 98)

Electrons and atoms

57 *Electrons*

The discovery of the electron was a landmark in physics and led to great technological advances.

■ Thermionic emission

The evacuated bulb in Figure 57.1 contains a small coil of wire, the **filament**, and a metal plate called the **anode** because it is connected to the positive of the 400 V d.c. power supply. The negative of the supply is joined to the filament which is also called the **cathode**. The filament is heated by current from a 6 V supply (a.c. or d.c.).

With the circuit as shown, the meter deflects, indicating current flow in the circuit containing the gap between anode and cathode. The current stops if *either* the 400 V supply is reversed to make the anode negative, *or* the filament is not heated.

This demonstration supports the view that negative charges, in the form of electrons, escape from the filament when it is hot because they have enough energy to get free from the metal surface. The process is known as **thermionic emission** and the bulb as a thermionic diode (since it has two electrodes). There is a certain minimum **threshold energy** (depending on the metal) which the electrons must have to escape. Also, the higher the temperature of the metal, the greater the number of electrons emitted. The electrons are attracted to the anode if it is positive and are able to reach it because there is a vacuum in the bulb.

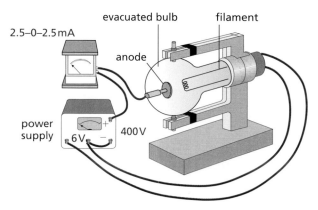

Figure 57.1 Demonstrating thermionic emission

■ Cathode rays

Beams of electrons moving at high speed are called **cathode rays**. Their properties can be studied using the 'Maltese cross tube', Figure 57.2.

Electrons emitted by the hot cathode are accelerated towards the anode but most pass through the hole in it and travel on along the tube. Those that miss the cross cause the screen to fluoresce with green or blue light and cast a shadow of the cross on it. The cathode rays evidently travel in straight lines.

If the N pole of a magnet is brought up to the neck of the tube, the rays (and the fluorescent shadow) move upwards. The rays are clearly deflected by a magnetic field and, using Fleming's left-hand rule (Chapter 53), we see that they behave like conventional current (positive charge flow) travelling from anode to cathode.

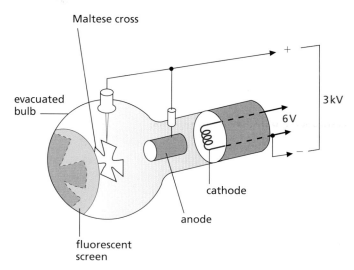

Figure 57.2 Maltese cross tube

There is also an optical shadow of the cross, due to light emitted by the cathode. This is unaffected by the magnet.

■ *Deflection of an electron beam*

a) By a magnetic field

In Figure 57.3 the evenly spaced crosses represent a uniform magnetic field (i.e. one of the same strength throughout the area shown) acting into and perpendicular to the paper. An electron beam entering the field at right angles to the field experiences a force due to the motor effect (Chapter 53), whose direction is given by Fleming's left-hand rule. This indicates that the force acts inwards at right angles to the direction of the beam and makes it follow a **circular** path as shown (the beam being treated as conventional current in the opposite direction).

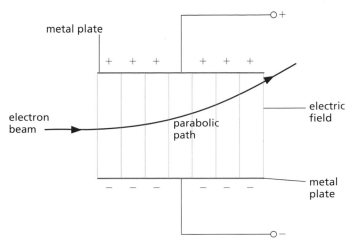

Figure 57.4 Path of an electron beam at right angles to an electric field

If an electron beam enters the field at right angles to it, the beam is attracted towards the positively charged plate and follows a **parabolic** path, as shown. In fact its behaviour is not unlike that of a projectile (Chapter 30) in which the horizontal and vertical motions can be treated separately.

c) Demonstration

The deflection tube in Figure 57.5 can be used to show the deflection of an electron beam in electric and magnetic fields. Electrons from a hot cathode strike a fluorescent screen S set at an angle. A p.d. applied across two horizontal metal plates Y_1Y_2 creates a *vertical* electric field which deflects the rays upwards if Y_1 is positive (as shown) and downwards if it is negative.

When current flows in the two coils X_1X_2 (in series) outside the tube, a *horizontal* magnetic field is produced across the tube. It can be used instead of a magnet to deflect the rays, or to cancel the deflection due to an electric field.

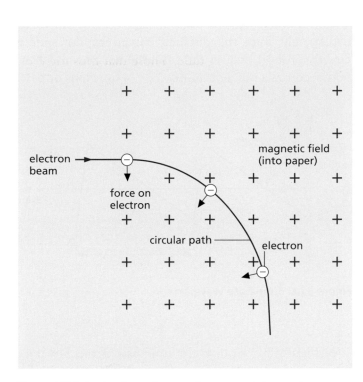

Figure 57.3 Path of an electron beam at right angles to a magnetic field

b) By an electric field

An electric field is a region where an electric charge experiences a force due to other charges. In Figure 57.4 the two metal plates behave like a capacitor which has been charged by connection to a voltage supply. If the charge is evenly spread over the plates, a uniform electric field is created between them and is represented by parallel, equally spaced lines.

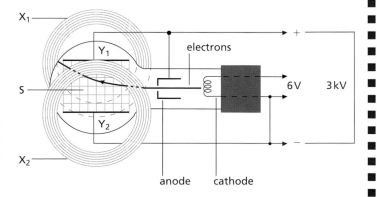

Figure 57.5 Deflection tube

Cathode ray oscilloscope (CRO)

The CRO is one of the most important scientific instruments ever to be developed. It contains, like a television and a computer monitor, a cathode ray tube which has three main parts, Figure 57.6.

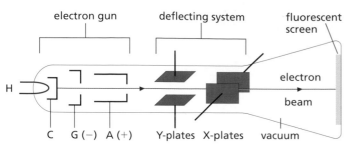

Figure 57.6 Main parts of a CRO

a) Electron gun

This consists of a **heater** H, a **cathode** C, another electrode called the **grid** G and two or three **anodes** A. G is at a negative voltage with respect to C and controls the number of electrons passing through its central hole from C to A; it is the **brilliance** or **brightness** control. The anodes are at high positive voltages relative to C; they accelerate the electrons along the highly evacuated tube and also **focus** them into a narrow beam.

b) Fluorescent screen

A bright spot of light is produced on the screen where the beam hits it.

c) Deflecting system

Beyond A are two pairs of deflecting plates to which p.ds can be applied. The **Y-plates** are horizontal but create a vertical electric field which deflects the beam vertically. The **X-plates** are vertical and deflect the beam horizontally.

The p.d. to create the electric field between the Y-plates is applied to the **Y-input** terminals (often marked 'high' and 'low') on the front of the CRO. The input is usually amplified by an amount which depends on the setting of the **Y-amp gain** control, before it is applied to the Y-plates. It can then be made large enough to give a suitable vertical deflection of the beam.

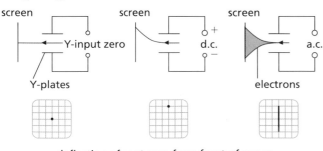

deflection of spot seen from front of screen

Figure 57.7 Deflection of the electron beam

In Figure 57.7a the p.d. between the Y-plates is zero as is the deflection. In b the d.c. input p.d. makes the upper plate positive and attracts the beam of negatively charged electrons upwards. In c the 50 Hz a.c. input makes the beam move up and down so rapidly that it produces a continuous vertical line (whose length increases if the Y-amp gain is turned up).

The p.d. applied to the X-plates is also via an amplifier, the X-amplifier, and can either be from an external source connected to the **X-input** terminal or, more commonly, from the **time base** circuit in the CRO.

The time base deflects the beam horizontally in the X-direction and makes the spot sweep across the screen from left to right at a steady speed determined by the setting of the time base controls (usually 'coarse' and 'fine'). It must then make the spot 'fly' back very rapidly to its starting point, ready for the next sweep. The p.d. from the time base should therefore have a sawtooth waveform like that in Figure 57.8. Since AB is a straight line, the distance moved by the spot is directly proportional to time and the horizontal deflection becomes a measure of time, i.e. a time axis or base.

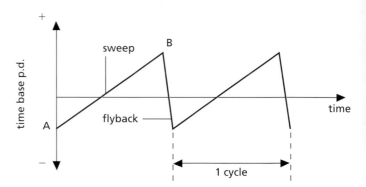

Figure 57.8 Time base waveform

In Figures 57.9a, b, c, the time base is on. For trace a the Y-input p.d. is zero, for b the Y-input is d.c. which makes the upper Y-plate positive. In both cases the spot traces out a horizontal line which appears to be continuous if the flyback is fast enough. For trace c the Y-input is a.c., i.e. the Y-plates are alternately positive and negative and the spot moves accordingly.

a Y-input zero   **b** Y-input d.c.   **c** Y-input a.c.

Figure 57.9 Deflection of the spot with time base on

■ Uses of the CRO

A small CRO is shown in Figure 57.10.

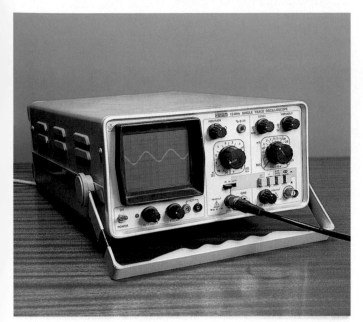

Figure 57.10 Single-beam CRO

a) Practical points

The **brilliance** or **intensity** control, which is sometimes the **on/off** switch as well, should be as low as possible when there is just a spot on the screen. Otherwise screen 'burn' occurs which damages the fluorescent material. If possible it is best to defocus the spot when not in use, or draw it into a line by running the time base.

When preparing the CRO for use, set the **brilliance**, **focus**, **X-** and **Y-shift** controls (which allow the spot to be moved 'manually' over the screen in the X and Y directions respectively) to their mid-positions. The **time base** and **Y-amp gain** controls can then be adjusted to suit the input.

When the **a.c./d.c. selector** switch is in the 'd.c.' (or 'direct') position, both d.c. and a.c. can pass to the Y-input. In the 'a.c.' (or 'via C') position, a capacitor blocks d.c. in the input but allows a.c. to pass.

b) Measuring p.ds

A CRO can be used as a d.c./a.c. voltmeter if the p.d. to be measured is connected across the Y-input terminals; **the deflection of the spot is proportional to the p.d.**

For example, if the **Y-amp gain** control is on, say, 1 V/div, a deflection of 1 vertical division on the screen graticule (like graph paper with squares for measuring deflections) would be given by a 1 V d.c. input. A line 1 division long (time base off) would be produced by an a.c. input of 1 V peak-to-peak, i.e. peak p.d. = 0.5 V.

c) Displaying waveforms

In this widely used role, the time base is on and the CRO acts as a 'graph-plotter' to show the waveform, i.e. the variation with time, of the p.d. applied to its Y-input. The displays in Figures 57.11a, b are of alternating p.ds with sine waveforms. For trace a, the time base frequency *equals* that of the input and one complete wave is obtained. For b it is *half* that of the input and two waves are formed. If the traces are obtained with the Y-amp gain control on, say, 0.5 V/div, the peak-to-peak voltage of the a.c. = 3 divs × 0.5 V/div = 1.5 V and the peak p.d. = 0.75 V.

Sound waveforms can be displayed if a microphone is connected to the Y-input terminals (see Chapter 16).

a

b

Figure 57.11 Alternating p.d. waveforms on the CRO

d) Measuring time intervals and frequency

These can be measured if the CRO has a calibrated time base. For example, when the time base is set on 10 ms/div, the spot takes 10 milliseconds to move 1 division horizontally across the screen graticule. If this is the time base setting for the waveform in Figure 57.11b then since 1 complete wave occupies 2 horizontal divisions, we can say

$$\text{time for 1 complete wave} = 2 \, \text{divs} \times 10 \, \text{ms/div}$$
$$= 20 \, \text{ms}$$
$$= 20/1000 = 1/50 \, \text{s}$$

∴   number of complete waves per second = 50

∴   frequency of a.c. applied to Y-input = 50 Hz

X-rays

X-rays are produced when high-speed electrons are stopped by matter.

a) Production

In an X-ray tube, Figure 57.12, electrons from a hot filament are accelerated across a vacuum to the anode by a large p.d. (up to 100 kV). The anode is a copper block with a 'target' of a high melting-point metal such as tungsten on which the electrons are focused by the concave cathode. The tube has a lead shield with a small exit for the X-rays.

The work done in transferring a charge Q through a p.d. V is

$$W = Q \times V \qquad \text{(see p. 203)}$$

This will equal the k.e. of the electrons reaching the anode if $Q =$ charge on an electron $= 1.6 \times 10^{-19}$ C and V is the accelerating p.d. Less than 1% of the k.e. of the electrons becomes X-ray energy; the rest heats the anode which has to be cooled.

High p.ds give short wavelength, very penetrating (**hard**) X-rays. Less penetrating (**soft**) rays, of longer wavelength, are obtained with lower p.ds. The absorption of X-rays by matter is greatest by materials of high density having a large number of outer electrons in their atoms, i.e. of high atomic number (Chapter 59). A more intense beam of rays is produced if the rate of emission of electrons is raised by increasing the filament current.

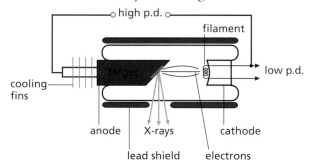

Figure 57.12 X-ray tube

b) Properties and nature

X-rays

(i) readily penetrate matter – up to 1 mm of lead,
(ii) are not deflected by electric or magnetic fields,
(iii) ionize a gas, making it a conductor, e.g. a charged electroscope discharges when X-rays pass through the surrounding air,
(iv) affect a photographic film,
(v) cause fluorescence,
(vi) give interference and diffraction effects.

These facts (and others) suggest that X-rays are electromagnetic waves of very short wavelength.

c) Uses

These were considered earlier (Chapter 14).

Photoelectric effect

Electrons are emitted by certain metals when electromagnetic radiation of small enough wavelength falls on them. The effect is called **photoelectric emission** and is given by zinc exposed to ultraviolet.

The photoelectric effect only occurs for a given metal if the frequency of the incident electromagnetic radiation exceeds a certain **threshold frequency**. We can explain this by assuming that

(i) all electromagnetic radiation is emitted and absorbed as packets of energy, called **photons**, and
(ii) the energy of a photon is directly proportional to its frequency.

Ultraviolet photons would therefore have more energy than light photons since UV has a higher frequency than light. The behaviour of zinc (and most other substances) in not giving photoelectric emission with light but with UV would therefore be explained: a photon of light has less than the minimum energy required to emit an electron.

The absorption of a photon by an atom results in the electron gaining energy and the photon disappearing. If the photon has more than the minimum amount of energy required to enable an electron to escape, the excess appears as k.e. of the emitted electron.

$$\frac{\text{energy}}{\text{of photon}} = \frac{\text{energy needed for}}{\text{electron to escape}} + \text{k.e. of electron}$$

The photoelectric effect is the process by which X-ray photons are absorbed by matter; in effect it causes **ionization** (Chapter 58) since electrons are ejected and positive ions remain. Photons not absorbed pass through with unchanged energy.

Waves or particles?

The wave theory of electromagnetic radiation can account for properties such as interference, diffraction and polarization which the photon theory cannot. On the other hand it does not explain the photoelectric effect which the photon theory does.

It would seem that electromagnetic radiation has a dual nature and has to be regarded as waves on some occasions and as 'particles' (photons) on others.

Questions

1 a In Figure 57.13a, to which terminals on the power supply must plates A and B be connected to deflect the cathode rays downwards?

b In Figure 57.13b, in which direction will the cathode rays be deflected?

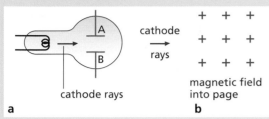

a   **b**

Figure 57.13

2 In an oscilloscope, a beam of electrons hits a screen. The screen has a fluorescent coating.

a The purpose of the fluorescent coating is to emit

   A infra-red radiation when electrons hit it
   B light when electrons hit it
   C microwaves when electrons hit it
   D ultraviolet radiation when electrons hit it

b The beam of electrons is produced by an electron gun. Figure 57.14 shows the principle of an electron gun.

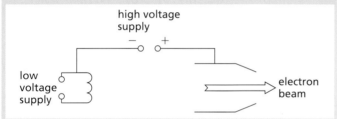

Figure 57.14

   (i) Copy the diagram and label the cathode (the part that emits electrons).
   (ii) Label the anode (the part that accelerates the electrons).
   (iii) How can the speed of the electrons in the beam be changed?
   (iv) The positive and negative connections to the high voltage supply are swapped over. State and explain what happens to the electron beam.

c An alternating voltage is applied to the input of an oscilloscope. Figure 57.15 shows the appearance of the screen. The sensitivity is set to 3 volts per cm.

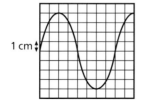

Figure 57.15

   (i) What is the maximum voltage?
   (ii) What is the minimum voltage?
   (iii) The frequency of the alternating voltage is doubled, but the value of the voltage is unchanged. On a similar grid draw the new oscilloscope display.

(London Foundation, June 98)

3 In an X-ray tube what is the effect on the X-ray beam of
  a increasing the accelerating voltage,
  b decreasing the filament current?

4 a What is the photoelectric effect?
  b Why do ultraviolet but not violet light photons give the photoelectric effect with zinc?

5 The apparatus shown in Figure 57.16 can be used to investigate the production of electrons from a hot filament.

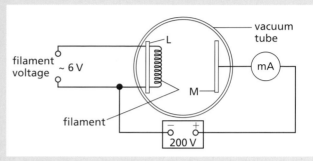

Figure 57.16

When the filament is glowing, a current is measured by the milliammeter.

a (i) Explain fully why there is a current between L and M.
   (ii) When the filament voltage is reduced, the current decreases. Explain why.
   (iii) State and explain what would happen to the current if air entered the vacuum tube.

b The charge on an electron is 1.6×10^{-19} C.
   (i) Calculate the kinetic energy of an electron arriving at plate M when the voltage between L and M is 200 V. Give your answer in joules.
   (ii) The number of electrons arriving at M is 2.0×10^{16} per second. Calculate the current in the vacuum tube.

(London Higher, June 99)

■ *Checklist*

After studying this chapter you should be able to

■ explain the terms **thermionic emission** and **cathode rays**,

■ describe experiments to show that cathode rays are deflected by magnetic and electric fields,

■ describe with the aid of diagrams the jobs done in a CRO by the electron gun, the X- and Y-plates, the fluorescent screen and the time base,

■ describe how the CRO is used to measure p.ds, to display waveforms and to measure time intervals and frequency,

■ describe with the aid of a diagram how X-rays are produced in an X-ray tube,

■ list six properties of X-rays which indicate that they are electromagnetic waves of short wavelength,

■ explain the term **photoelectric emission**,

■ explain the term **threshold frequency** and why this leads to the idea of **photons** of electromagnetic energy.

58 Radioactivity

Ionizing effect of radiation
Geiger–Müller (GM) tube
Alpha, beta, gamma rays
Particle tracks

Radioactive decay: half-life
Uses of radioactivity
Dangers and safety

The discovery of radioactivity in 1896 by the French scientist Becquerel was accidental. He found that uranium compounds emitted radiation which (i) affected a photographic plate even when wrapped in black paper and (ii) ionized a gas. Soon afterwards Marie Curie discovered the radioactive element radium. We now know that radioactivity arises from unstable nuclei (Chapter 59) which may occur naturally or be produced in reactors. Radioactive materials are widely used in industry, medicine and research.

We are all exposed to natural **background radiation** caused partly by radioactive materials in rocks, the air and our bodies, and partly by cosmic rays from outer space (see p. 269).

Ionizing effect of radiation

A charged electroscope discharges when a lighted match or a radium source (**held in forceps**) is brought near the cap, Figures 58.1a, b.

In the first case the flame knocks electrons out of surrounding air molecules leaving them as positively charged air **ions**, i.e. air molecules which have lost one or more electrons, Figure 58.2; in the second case radiation causes the same effect, called **ionization**. The positive air ions are attracted to the cap if it is negatively charged; if it is positively charged the electrons are attracted. As a result in either case the charge on the electroscope is neutralized, i.e. it loses its charge.

Figure 58.2 Ionization

Geiger–Müller (GM) tube

The ionizing effect is used to detect radiation.

When radiation enters a GM tube, Figure 58.3, either through a thin end-window made of mica, or, if it is very penetrating, through the wall, it creates argon ions and electrons. These are accelerated towards the electrodes and cause more ionization by colliding with other argon atoms.

On reaching the electrodes, the ions produce a current pulse which is amplified and fed either to a **scaler** or a **ratemeter**. A scaler counts the pulses and shows the total received in a certain time. A ratemeter gives the counts per second (or minute), or **count-rate**, directly. It usually has a loudspeaker which gives a 'click' for each pulse.

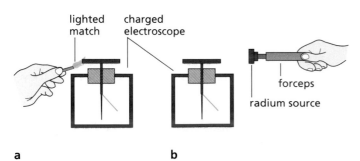

a   **b**

264 **Figure 58.1**

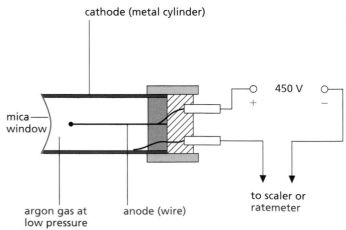

Figure 58.3 GM tube

■ *Alpha, beta, gamma rays*

Experiments to study the penetrating power, ionizing ability and behaviour of radiation in magnetic and electric fields, show that a radioactive substance emits one or more of three types of radiation – called alpha (α), beta (β⁻ or β⁺) and gamma (γ) rays.

Penetrating power can be investigated as in Figure 58.4 by observing the effect on the count-rate of placing in turn between the GM tube and the lead sheet,

(i) a sheet of thick paper (the radium source, lead and tube must be close together for this part),
(ii) a sheet of aluminium 2 mm thick,
(iii) a sheet of lead 2 cm thick.

Radium (Ra-226) emits α, β⁻ and γ-rays. Other sources can be tried, e.g. americium, strontium and cobalt.

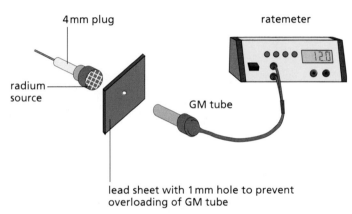

Figure 58.4 Investigating the penetrating power of radiation

a) Alpha rays

These are stopped by a thick sheet of paper and have a range in air of only a few centimetres since they cause intense ionization in a gas due to frequent collisions with gas molecules. They are deflected by electric and *strong* magnetic fields in a direction and by an amount which suggests they are helium atoms minus two electrons, i.e. **helium ions with a double positive charge**. From a particular substance, they are all emitted with the same speed (about 1/20th of that of light).

Americium (Am-241) is a pure α source.

b) Beta rays

These are stopped by a few millimetres of aluminium and some have a range in air of several metres. Their ionizing power is much less than that of α-particles. As well as being deflected by electric fields, they are more easily deflected by magnetic fields. Measurements show that β⁻-rays are streams of **high-energy electrons**, like cathode rays, emitted with a range of speeds up to that of light.

Strontium (Sr-90) emits β⁻-rays only.

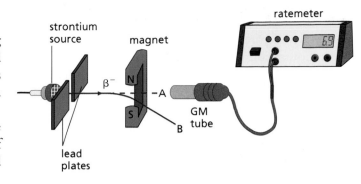

Figure 58.5 Demonstrating magnetic deflection of β⁻-particles

The magnetic deflection of β⁻-particles can be shown as in Figure 58.5. With the GM tube at A and without the magnet, the count-rate is noted. Inserting the magnet reduces the count-rate but it increases again when the GM tube is moved to B.

c) Gamma rays

These are the most penetrating and are stopped only by many centimetres of lead. They ionize a gas even less than β-particles and are not deflected by electric and magnetic fields. They give interference and diffraction effects and are **electromagnetic radiation** travelling at the speed of light. Their wavelengths are those of very short X-rays, from which they differ only because they arise in atomic nuclei whereas X-rays come from energy changes in the electrons outside the nucleus.

Cobalt (Co-60) emits γ and β⁻ but can be covered with aluminium to provide pure γ.

A GM tube detects β-particles and γ-photons and energetic α-particles; a charged electroscope detects α only. All three types of rays cause fluorescence.

The behaviour of the three kinds of radiation in a magnetic field is summarized in Figure 58.6. The deflections (not to scale) are found from Fleming's left-hand rule, taking negative charge moving to the right as equivalent to positive (conventional) current to the left.

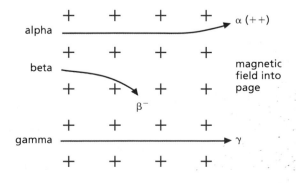

Figure 58.6 Deflection of α, β and γ-rays in a magnetic field

■ *Particle tracks*

The paths of particles of radiation were first shown up by the ionization they produced in devices called cloud chambers. When air containing vapour, e.g. alcohol, is cooled enough, saturation occurs. If ionizing radiation passes through the air, further cooling causes the saturated vapour to condense on the air ions created. The resulting white line of tiny liquid drops shows up as a track when illuminated.

In a **diffusion cloud chamber**, α-particles gave straight, thick tracks, Figure 58.7a. Very fast β-particles produced thin, straight tracks while slower ones gave short, twisted, thicker tracks, Figure 58.7b. γ-rays eject electrons from air molecules; the ejected electrons behaved like β⁻-particles in the cloud chamber and produced their own tracks spreading out from the γ-rays.

The **bubble chamber**, in which the radiation leaves a trail of bubbles in liquid hydrogen, has now replaced the cloud chamber in research work. The higher density of atoms in the liquid gives better defined tracks, Figure 58.8, than obtained in a cloud chamber. A magnetic field is usually applied across the bubble chamber which causes charged particles to move in circular paths; the sign of the charge can be deduced from the way the path curves.

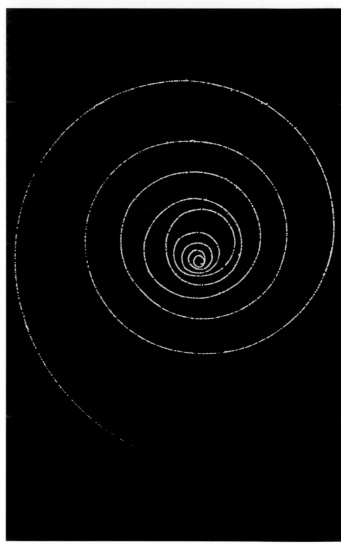

Figure 58.8 Charged particle tracks in a bubble chamber

a α-particles

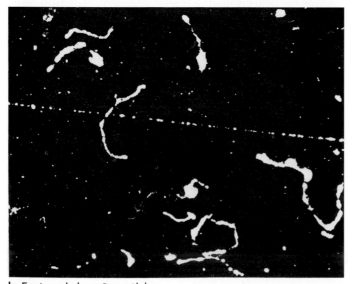

b Fast and slow β-particles

Figure 58.7 Tracks in a cloud chamber

■ *Radioactive decay: half-life*

Radioactive atoms have unstable nuclei and 'decay' into atoms of different elements with more stable nuclei when they emit α or β-particles. These changes are spontaneous and cannot be controlled; also, it does not matter whether the material is pure or combined chemically with something else.

a) Half-life

The **rate of decay** is unaffected by temperature but every radioactive element has its own definite decay rate, expressed by its **half-life**. This is the **average time for half the atoms in a given sample to decay**. It is difficult to know when a substance has lost all its radioactivity, but the time for its activity to fall to half its value can be found more easily.

b) Decay curve

The average number of disintegrations (i.e. decaying atoms) per second of a sample is its **activity**. If it is measured at different times (e.g. by finding the count-rate using a GM tube and ratemeter), a decay curve of activity against time can be plotted. The ideal one in Figure 58.9 shows that the activity decreases by the *same* fraction in successive equal time intervals. It falls from 80 to 40 disintegrations per second in 10 minutes, from 40 to 20 in the next 10 minutes, from 20 to 10 in the third 10 minutes and so on. The half-life is 10 minutes.

Half-lives vary from millionths of a second to millions of years. For radium it is 1600 years.

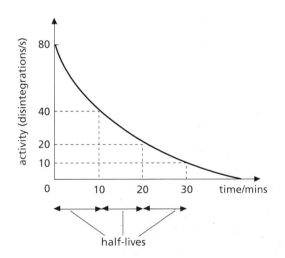

Figure 58.9 Decay curve

c) Experiment

The half-life of the α-emitting gas **thoron** can be found as in Figure 58.10. The thoron bottle is squeezed three or four times to transfer some thoron to the flask, Figure 58.10a. The clips are then closed, the bottle removed and the stopper replaced by a GM tube so that it seals the top, b.

a

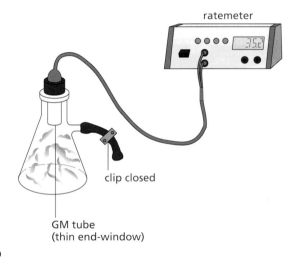

b

Figure 58.10

When the ratemeter reading has reached its maximum and started to fall, the count-rate is noted every 15 s for 2 minutes and then every 60 s for the next few minutes. (The GM tube is left in the flask for at least 1 hour until the radioactivity has decayed.)

A measure of the background radiation is obtained by recording the counts for a period (say 10 minutes) at a position well away from the thoron equipment. The count-rates in the thoron decay experiment are then corrected by subtracting the average background count-rate from each reading (as in question 2, p. 270). A graph of the corrected count-rate against time is plotted and the half-life (52 s) estimated from it.

d) Random nature

During the previous experiment it becomes evident that the count-rate varies irregularly: the loudspeaker of the ratemeter 'clicks' erratically, not at a steady rate. This is because radioactive decay is a **random** process, in that it is a matter of pure chance whether or not a particular nucleus will decay during a certain period of time. All we can say is that about half the nuclei in a sample will decay during the half-life. We cannot say which nuclei these will be, nor can we influence the process in any way.

■ Fundamental particles

Over the last 50 years particle accelerators such as the super proton accelerator at CERN shown in Figure 59.9 have led to the discovery of a host of subatomic particles. The neutrino, antineutrino and the positron were introduced in our discussion of beta decay above; together with the electron these are thought to be fundamental particles of matter which cannot be further divided; more generally they are called **leptons**.

Figure 59.9 View inside the cylindrical magnets which guide the particle beams in the CERN particle accelerator

Protons and neutrons are no longer thought to be fundamental (indivisible) particles; they have been found to be made up of smaller particles called **quarks** which are held together by 'gluons'. **Leptons and quarks are fundamental particles from which all matter is composed.**

Quarks have non-integral charge, for example the 'up' quark (u) has charge $+2/3$ of the electron charge, while the 'down' quark (d) has charge $-1/3$ of the electron charge. A proton is composed of two 'up' quarks and one down quark (uud) giving an overall charge of $+1e$, while a neutron is made up of one 'up' quark and two 'down' quarks (udd), Figure 59.10.

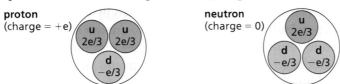

Figure 59.10 Quark structure of proton and neutron

We can now describe beta decay in terms of quarks:

in β^- decay a 'down' quark changes to an 'up' quark,

$$udd \rightarrow uud + {}_{-1}^{0}e$$

while in β^+ decay an 'up' quark changes to a 'down' quark,

$$uud \rightarrow udd + {}_{+1}^{0}e$$

Check that the charges on each side of the reactions balance.

■ Nuclear energy

a) $E = mc^2$

Einstein predicted that if the energy of a body changes by an amount E, its mass changes by an amount m given by the equation

$$E = mc^2$$

where c is the speed of light ($3 \times 10^8 \, \text{m/s}$). The implication is that any reaction in which there is a decrease of mass, called a **mass defect**, is a source of energy. The energy and mass changes in physical and chemical changes are very small; those in some nuclear reactions, e.g. radioactive decay, are millions of times greater. It appears that mass (matter) is a very concentrated form of energy.

b) Fission

The heavy metal uranium is a mixture of isotopes of which ${}^{235}_{92}\text{U}$, called uranium-235, is the most important. Some atoms of this isotope decay quite naturally, emitting high-speed neutrons. If one of these hits the nucleus of a neighbouring uranium-235 atom (being uncharged the neutron is not repelled by the nucleus), this may break (**fission**) into two nearly equal radioactive nuclei, often of barium and krypton, with the production of two or three more neutrons:

$$\underset{\text{neutron}}{{}^{235}_{92}\text{U} + {}^{1}_{0}\text{n}} \rightarrow \underset{\text{fission fragments}}{{}^{144}_{56}\text{Ba} + {}^{90}_{36}\text{Kr}} + \underset{\text{neutrons}}{2{}^{1}_{0}\text{n}}$$

The mass defect is large and appears mostly as k.e. of the fission fragments. These fly apart at great speed, colliding with surrounding atoms and raising their average k.e., i.e. their temperature, so producing heat.

If the fission neutrons split other uranium-235 nuclei, a **chain reaction** is set up, Figure 59.11. In practice some fission neutrons are lost by escaping from the surface of the uranium before this happens. The ratio of those causing fission to those escaping increases as the mass of uranium-235 increases. This must exceed a certain **critical** value to sustain the chain reaction.

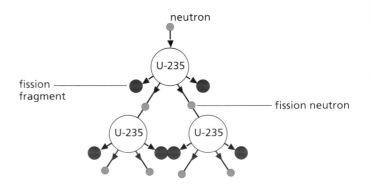

Figure 59.11 Chain reaction

c) Nuclear reactor

In a nuclear power station heat from a nuclear reactor produces the steam for the turbines. Figure 59.12 is a simplified diagram of one type of reactor.

The chain reaction occurs at a steady rate which is controlled by inserting or withdrawing neutron-absorbing rods of boron among the uranium rods. The graphite core is called the **moderator** and slows down the fission neutrons; fission of uranium-235 occurs more readily with slow than with fast neutrons. Carbon dioxide gas is pumped through the core and carries off heat to the **heat exchanger** where steam is produced. The concrete shield gives workers protection from γ-rays and escaping neutrons. The radioactive fission fragments must be removed periodically if the nuclear fuel is to be used efficiently.

In an **atomic bomb** an increasing uncontrolled chain reaction occurs when two pieces of uranium-235 come together and exceed the critical mass.

d) Fusion

The union of light nuclei into heavier ones can also lead to a loss of mass and, as a result, the release of energy. At present, research is being done on the controlled fusion of isotopes of hydrogen (deuterium and tritium) to give helium. Temperatures of over 100 million °C are required. Fusion is believed to be the source of the Sun's energy.

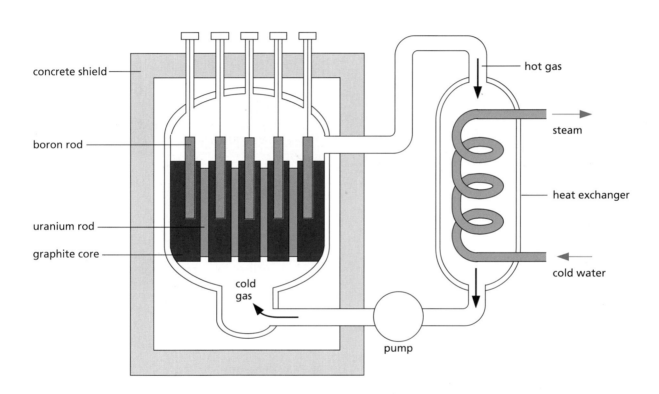

Figure 59.12 Nuclear reactor

Questions

1 Nuclear fission is used in nuclear power stations for the production of electricity.

  a What is meant by **nuclear fission**?
  b What type of energy is released in nuclear fission?
  c Compile a table giving *two* advantages and *two* disadvantages of having nuclear power stations.

Advantages	Disadvantages

  d Complete the passage using words from the box.

> generator   heat exchanger   reactor   steam

In a nuclear power station, nuclear fission takes place in the .... Energy is taken from the nuclear fuel by the coolant and carried to the ...... where it is used to produce ...... which drives the turbines. These turn the ...... to produce electricity.

(London Foundation, June 99)

2 The diagrams in Figure 59.13 represent three atoms, **A**, **B** and **C**.

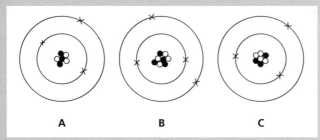

Figure 59.13

  a Two of the atoms are from the *same* element.
    (i) Which of **A**, **B** and **C** is an atom of a different element?
    (ii) Give *one* reason for your answer.
  b Two of these atoms are isotopes of the same element.
    (i) Which two are isotopes of the same element?
    (ii) Explain your answer.
  c Which of the particles ○, X and ●, shown in the diagrams
    (i) has a positive charge?
    (ii) has no charge?
    (iii) has the smallest mass?
  d Using the same symbols as those used in the atom diagrams, draw an alpha particle.

(AQA (NEAB) Higher, June 99)

3 The graph in Figure 59.14 shows data about the number of neutrons and protons in stable nuclei.

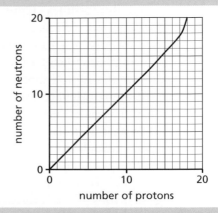

Figure 59.14

An atom of phosphorus-32 has 17 neutrons.

  a Calculate the number of protons in the nucleus of phosphorus-32.
  b Copy the graph and mark an X to show the position of phosphorus-32.
  c How can you tell from the graph that phosphorus-32 is radioactive?
  d Figure 59.15 shows the changes which take place in the nucleus during β^- decay.

Figure 59.15

    (i) Give the names of the particles **A**, **B** and **C**.
    (ii) Explain why particles **A** and **B** are *not* fundamental particles.
    (iii) By calculating the n/p ratios, show that when phosphorus-32 undergoes β^- decay, the resultant isotope is stable. *(London Higher, June 99)*

4 Evidence for the structure of the atom comes from alpha particle scattering. Figure 59.16 represents alpha particles from a radioactive source being directed at thin gold foil.

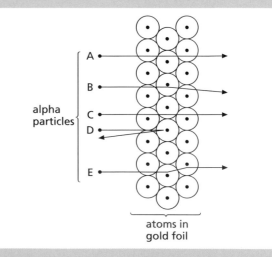

Figure 59.16

a (i) Explain why particle B is deflected but particles A and C are not.

(ii) What does the deflection of particle D show about the charge on the nucleus? Explain your answer.

(iii) Only a very small number of particles are deflected in the same way as particle D. What does this show about the structure of a gold atom?

b Suggest the likely result of an experiment using thick gold foil instead of thin gold foil.

c The results of alpha particle scattering experiments led to an atomic model that describes the atom as being made up of protons, neutrons and electrons.

(i) Explain why the electron is described as a **fundamental** particle.

Protons and neutrons are each made up of quarks. The charge on an up quark is equal to two-thirds that on a positron, $+e$. The charge on a down quark is equal to one-third that on an electron, $-e$.

(ii) Figure 59.17 represents the decay of a neutron.

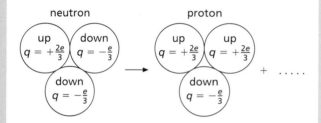

Figure 59.17

Write the name *and* symbol of the particle on the right hand side to complete the decay equation.

(iii) Figure 59.18 represents the decay of a proton.

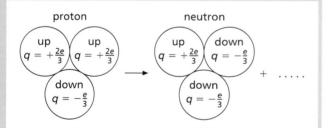

Figure 59.18

Write the name *and* symbol of the particle on the right hand side to complete the decay equation.

(*London Higher, June 98*)

■ *Checklist*

After studying this chapter you should be able to

- ■ describe how Rutherford and Bohr contributed to views about the structure of the atom,
- ■ describe the Geiger–Marsden experiment which established the nuclear model of the atom,
- ■ recall the charge, relative mass and location in the atom of protons, neutrons and electrons,
- ■ define the terms **proton number** (Z), **neutron number** (N) and **nucleon number** (A) and use the equation $A = Z + N$,
- ■ explain the terms **isotope** and **nuclide** and use symbols to represent them, e.g. $^{35}_{17}\text{Cl}$,
- ■ recall the effect of N/Z ratios on nuclear stability,
- ■ write equations for radioactive decay and interpret them,
- ■ outline the Schrödinger model of the atom and explain in terms of **energy levels** how line spectra are produced,
- ■ identify leptons and quarks as fundamental particles and recall their properties,
- ■ connect the release of energy in a nuclear reaction with a change of mass according to the equation $E = mc^2$,
- ■ describe the process of **fission**,
- ■ describe a nuclear reactor,
- ■ outline the process of **fusion**.

60 Electronics and control

Electronics is being used more and more in our homes, factories, offices, schools, banks, shops and hospitals. The development of semiconductor devices such as transistors and integrated circuits ('chips') has given us, among other things, automatic banking machines, laptop computers, programmable control devices, robots, computer games, digital cameras, Figure 60.1a, and heart pacemakers, Figure 60.1b.

Figure 60.1a Digital camera

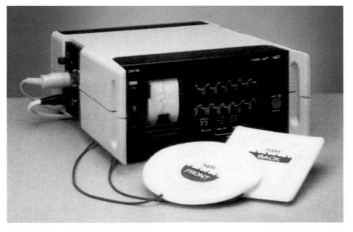

Figure 60.1b Heart pacemaker

■ *Semiconductors*

Semiconductors, of which silicon and germanium are the two best known, are insulators if they are very pure, especially at low temperatures. Their conductivity increases at higher temperatures. This is the opposite to the case with metals. In metals the number of electrons available for conduction, i.e. to carry current, is fixed. As the temperature increases the resistance *increases* due to increased vibration of the atoms making electron flow more difficult. In semiconductors (and carbon) this effect is more than offset by the 'freeing' of more charges for conduction as the temperature increases. As a result the resistance *decreases* at higher temperatures.

The conductivity of semiconductors can be greatly increased by adding tiny but controlled amounts of certain other substances (called 'impurities') by a process known as **doping**. They can then be used to make diodes, transistors and integrated circuits.

■ *Semiconductor diode*

A diode is a two-terminal, one-way device which lets current pass through it in one direction only. One is shown in Figure 60.2 with its symbol. (You will also come across the symbol without its outer circle.) The wire nearest the band is the **cathode** and the one at the other end is the **anode**.

The diode conducts when the anode goes to the + terminal of the voltage supply and the cathode to the − terminal, Figure 60.3a. It is then **forward biased**; its resistance is small and conventional current passes in the direction of the arrow on its symbol. If the connections are the other way round, it does not conduct; its resistance is large and it is **reverse biased**, Figure 60.3b.

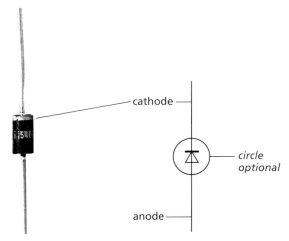

Figure 60.2 A diode and its symbol

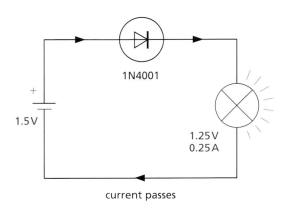

a Forward biased

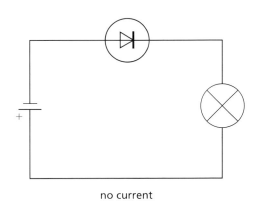

b Reverse biased

Figure 60.3 Action of a diode

The lamp in the circuit shows when the diode conducts by lighting up. It also acts as a resistor to limit the current when the diode is forward biased. Otherwise the diode might overheat and be damaged.

■ *Transistor*

Transistors are the small semiconductor devices which have revolutionized electronics. They are made both as separate components, like those in Figure 60.4a in their cases, and also as parts of **integrated circuits** (ICs) in which millions may be 'etched' on a 'chip' of silicon, Figure 60.4b.

Transistors have three connections called the **base** (B), the **collector** (C) and the **emitter** (E). In the transistor symbol shown in Figure 60.5, the arrow indicates the direction in which conventional current flows through it when C and B are connected to battery +, and E to battery −. Again, the outer circle of the symbol is not always included.

Figure 60.4a Transistor components

Figure 60.4b Integrated circuits which may each contain millions of transistors

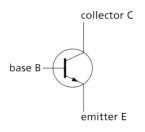

Figure 60.5 Symbol for a transistor

■ *Logic gates*

Logic gates are switching circuits used in computers and other electronic systems. They 'open' and give a 'high' output voltage, i.e. a signal (e.g. 5 V), depending on the combination of voltages at their inputs, of which there is usually more than one.

There are five basic types, all made from transistors in integrated circuit form. The behaviour of each is described by a **truth table** showing what the output is for all possible inputs. 'High' (e.g. 5 V) and 'low' (e.g. near 0 V) outputs and inputs are represented by 1 and 0 respectively and are referred to as **logic levels** 1 and 0.

a) NOT gate or inverter

This is the simplest gate, with one input and one output. It produces a 'high' output if the input is 'low', i.e. NOT high, and vice versa. Whatever the input, the gate inverts it. The symbol and truth table are given in Figure 60.18.

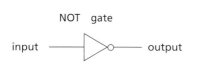

input	output
0	1
1	0

Figure 60.18 NOT gate symbol and truth table

b) OR, NOR, AND, NAND gates

All these have two or more inputs and one output. The truth tables and symbols for 2-input gates are shown in Figure 60.19. Try to remember the following.

> **OR:** output is 1 if input A **OR** input B **OR** both are 1
> **NOR:** output is 1 if neither input A **NOR** input B is 1
> **AND:** output is 1 if input A **AND** input B are 1
> **NAND:** output is 1 if input A **AND** input B are **NOT** both 1

Note from the truth tables that the outputs of the NOR and NAND gates are those of the OR and AND gates respectively inverted. They have a small circle at the output end of their symbols to show this inversion.

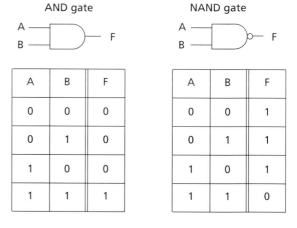

OR gate

A	B	F
0	0	0
0	1	1
1	0	1
1	1	1

NOR gate

A	B	F
0	0	1
0	1	0
1	0	0
1	1	0

AND gate

A	B	F
0	0	0
0	1	0
1	0	0
1	1	1

NAND gate

A	B	F
0	0	1
0	1	1
1	0	1
1	1	0

Figure 60.19 Symbols and truth tables for 2-input gates

c) Testing logic gates

The truth tables for the various gates can be conveniently checked by having the logic gate IC mounted on a small board with sockets for the power supply, inputs A and B and output F, Figure 60.20. A 'high' input (i.e. logic level 1) is obtained by connecting the input socket to the positive of the power supply, e.g. +5 V, and a 'low' one (i.e. logic level 0) to 0 V.

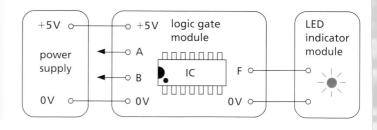

Figure 60.20 Modules for testing logic gates

The output can be detected using an indicator module containing an LED which lights up for a 1 and stays off for a 0.

■ *Logic gate control systems*

Logic gates can be used as processors in electronic control systems. Many of these can be demonstrated by connecting together commercial modules like those in Figure 60.22b below.

a) Security system

The block diagram for a simple system that might be used by a jeweller to protect an expensive clock is shown in Figure 60.21. The clock sits on a push switch which sends a 1 to the NOT gate unless the clock is lifted when a 0 is sent. In that case the output from the NOT gate is a 1 which rings the bell.

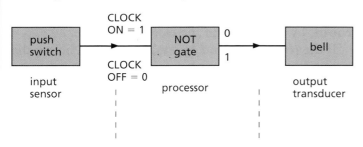

Figure 60.21 Simple alarm system

b) Safety system for a machine operator

The system could prevent a machine (e.g. an electric motor) being switched on before another switch had been operated by a protective safety guard being in the correct position. In Figure 60.22a, when switches A *and* B are pressed they supply a 1 to each input of the AND gate which can then start the motor. Figure 60.22b shows the system built from modules with a transducer driver included to supply the large current required by the motor.

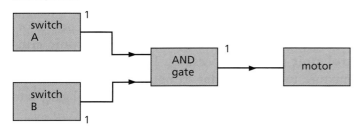

Figure 60.22a Safety system for controlling a motor

Figure 60.22b Modules for demonstrating the safety system

c) Heater control system

The heater control has to switch on the heating system when it is

(i) **cold**, i.e. the temperature is below a certain value and the output from the temperature sensor is 0, and
(ii) **daylight**, i.e. the light sensor output is 1.

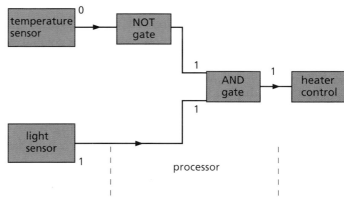

Figure 60.23 Heater control system

With these outputs from the sensors applied to the processor in Figure 60.23, the AND gate has two 1 inputs. The output from the AND gate is then 1 and will turn on the heater control. Any other combination of sensor outputs produces a 0 output from the AND gate, as you can check.

d) Street lights

A system is required which allows the street lights either to be turned on manually by a switch at any time, or automatically by a light sensor when it is dark. The arrangement in Figure 60.24 achieves this since the OR gate gives a 1 output when either or both of its inputs are 1.

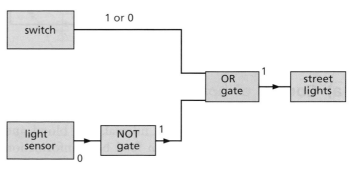

Figure 60.24 Control system with manual override

b The circuit diagram in Figure 60.50 shows one design for a burglar alarm. The alarm is activated when a burglar steps onto a pressure switch, S, hidden under a door mat.

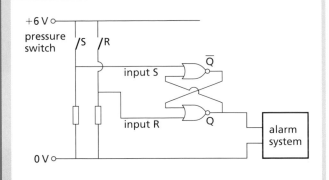

Figure 60.50

(i) Copy and complete the truth table to show what happens to the outputs Q and Q̄ when the burglar first steps onto and then off the door mat.

	S	R	Q	Q̄
burglar steps onto the door mat	1	0		
burglar steps off the door mat	0	0		

(ii) Explain in terms of the outputs Q and Q̄, the effect on the alarm if the switch R is momentarily closed after the alarm has been activated.

(iii) Why is this combination of two NOR gates called a bi-stable latch circuit?

(AQA (SEG) Higher, Summer 99)

■ *Checklist*

After studying this chapter you should be able to

■ explain the term **doping** of semiconductors,

■ explain what is meant by a diode being **forward biased** and **reverse biased**,

■ describe the action of a transistor with the aid of a circuit diagram,

■ describe the action of an **LDR**, a **thermistor**, an **LED** and a **photodiode**,

■ describe how a transistor can be used as a switch,

■ recall the functions of the **input sensor**, **processor** and **output transducer** in an electronic system and give some examples,

■ explain the operation of light-, temperature- and time-operated transistor switching circuits with the aid of diagrams,

■ describe the action of **NOT**, **OR**, **NOR**, **AND** and **NAND** logic gates and recall their truth tables,

■ design and draw block diagrams of logic control systems for given requirements,

■ explain with the aid of a sketch the term **feedback** in electronic systems and how it can be positive or negative, and give examples of the use of feedback,

■ describe the action of an **SR bistable** and explain how it can be used as a **latch**,

■ distinguish between **analogue** and **digital** circuits and devices,

■ show an appreciation of the impact of electronics on society.

61 Telecommunications

Telecommunications is concerned with sending and receiving information over a distance. In the broadest sense, information can be in the form of written or spoken words, numbers, diagrams, pictures, music or computer data.

■ *Early history*

The earliest line-of-sight methods involved sending smoke signals from hilltop to hilltop. It was the fore-runner of flag signalling by semaphore and ship-to-ship signalling by Aldis lamp.

The mid-19th century saw the invention of the electric telegraph by **Wheatstone**, Figure 61.1a, in which messages were sent in Morse code as currents in cables. Shortly afterwards **Bell** developed the telephone, Figure 61.1b, making it possible to transmit speech electrically. The use of electrical signals increased the speed of communications substantially.

Figure 61.1a Telegraph transmitting key c.1890

Figure 61.1b Early telephone c.1890

Practical work

Simple radio receiver

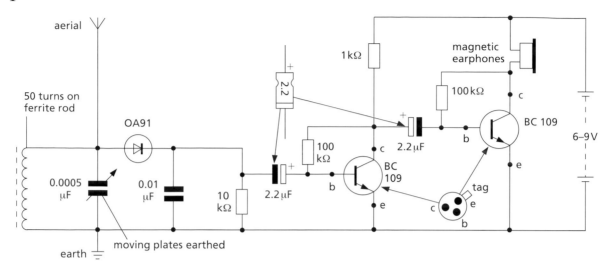

Figure 61.12 Simple radio receiving circuit

Connect the circuit of Figure 61.12 on an S-DeC, Figure 61.13. The transistor leads can be lengthened and connections made to the 'tags' on the variable capacitor as in Figure 60.13b, p. 284. The resistor colour code is given on p. 210.

For an aerial, support a length of wire (e.g. 10 m) as high as you can; making an earth connection to a water tap will improve reception, Figure 61.14.

You should be able to tune in to one or two stations (depending on your location) by altering the variable capacitor.

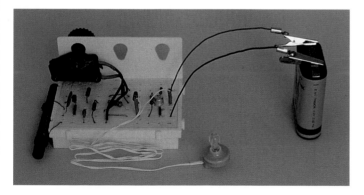

Figure 61.13 Radio receiving circuit built on an S-DeC

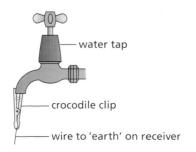

Figure 61.14 Connect the aerial to earth via a tap

■ *Television*

a) Black and white

A television receiver is basically a CRO with two time bases. The horizontal or **line time base** acts as in the CRO. The vertical or **frame time base** operates at the same time and draws the spot at a much slower rate down to the bottom of the screen and then returns it almost at once to the top. The spot thus 'draws' a series of parallel lines of light (625) which covers the screen, Figure 61.15, and is called a **raster**.

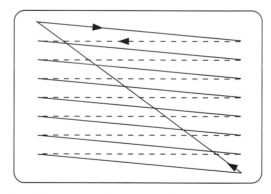

Figure 61.15 Pattern of lines (raster) on a TV screen

A picture is produced by the incoming signal altering the number of electrons which travel from the electron gun to the screen. The greater the number the brighter the spot. The brightness of the spot varies from white through grey to black as it sweeps across the screen. A complete picture appears every 1/25 s, but because of the persistence of vision we see the picture as continuous. If each picture is just slightly different from its predecessor, the resultant effect is that of a 'movie' and not a sequence of 'stills'.

b) Colour

One type of colour television has three electron guns and the screen is coated with about a million tiny light-emitting 'dots' arranged in triangles. One 'dot' in each triangle emits red light when hit by electrons, another green light and the third blue light.

As the three electron beams scan the screen, an accurately placed 'shadow mask' consisting of a perforated metal plate with about one-third of a million holes ensures that each beam strikes only dots of one 'colour', e.g. electrons from the 'red' gun strike only 'red' dots, Figure 61.16.

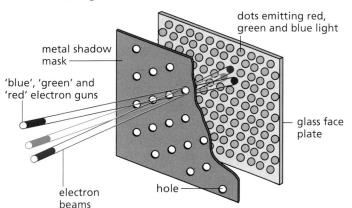

Figure 61.16 Colour TV screen

When a triangle of dots is struck it may be that the red and green electron beams are intense but not the blue. The triangle will emit red and green light strongly and appear yellowish (see Chapter 9). The triangles of dots are struck in turn, and since the dots are so small and the scanning so fast, we see a continuous colour picture.

c) Energy transfers

Electrons emitted by an electron gun in the cathode ray tube of a television receiver or a CRO are accelerated towards the screen by the high p.d. which is applied across the tube. If this is V and an electron has charge e and mass m, then an amount of electrical energy eV (from $W = QV$, Chapter 45) is transferred to kinetic energy of the electron. Therefore

$$\text{k.e.} = \tfrac{1}{2}mv^2 = eV$$

where v is the final velocity of the electron. This k.e. is transferred to heat and light when the electron hits the screen.

We can calculate the k.e. If we take $e = 1.6 \times 10^{-19}\,\text{C}$ and $V = 25\,\text{kV}$ then

$$\text{k.e.} = eV = 1.6 \times 10^{-19}\,\text{C} \times 25 \times 10^3\,\text{V}$$
$$= 4.0 \times 10^{-15}\,\text{J}$$

■ *Optical fibre communication systems*

For both local and long-distance communication, copper telephone cables carrying electric currents are being replaced by optical fibres. These are very thin fibres of very pure glass in which 'light' is trapped by total internal reflection (Chapter 6) and used to carry information.

a) Outline of system

A simplified block diagram of a communication system is shown in Figure 61.17 and an optical cable (showing fibres inside) in Figure 61.18. Electrical signals from the **input transducer** (representing the information to be transmitted) are modulated by a process called **pulse code modulation** in a **coder**. This produces a stream of equivalent digital electrical pulses which are changed by the **optical transmitter** into 'light' pulses for transmission by the optical fibre. The transmitter is either an LED or a miniature diode laser (p. 283). The 'light' used is infrared radiation because it suffers less absorption in the glass.

At the receiving end the **optical receiver** is a **photodiode** (a diode that conducts when light falls on it). This converts the incoming infrared signals into the corresponding electrical ones before they are processed by the **decoder** for passage to the **output transducer**.

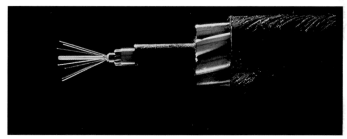

Figure 61.18 Optical cable

b) Advantages

A digital optical fibre system has important advantages over other communication systems.

(i) It has a higher information-carrying capacity, i.e. more information can be sent in a given time due to the higher frequency of light compared to that of radio waves or to the frequencies of electrical signals sent along copper cables. About 30 000 telephones calls can be carried at once on a pair of fibres, whereas a copper cable can carry only about 2000.

(ii) The amplitude of a signal falls off less quickly than with the electrical signal in a copper cable, i.e. there is less **attenuation**. This means that greater distances can be covered without amplifiers than with copper cables.

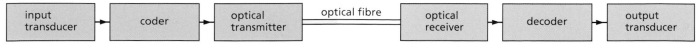

Figure 61.17 Optical fibre communication system

Questions

1 List the limitations of sending information by
   a a line-of-sight method,
   b telegraphy,
   c telephony.

2 a Figure 61.23 represents a radio.

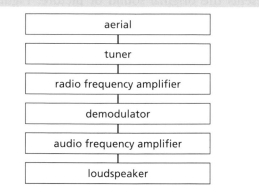

Figure 61.23

Which of the blocks in the diagram:
(i)   detects radio waves;
(ii) increases the voltage of the sound signal;
(iii) produces sound when there is a varying current
in it?

b Figure 61.24 represents a radio wave carrying a
sound signal.

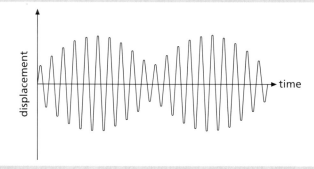

Figure 61.24

(i) What is happening to the amplitude of the radio
wave?
(ii) Use similar axes to sketch the sound signal that
the radio wave is carrying.

(*London Foundation, June 98*)

3 Figure 61.25 shows the circuit of a simple radio receiver.
   a State the names of the parts labelled A, B, C.
   b What is the purpose of the part labelled D?

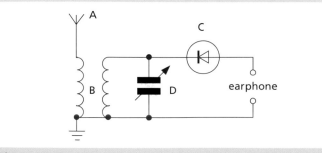

Figure 61.25

4 a The names of five basic blocks used in
communications systems are shown in Figure 61.26.
The uses of these blocks are shown on the right. The
uses are *not* in the correct order.
   Copy and complete the diagram by drawing a line
between each block and its correct use. One has been
done for you.

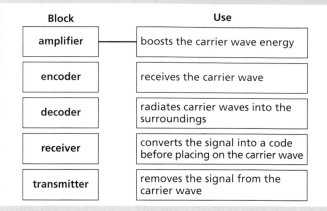

Figure 61.26

b People can communicate by talking and listening.
Different parts of the body have different functions
in this method of communication. Complete the
table using words from the box.

| brain | mouth | outer ear | voice box |

Function	Part
encoder	voice box
decoder	
receiver	
transmitter	

c State and explain *two* disadvantages of using talking
and listening as a method of communication.
(*London Foundation, June 99*)

5 Discuss the way in which information is stored on
   a an audio CD,
   b audio tape.

6 Figure 61.27 shows three ways in which radio waves
can travel.

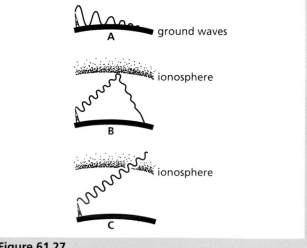

Figure 61.27

The ground waves in diagram **A** follow the Earth's curvature. The radio waves shown in diagram **B** are reflected by the ionosphere. The radio waves shown in diagram **C** are able to travel through the ionosphere.

a Write down the names of the waves shown in diagrams **B** and **C**.

b Explain how the diagrams show that each wave is attenuated as it travels.

c Radio 4 is broadcast from England. A motorist driving away from England through France can receive the Radio 4 long wave signal on her car radio.
(i) Which of the diagrams shows the way in which long wave signals travel?
(ii) How does attenuation affect the signal she receives as she continues to drive south?
(iii) Explain why she is unable to receive the Radio 4 VHF signal.

(London Higher, June 98)

7 In recent years there has been a rapid growth in the number and use of mobile telephones. Mobile telephones use radio waves for transmitting speech. They have to use frequencies that are not already used by radio stations.

a Radio waves used for mobile telephones have a typical wavelength of 0.30 m. Calculate the frequency of these radio waves, given that their speed is 3.0×10^8 m/s.

b Mobile telephones can be used to communicate throughout Europe using satellite links. A set of three satellites, each in an elliptical orbit, is used to give 24 hour coverage. Figure 61.28 shows the orbit of one of these satellites.

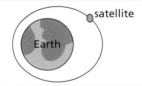

Figure 61.28

(i) Copy the diagram and draw an arrow to show the gravitational force acting on the satellite.
(ii) Describe how the size of this force changes as the satellite makes one orbit of the Earth.
(iii) Place an M on your diagram where the acceleration of the satellite is greatest.

c Figure 61.29 shows how a dish aerial is used to focus waves and transmit them to the satellite. Focusing the radio waves minimises the effects of diffraction.

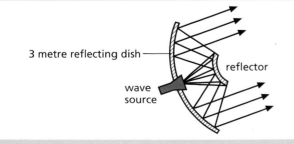

Figure 61.29

(i) Explain why it is important to minimise the effects of diffraction.
(ii) What *two* factors affect the amount of diffraction that takes place when a wave passes through an opening?
(iii) The waves used for satellite transmission have a much shorter wavelength than the 0.30 m used by the mobile telephones. Suggest why a wavelength of 0.30 m is unsuitable for satellite transmission.

d Transatlantic telephone calls can be carried either using satellite links or by optical fibres on the seabed. Describe how optical fibres can be used to carry telephone calls. Suggest *one* advantage and *one* disadvantage of using optical fibres rather than satellite links.

(London Higher, June 98)

■ *Checklist*

After studying this chapter you should be able to

■ describe the early history of telecommunications,

■ explain how information can be represented electrically in digital and analogue form,

■ describe with the aid of a diagram the main parts of a communication system,

■ recall that radio waves are a form of electro-magnetic radiation, their range of frequency and wavelength, and the various uses of the different bands,

■ recall the ways in which radio waves travel,

■ outline a radio system using amplitude modulation,

■ recall how electrical oscillations are produced,

■ explain the action of the tuning and detector circuits in a radio receiver,

■ outline how black and white and colour television receivers work,

■ recall that in a cathode ray tube the k.e. of an electron of charge e, mass m, accelerated through a p.d. V to velocity v is given by $\frac{1}{2}mv^2 = eV$,

■ give an outline of an optical fibre communication system and state its advantages over other systems,

■ outline the process of pulse code modulation,

■ describe how geostationary communication satellites operate,

■ recall some ways in which information can be stored.

Electrons and atoms
Additional questions

Electrons

1 One type of cathode ray oscilloscope (CRO) used in a laboratory is shown.

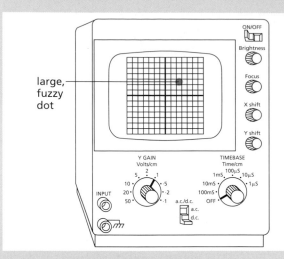

a When first switched on, the screen shows a large, fuzzy dot. Explain how the CRO should be adjusted to give a dot which is:
- small and clear;
- at the centre of the screen

b A battery is now connected to the input. The dot moves a small distance up the screen.

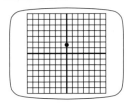

Explain what you would do to measure the voltage of the battery.

c The battery is now replaced with an a.c. supply and the CRO adjusted to give the display below. The Y-gain control stays set at 1 V/cm.

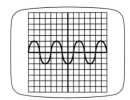

(i) How is the CRO adjusted to give this display?
(ii) On a copy of the diagram above show how the display would look if the Y-gain is changed to 0.5 V/cm. All the other controls and the input stay the same.

(*SEG Foundation, Summer 98*)

2 The diagram below shows an electron beam. The beam is deflected as it passes between two metal plates.

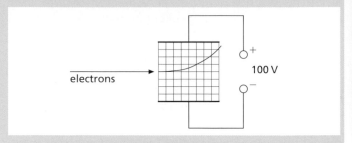

a How can you tell from the diagram that the electrons have a negative charge?
b On copies of the diagrams below show the paths of the electron beam when the voltage between the metal plates is changed as shown.

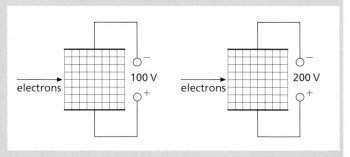

(*London Foundation, June 98*)

Radioactivity and atomic structure

3 Atomic nuclei contain both protons and neutrons. $^{11}_{6}C$, $^{12}_{6}C$ and $^{14}_{6}C$ are all forms of carbon. The nucleus of $^{11}_{6}C$ consists of 6 protons and 5 neutrons.
a (i) Write down the number of protons and neutrons in $^{12}_{6}C$:
$^{12}_{6}C$ contains ...... protons and ...... neutrons.
(ii) Explain why $^{14}_{6}C$ is described as 'neutron-rich'.
(iii) What do the nuclei of $^{11}_{6}C$, $^{12}_{6}C$ and $^{14}_{6}C$ all have in common?
b The following graph can be used to show the numbers of neutrons and protons in the lighter nuclei. The crosses represent the nuclei of $^{12}_{6}C$ and $^{14}_{6}C$.

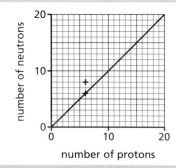

For a nucleus to be stable, a plot of its number of neutrons against its number of protons must lie on or close to the diagonal line.

(i) Copy the graph and mark X to show the position of $^{11}_{6}C$.

(ii) What is the condition for a nucleus with less than 10 protons to be stable?

(iii) Which form of carbon is the most stable? Explain your answer.

c The diagram below represents the decay of a nucleus of $^{14}_{6}C$.

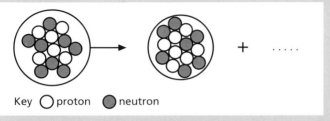

Key ○ proton ● neutron

(i) Use the diagram to write down what happens to the numbers of protons and neutrons when $^{14}_{6}C$ decays.

(ii) What particle is missing from the right hand side of the diagram?

(London Foundation, June 98)

4 a The graph shows how a sample of barium-143, a radioactive *isotope* with a short *half-life*, decays with time.

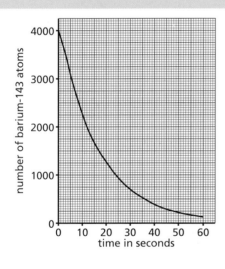

(i) What is meant by the term *isotope*?

(ii) What is meant by the term *half-life*?

(iii) Use the graph to find the half-life of barium-143.

b Humans take in the radioactive isotope carbon-14 from their food. After their death, the proportion of carbon-14 in their bones can be used to tell how long it is since they died. Carbon-14 has a half-life of 5700 years.

(i) A bone in a living human contains 80 units of carbon-14. An identical bone taken from a skeleton found in an ancient burial ground contains 5 units of carbon-14. Calculate the age of the skeleton. Show clearly how you work out your answer.

(ii) Why is carbon-14 unsuitable for dating a skeleton believed to be about 150 years old?

c The increased industrial use of radioactive materials is leading to increased amounts of radioactive waste. Some people suggest that radioactive liquid waste can be mixed with water and then safely dumped at sea. Do you agree with this suggestion? Explain the reason for your answer.

(SEG Higher, Summer 98)

5 The diagram shows the variation in background radiation in England, Scotland and Wales.

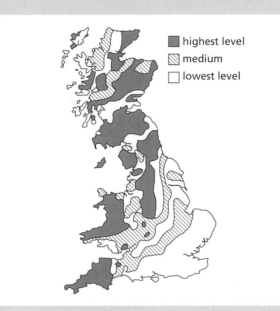

■ highest level
▨ medium
□ lowest level

a The background radiation is calculated by finding the average value of a large number of readings. Suggest why this method is used.

b The high levels of radiation in some parts of Britain are caused by radon gas escaping from underground rocks such as granite.

Radium-224 ($^{224}_{88}Ra$) decays to form radon-220 ($^{220}_{86}Rn$).

(i) What particle is emitted when radium-224 decays?

(ii) Radon-220 then decays to polonium by emitting an alpha particle. Complete the decay equation for radon-220.

$$^{220}_{86}Rn \rightarrow \quad Po \quad + \quad He$$

(iii) The half-lives of these isotopes are given in the table.

Isotope	Half-life
radium-224	3.6 days
radon-220	52 seconds

A sample of radium-224 decays at the rate of 360 nuclei per second. The number of radon-220 nuclei is growing at less than 360 per second. Suggest a reason for this.

(iv) Radon-220 has a short half-life and it emits the least penetrative of the three main types of radioactive emission. Explain why the presence of radon gas in buildings is a health hazard.

(London Higher, June 99)

Electronics and control

6 a The following is a list of devices which may be used in electrical circuits.

> capacitor   diode   microphone
>
> multimeter   transistor

Using each device only once, name the device which
(i) allows current to flow in one direction only;
(ii) can store charge;
(iii) can act as an electronic switch;
(iv) can be used to measure voltage;
(v) can act as a sound sensor.

b The diagram below shows a simple electronic bicycle alarm system. There are two switches in the system:

- a key operated switch which the rider turns on when the bicycle is parked;
- a pressure switch fitted in the saddle.

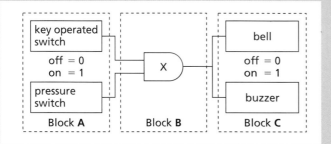

(i) The alarm system consists of three blocks, **A**, **B** and **C**. Use the following words to name each block.

control circuit   output sensors   input sensors

(ii) What sort of logic gate is gate X?
(iii) Copy and complete the truth table for the system.

Key operated switch	Pressure switch	Buzzer	Bell
1	1		
1	0		
0	1		
0	0		

(iv) What conditions will cause the alarm to go off?
(SEG Foundation, Summer 98)

7 The diagram shows part of an alarm system used to protect CD players on display in a shop. Each CD player is placed on two pressure switches. When a pressure switch is pressed down it provides a high output (1). The alarm bell sounds if the CD player is lifted up.

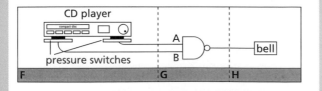

a This electronic alarm system consists of three parts **F**, **G** and **H**. Write the name of each of these parts of the system. Choose the *best* word from the list below.

> input   output   processor
>
> relay   supply   thermistor

b The system contains a NAND logic gate.
(i) Write the truth table for this logic gate, started below.

Input A	Input B	Output
0	0	

(ii) The NAND gate can be replaced by a combination of two other logic gates to give the same effect. Copy the diagram below and draw this combination in the box labelled **G**. Label the gate.

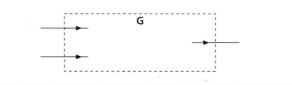

(iii) Somebody suggests replacing the NAND gate with a NOR gate. Describe how this would change the way the system works.

(OCR Higher, June 99)

8 The diagram below shows an electronic system installed in a greenhouse.

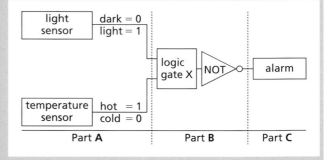

The alarm is designed to operate when there is an output from the NOT gate.

a (i) Which part, **A**, **B** or **C** of the system acts as the **processor**?
(ii) Which part, **A**, **B** or **C** of the system is **controlled** by the processor?

b The alarm should be activated if the temperature in the greenhouse drops during the night. Which type of logic gate should be used at X?

c Name a device suitable for use as
(i) the light sensor,
(ii) the temperature sensor.

d The following graph shows how the resistance of the temperature sensor changes with temperature. Use the graph to find the resistance of the temperature sensor at 33 °C.

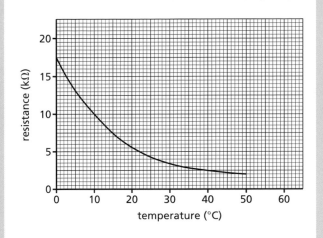

e The temperature sensor is connected in the circuit below.

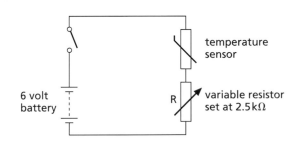

(i) What is the total resistance in the circuit when the temperature is 33 °C?
(ii) Calculate the current in the circuit when the temperature is 33 °C.
f Resistor R is connected to an electronic circuit as shown below.

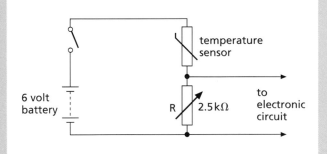

The electronic circuit is switched on when the potential difference across R is 1.4 V or more.
(i) Calculate the output voltage when the temperature is 33 °C.
(ii) Will the electronic circuit be ON when the temperature is 33 °C?
(iii) Calculate the output voltage when the temperature is 0 °C.
(iv) Will the electronic circuit be ON when the temperature is 0 °C?
(v) Suggest a use for this circuit.
(AQA (NEAB) Higher, June 98)

9 The diagram below shows the logic levels for an SR bistable in its RESET state.
Redraw the diagram with the new logic levels when the S input is changed to 0. Explain the changes.
Hint. You need to refer in your explanation to the truth table of a NAND gate.

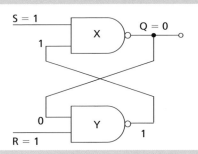

Telecommunications

10 This question is about communications.
a Communications use signals to carry a message. Finish the table by writing the type of signal. Choose the *best* words from this list.

electrical   heat   light   sound

Method of communication	What carries the signal	Type of signal
voice	air	
telephone	wire	
telephone	optical cable	

b A telephone receiver has a microphone and an earphone. The diagram shows a simple version of a telephone link.

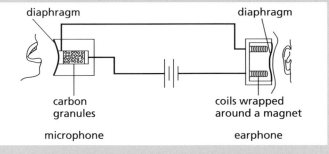

Finish the sentences by choosing the *best* words from this list. Each word may be used once, more than once or not at all.

decreases   increases   magnetic field

sound   spin   vibrate

Sound waves from the voice make the microphone diaphragm ....
When the diaphragm moves in, the carbon granules are squashed.
The resistance of the granules decreases, so the current in the circuit ....
The changing current causes a changing ...... in the earphone coils.
This makes the diaphragm ...... which creates sound.
(OCR Foundation, June 99)

11 a A dynamo is used to power a lamp. The cathode ray oscilloscope (CRO) is connected to show the voltage across the lamp. The trace first obtained is shown in the diagram on the CRO.

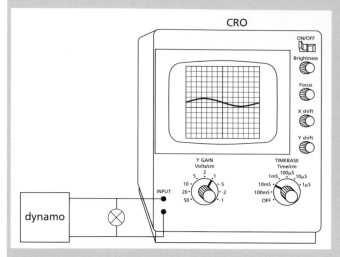

(i) Which control would be adjusted to increase the height of the trace?

(ii) Which control would be adjusted so that the trace shows more than one complete wave?

b The diagram shows the trace obtained after suitable adjustments have been made.

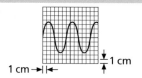

1 cm →|←   ⊥ 1 cm

If the Y-gain is set at 2 volts/cm, what is the peak voltage across the lamp?

c A diode is added to the circuit. The new trace seen on the CRO is shown.

What name is given to this type of trace?

d A capacitor is then added to smooth the voltage across the lamp.

(i) Sketch the diagram below and, using the correct circuit symbol, show where the capacitor should be connected in the circuit.

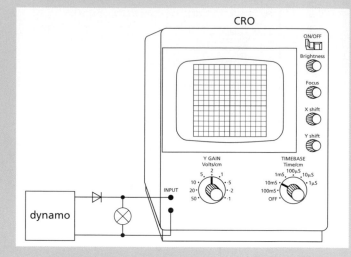

(ii) Explain how the capacitor achieves this smoothing effect.

(iii) On the screen of the CRO in your sketch, draw the trace which would now be seen.

e The lighting circuit for a bicycle may include both a dynamo and a battery.

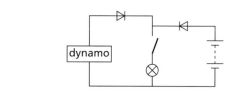

Why are *both* diodes needed in the circuit?

(*AQA (SEG) Foundation, Summer 99*)

12 An electron, charge e and mass m, is accelerated in a cathode ray tube by a p.d. of 1000 V. Calculate

a the kinetic energy gained by the electron,

b the speed it acquires.

$(e = 1.6 \times 10^{-19} \, C, \, m = 9.1 \times 10^{-31} \, kg)$

13 a Describe a geostationary satellite orbit.

b State the conditions for a satellite to remain in a geostationary orbit.

c Explain why a geostationary orbit is necessary for effective satellite communication.

Earth and space physics

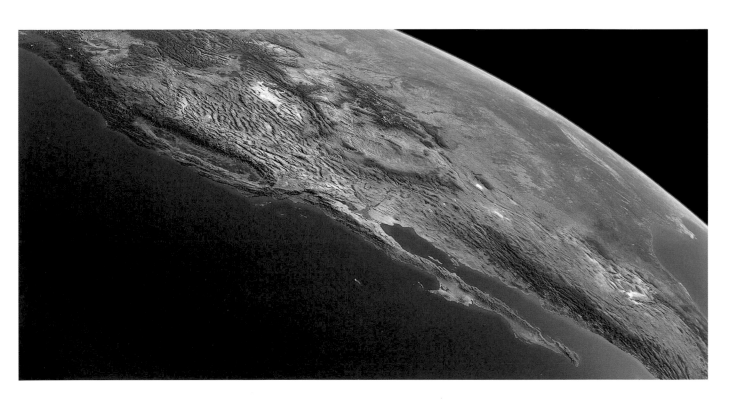

■ *Earthquakes, volcanoes, tidal waves*

a) Earthquakes

Large earthquakes are due to the movement of tectonic plates along the boundaries where they meet; many are the result of oceanic plates descending into the mantle. They also occur in mid-ocean ridges and rift valleys.

When two plates slide past each other, the rocks become stressed, eventually break and energy is released as shock waves. For example, earthquakes occur along the San Andreas Fault in California, Figure 62.9, which marks the boundary between the North American and Pacific Plates (Figure 62.6). The shock waves result in earth tremors or even horizontal or vertical shifts of several metres in the rocks.

Smaller earthquakes occur within plates to relieve local stresses.

While major earthquakes can cause great damage to property and tragic loss of human life, the effects are less the deeper the origin of the earthquake, the firmer the rocks are at the surface and the more substantial the buildings affected, e.g. if they have foundations that are larger than the wavelengths of the tremors that might strike them.

The **Richter Scale** gives an indication of the total amount of energy released by an earthquake; it is an open-ended scale of magnitudes but earthquakes above magnitude 9 are extremely rare. One rated of magnitude 2.0, which is ten times greater than one of 1.0, would scarcely be noticed. One of 3.0 is one hundred times greater than one of 1.0, and so on. The earthquake of 2001 in Gujerat, India, had magnitude 7.9, Figure 62.10, and killed over 30 000 people. Altogether there are more than one million earthquakes annually but fortunately only about one of this magnitude per year.

Britain is far removed from plate boundaries and so earthquakes today are rare but mild ones occur occasionally. They arise from slight slipping of rock faults caused by earthquakes that happened millions of years ago. An example is the Great Glen Fault which separates Scotland into two parts and is now occupied by three lochs and the Caledonian Canal.

Figure 62.9 Aerial view of the San Andreas Fault, California

Figure 62.10 Devastation after the Indian earthquake of January 2001

b) Volcanoes

The fact that many of the main volcanic regions occur along tectonic plate boundaries suggests that volcanic eruptions might arise from the same forces that cause earthquakes and the formation of mountains, i.e. moving plates. When eruptions occur, hot magma, being less dense, rises up through the crust and is ejected as **lava**. The eruption is accompanied by a huge explosion due to water near the surface being turned into steam; clouds of ash and gases, e.g. poisonous sulphur dioxide, are also ejected and a crater is blown.

There are about 500 active volcanoes on Earth and more under the oceans. They are common round the edges of the Pacific Ocean. The eruption of Mount St Helen's on the north-west coast of the USA in 1980, Figure 62.11, blew away half the mountainside and caused loss of life despite being predicted from events before the eruption such as earth tremors, swelling of the mountainside and the presence of certain gases at its top.

When lava solidifies in the crater of a volcano, the volcano becomes dormant.

Figure 62.11 Eruption of Mount St Helen's, Washington State

c) Tidal waves (tsunamis)

These are huge waves that occur as a result of an earthquake or volcanic eruption under water. They have enormous energy and travel at speeds up to 700 km/h.

■ *The Earth and its rocks*

The study of rocks is called **geology** and it is a subject which has told us much about the history of the Earth. Rocks can be put into one of three groups depending on their origin and how they were formed.

a) Igneous rocks

These formed when magma from below the Earth's surface cooled and solidified, either in the Earth's crust or on its surface if it was ejected through the crust as lava in a volcanic eruption. Examples are **granite** and **basalt**. They are the bedrock of large parts of most continental and oceanic crust and when not covered by sediment or the sea, they are exposed at the Earth's surface, Figures 62.12a, b. They contain no fossils.

Figure 62.12a Granite rocks on Dartmoor

Figure 62.12b Basalt rocks in the Giant's Causeway, Northern Ireland

b) Sedimentary rocks

These are the result of the weathering (i.e. breaking down) of older rocks over millions of years to form sediments. Sediments consisting of loose pieces of rock, gravel, sand and mud were produced and in many cases carried from their place of origin to the sea-bed. There they were deposited in layers (strata), one on top of the other. The transporting agents responsible for the erosion (i.e. eating away) of the rocks were rain, wind, rivers and glaciers; they moved small-particle sediments farther than large-particle ones.

The lower layers of sediment were pressed together tightly, cemented and in time hardened into sedimentary rock. In some instances movements inside the Earth made the layers bend and fold, like those shown in Figure 62.13, often creating faults and fractures in the rock.

Figure 62.14 Slate quarry in the Lake District

Figure 62.13 Folded sedimentary rocks in Dorset

Examples of sedimentary rocks are **sandstone**, **limestone**, **clay** and **coal**. They form about three-quarters of the rocks on or very close to the Earth's surface. Many deposits, e.g. of limestone and coal, contain fossils from the remains of animals and plants.

c) Metamorphic rocks

These were formed when either igneous or sedimentary rocks were subjected to very high temperatures and/or pressures inside the Earth; examples are **slate**, formed from clay, and **marble**, formed from limestone. Metamorphic rocks can be exposed at ground level if weathering or quarrying removes the material covering them, Figure 62.14. They are often found in volcanically active regions. They rarely have fossils.

Rocks are important in everyday life. They break down to give the **soil** in which our food is grown. Some are used to make **building materials** such as stone, bricks, concrete and glass. Others contain **ores** that provide metals like iron and copper that are essential in industry. All contain **minerals**, i.e. naturally occurring crystalline inorganic substances; for example, granite is mostly feldspar, Figure 62.15. There are over 2000 different minerals, each with its characteristic physical and chemical properties. Many minerals are silicates, containing oxygen and silicon; these make up most of the Earth's crust and are called 'rock-making' minerals.

Figure 62.15 Granite showing large pink crystals of feldspar

Weathering and soil

Weathering is the process which occurs when rocks are broken down into smaller pieces by the action of wind, water, frost, sun, etc. For example, temperature changes cause the outer layers of rocks to expand during the day and to contract at night. Internal stress is created, leading to cracking and disintegration. Similarly, if rainwater freezes in rock cracks, it expands as it changes to ice, making the cracks wider and deeper and causing bits to break off.

Ultimately the sediment particles produced by weathering become fine enough to be classed as soil in which plants can grow. When the plants die, they decompose to form a dark sticky substance called **humus** which plays a vital part in releasing chemicals from minerals in the soil. These are then absorbed in solution by plant roots.

There are three layers in most soils, as shown by the **soil profile** in Figure 62.16. The top layer contains the smallest particles and humus. The middle layer or sub-soil consists of larger particles and rests on the bottom layer of weathered parent rock.

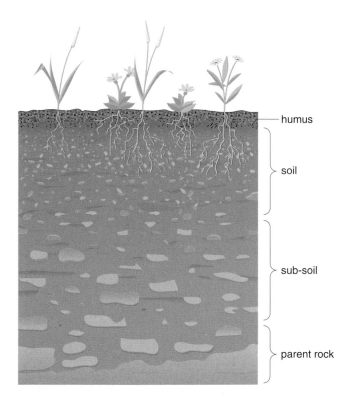

Figure 62.16 Soil profile

Particle sizes vary in different soils. In clay they are very small and the spaces between them soon fill up with water and the soil becomes waterlogged. In sand, particles and spaces are larger and the soil drains well.

Questions

1 Which of these statements is *incorrect*?
   a P waves can travel through solid material.
   b P waves can travel through liquid material.
   c S waves can travel through solid material.
   d S waves can travel through liquid material.

2 a State *three* pieces of evidence which support the theory of continental drift.
   b What causes continental drift?

3 Figure 62.17 represents the structure of the Earth.

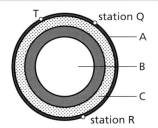

Figure 62.17

   a Name the parts A to C.
   b An earthquake occurs at the point T on the Earth's surface. Two types of shock wave are produced by the earthquake, P waves and S waves. Describe *two* similarities and *two* differences between P waves and S waves as they travel through the Earth.
   c State whether P waves or S waves or both will reach
      (i) Station Q, (ii) Station R.
      (AQA (NEAB) Higher, June 99)

Checklist

After studying this chapter you should be able to

■ describe the nature of the Earth's layered structure and state from what the evidence for this structure comes,
■ recall the properties of the three types of **seismic wave** and describe how they are transmitted through the Earth,
■ understand how to locate the epicentre of an earthquake,
■ recall that the Earth's lithosphere consists of large slabs of rock called **tectonic plates** whose relative motion can account for the drift, shape and rock records of the continents,
■ explain the formation of mountain ranges, island chains, mid-ocean ridges and rift valleys in terms of moving tectonic plates,
■ use the theory of plate tectonics to explain how earthquakes and volcanoes arise at plate boundaries,
■ recall the three main groups of rocks and their properties,
■ explain how soil is produced by weathering.

63 The Solar System

Members of the Solar System

The **Solar System** consists of the Sun and the nine planets moving round it in elliptical orbits, i.e. slightly flattened circles, Figure 63.1. It also includes the asteroids, comets and the moons that travel round most of the planets.

The **four inner planets**, Mercury, Venus, Earth and Mars are all small, of similar size, solid and rocky, probably with a layered structure.

The **four outer planets**, Jupiter, Saturn, Uranus and Neptune are much larger and colder and consist mainly of gases. The outermost planet, Pluto, is quite small and thought to be made of ice. It has a more strongly elliptical orbit which is tilted slightly compared with those of the other planets.

The **asteroids** consist of pieces of rock of various sizes which orbit between Mars and Jupiter; their density is similar to that of the inner planets. If they enter the Earth's atmosphere they burn up and fall to Earth as meteors or shooting stars. **Comets** consist of dust embedded in ice made from water and methane and are sometimes called 'dirty snowballs'. Their density is similar to that of the outer planets and they have large elliptical orbits (see p. 328).

The **Sun**, being a star, produces its own light which takes 8 minutes to reach us; the planets and our Moon, on the other hand, are seen from the Earth by reflected solar light. The Sun has a surface temperature of about 6000 °C and in its central core, where the temperature must be many millions of °C, nuclear fusion occurs. This results in hydrogen being changed to helium and the whole range of electromagnetic radiation from gamma rays to radio waves is emitted. It is a 'yellow dwarf' star (see Chapter 64) estimated to be about halfway through its lifetime of 10 000 million years.

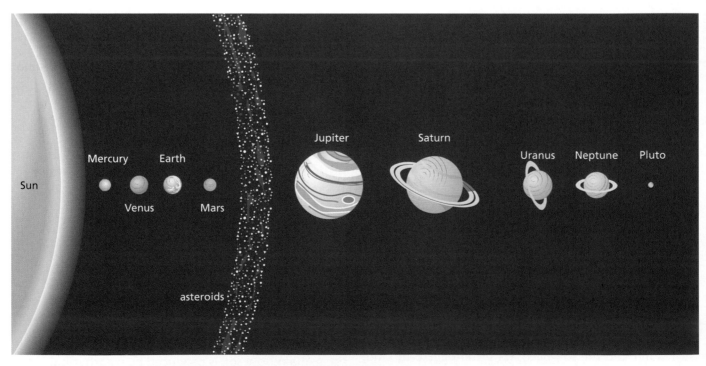

Figure 63.1 The Solar System (distances from the Sun not to scale)

Motion of the Earth

The occurrence of certain natural events is readily explained by the Earth's motion.

a) Day and night

These are caused by the Earth spinning on its axis (i.e. about the line through its north and south poles) and making one complete revolution every 24 hours. This creates day for the half of the Earth's surface facing the Sun and night for the other half.

b) The seasons

Two factors are responsible for these. The first is the motion of the Earth round the Sun once every 365.24 days, i.e. in 1 year, and the second is the tilt of the Earth's axis (at 23.5°) to the plane of its path round the Sun. Figure 63.2 shows the tilted Earth in four different positions of its orbit.

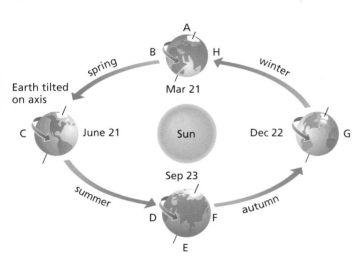

Figure 63.2 Seasons for the northern hemisphere

Over part BCD of the orbit, the northern hemisphere is tilted towards the Sun and so it is spring and summer with the hours of daylight being greater than those of darkness. The southern hemisphere is tilted away from the Sun and is having autumn and winter with shorter days than nights. The northern hemisphere receives more solar radiation and the weather is consequently warmer.

Over FGH the situation is reversed. The southern hemisphere is tilted towards the Sun, while the northern hemisphere is tilted away from it and experiences autumn and winter.

At C the northern hemisphere has its longest day, while the southern hemisphere has its shortest, usually on 21 June. At G the opposite is true and occurs about 22 December.

At A and E night and day are equal in both hemispheres. These are the **equinoxes**, often 21 March and 23 September.

c) Rising and setting of the Sun

The Earth's rotation on its axis causes the Sun to have an apparent daily journey from east to west. It rises exactly in the east and sets exactly in the west only at the equinoxes. In the northern hemisphere in summer it rises north of east and sets north of west. In winter it rises and sets south of these points.

Each day the Sun is highest above the horizon at noon and directly due south in the northern hemisphere; this height itself is greatest and the daylight hours longest about 21 June. Thereafter the Sun's noon height slowly decreases and near 22 December it is lowest and the number of daylight hours is smallest, Figure 63.3.

In the southern hemisphere the Sun is due north at noon.

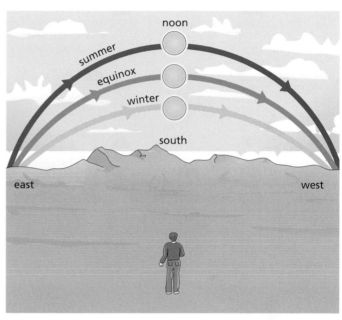

Figure 63.3 Rising and setting of the Sun (in the northern hemisphere)

Motion of the Moon

The Moon is a satellite of the Earth and travels round it in an approximately circular orbit once every 27.3 days at an average distance away of about 400 000 km or 240 000 miles. It also revolves on its own axis in 27.3 days and so always has the same side facing the Earth, hence we never see the 'dark side of the Moon'. We see the Moon by reflected sunlight since it does not produce its own light. It does not have an atmosphere. It does have a gravitational field due to its mass, but the field strength is only one-sixth of that on Earth. Hence the astronauts who walked on the Moon moved in a 'springy' fashion but did not fly off into space.

a) Phases of the Moon

The Moon's appearance from the Earth changes during its monthly journey; it has different phases. In Figure 63.4 the outer circle shows that exactly half of it is always illuminated by the Sun. What it looks like from the Earth in its various positions is shown inside this. In the New phase, the Moon is between the Sun and the Earth and the side facing the Earth, being unlit, is not visible from the Earth. A thin 'new' crescent appears along one edge as it travels in its orbit, gradually increasing until at the First Quarter phase, half of the Moon's face can be seen. At Full Moon it is on the opposite side of the Earth from the Sun and appears as a complete circle. Thereafter it wanes through the Last Quarter until only the 'old' crescent can be seen.

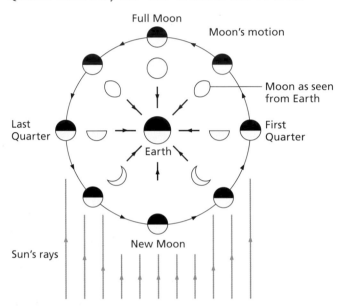

Figure 63.4 Phases of the Moon

Figure 63.5 The surface of the Moon partially illuminated as seen from Earth

b) Eclipse of the Sun

There is an eclipse of the Sun by the Moon (a **solar eclipse**) when the Sun, Moon and Earth are in a straight line. Solar eclipses can only occur at the New Moon phase. Anyone at B in Figure 63.6a sees a **total** eclipse of the Sun (i.e. they can't see the Sun at all). Anyone at A sees a **partial** eclipse (i.e. part of the Sun is still visible).

Sometimes the Moon is farther from the Earth (since its orbit is not a perfect circle), Figure 63.6b, and then while those at A would still see a partial eclipse, anyone at C sees an **annular** eclipse (i.e. only the central region of the Sun is hidden).

A total eclipse seen from one place may last for up to 7 minutes. During this time, although it is day, the sky is dark, stars are visible, the temperature falls and birds stop singing.

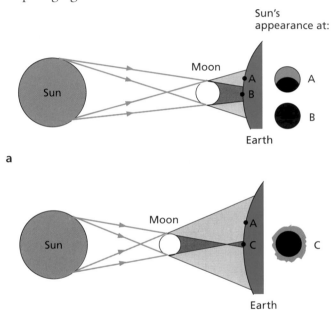

Figure 63.6 Eclipse of the Sun

c) Eclipse of the Moon

A **lunar eclipse** occurs when the Moon passes into the Earth's shadow, i.e. the Earth comes between the Sun and the Moon, Figure 63.7. Lunar eclipses can only occur at the Full Moon phase.

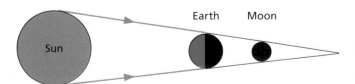

Figure 63.7 Eclipse of the Moon

d) Rising and setting of the Moon

Like the Sun, the Moon seems to have a daily trip across the sky from the east where it rises to the west where it sets, due to the Earth's rotation on its axis.

More about the planets

a) Data

Facts and figures about the Sun, Earth, Moon and the planets are listed in Tables 63.1 and 63.2; times are given in Earth hours (h), days (d) or years (y). As well as being of general interest this data indicates (i) factors that affect conditions on the surface of the planets, and (ii) some of the environmental problems that a visit or attempted colonization would encounter!

Table 63.1 Data for the Sun, Earth and Moon

	Mass /kg	Radius /m	Density /kg/m³	Surface gravity /N/kg	Orbital period
Sun	2.0×10^{30}	7.0×10^{8}	1410	274	
Earth	6.0×10^{24}	6.4×10^{6}	5520	9.8	365 days (around Sun)
Moon	7.4×10^{22}	1.7×10^{6}	3340	1.7	27 days (around Earth)

Table 63.2 Data for the planets

Planet	Av. distance from Sun /million km	Orbit time round Sun /days or yrs	Surface temp. /°C	Spin time about axis /hrs or days	Diameter /thousand km	Relative mass /Earth = 1.0	Surface gravity /N/kg	No. of moons
Mercury	58	88 d	350	58.5 d	4.8	0.05	3.6	0
Venus	108	225 d	460	243 d	12.2	0.81	8.7	0
Earth	150	365 d	20	23.9 h	12.8	1.0	9.8	1
Mars	228	687 d	−23	24.6 h	6.8	0.11	3.7	2
Jupiter	778	11.9 y	−120	10 h	143	318.0	25.9	16
Saturn	1430	29.5 y	−180	10.6 h	120	95.0	11.3	18
Uranus	2870	84 y	−210	17.2 h	51	14.0	10.4	17
Neptune	4500	165 y	−220	16 h	50	17.5	14.0	8
Pluto	5900	248 y	−230	6.3 d	2.3	0.003	?	1

b) Some features

A planet's year (i.e. orbit time round the Sun) increases with distance from the Sun. The orbital speed decreases with distance; Neptune travels much more slowly than Mercury. Surface temperatures decrease markedly with distance from the Sun, with one exception. **Venus** has a high surface temperature (460 °C) due to its dense atmosphere of carbon dioxide acting as a heat trap (i.e. the greenhouse effect). Its very slow 'spin time' of 243 Earth days means its day is longer than its year of 225 Earth days!

Mercury, also with a slow 'spin time' (58.5 days) has practically no atmosphere and so while its noon temperature is 350 °C, at night it falls to −170 °C. Its surfaces are also exposed to long periods of heat or cold because of its slow rotation.

Mars is the Earth's nearest neighbour. It is colder than the Earth, temperatures on its equator seldom exceeding 0 °C even in summer. Its atmosphere is very thin and consists mostly of carbon dioxide with traces of water vapour and oxygen. Its axis is tilted at an angle of 24° and so it has seasons but these are longer than on Earth. There is now no liquid water on the surface but it has polar ice-caps of water ice and solid carbon dioxide ('dry ice'). In some parts there are large extinct volcanoes and evidence such as gorges of torrential floods in the distant past, Figure 63.8a. It is now a comparatively inactive planet though high winds do blow at times causing dust storms.

Figure 63.8a The surface of Mars as seen by the Mars Global Surveyor, showing a vast canyon 6000 km long and layered rocks indicating a geologically active history

Jupiter is by far the largest planet in the Solar System. It is a gaseous planet and is noted for its Great Red Spot, Figure 63.8b, which is a massive swirling storm that has been visible from Earth for over 200 years.

Figure 63.8b Jupiter's Great Red Spot pictured by the Voyager 2 space probe

Saturn has rings made up of ice particles, Figure 63.8c, which are clearly visible through a telescope. Like Jupiter, it has an ever-changing very turbulent atmosphere of hydrogen, helium, ammonia and methane gas.

Uranus and **Neptune** have methane in their atmosphere as well as hydrogen and helium. The four outer 'gas giants' are able to retain these lighter gases in their atmospheres, unlike the Earth, because of the greater gravitational attraction their large masses exert.

The **Voyager** and other unmanned space missions have obtained much new information about the outer planets and their many moons as they flew past them. Uranus and Neptune were found to have rings, but not so large as Saturn's. Signs of geological activity on the planets can be detected by direct observations and, in some cases, from the emission of infrared radiation.

Figure 63.8c Saturn's rings as seen by the Voyager 1 probe

Less is known about **Pluto**. It is believed to be smaller than our Moon and to have an 'atmosphere' of frozen methane. It has its own moon, Charon.

Observing the planets

In addition to the Sun and Moon, five other fairly bright objects can be seen without a telescope moving among the stars. They are the planets (or 'wanderers') Mercury, Venus, Mars, Jupiter and Saturn.

Like the Sun and Moon, the planets seem to rise in the east and set in the west but sometimes their movements appear to lack the regularity associated with other heavenly bodies and were a puzzle to early astronomers. Their positions against the background of the stars depend on where they and the Earth are as they orbit the Sun.

Venus and Jupiter are the two brightest planets. Venus appears periodically as a very bright evening 'star'. It is first visible just after sunset, close to the Sun, and sets shortly after sunset. It gradually moves eastwards from the Sun on subsequent days and sets later in the evening, Figure 63.9a. Some time later it appears as a morning star, rising at first just before the Sun and then earlier from day to day as its westward motion away from the Sun increases, Figure 63.9b. Jupiter is visible for several months each year.

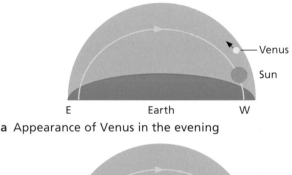

a Appearance of Venus in the evening

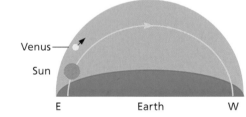

b Appearance of Venus in the morning

Figure 63.9

Mars is outshone only by Venus and Jupiter and is visible from Earth as an orange-red planet for several months every year.

Mercury is difficult to see from the Earth because it is close to the Sun. It appears briefly, low in the sky after sunset and before dawn.

Maps of the sky are published monthly in some national newspapers and show where to look for different planets. Using even a small telescope or binoculars improves the details that can be seen, e.g. Jupiter's moons and Saturn's rings.

Warning: Looking directly at the Sun through a telescope or binoculars can cause blindness. The only safe way to view the Sun is to project its image through a telescope onto a piece of card.

Gravity and satellites

Newton proposed that all objects in the Universe having mass attracted each other with a force called gravity. The greater the mass of each object and the smaller their distance apart, the greater is the force. In fact, halving the distance quadruples the force. Earlier we saw that it is the force of gravity arising from the mass of the Earth that causes a body to fall with an acceleration $g = 9.8\,\text{m/s}^2$ (Chapter 30). The Moon has a smaller mass than the Earth and on it $g = 1.6\,\text{m/s}^2$.

To keep a body moving in a circular path requires a centripetal force, $F = mv^2/r$, as we saw previously (Chapter 34). The centripetal force needed increases if

(i) the mass m of the body increases,
(ii) the speed v of the body increases, and
(iii) the radius r of the orbit decreases.

In the case of the planets orbiting the Sun in near-circular paths, it is the force of gravity between the planet and the Sun which provides the necessary centripetal force. The Moon is similarly kept in a circular orbit round the Earth by the force of gravity between it and the Earth. The further a planet is from the Sun, the lower its speed and the longer the time it takes to make a complete orbit.

Similarly, a high-level geostationary communication satellite at 36 000 km experiences a smaller gravitational force than a low-level polar one orbiting at 850 km (Chapter 34) since it is at a greater distance from the Earth. Its speed is less, so it takes longer to make a complete orbit of the Earth, 24 hours compared with 100 minutes.

Space programmes

There have been many beneficial spin-offs from space programmes. They have been the reason for much of the miniaturization of electronic components and systems, and they have led to the development of new materials. The technology has enabled satellites to be put into orbit to assist air and sea navigation, to improve the reliability of worldwide telecommunications, to make weather forecasting more accurate, to allow astronomical observations to be made unaffected by the Earth's atmosphere and to permit surveillance and monitoring of the Earth's surface for a variety of reasons.

In the realm of space travel one of the most notable feats occurred on 20 July 1969 when the American lunarnauts Edwin Aldrin and Neil Armstrong landed on the Moon with their Lunar Excursion Module.

The exploration of space, even by unmanned vehicles, is very costly and the benefits have to be weighed against this. Huge amounts of fuel are needed for rockets and their payloads to leave the Earth and reach the speeds required on their long journeys.

Limitations on manned space travel are set by the unfriendliness of the environments of possible destinations and by the difficulties of maintaining the conditions to sustain life, such as the supply of oxygen, water, food, warmth and sanitation. Despite these difficulties and the problems of existing in zero gravity, the Russian cosmonaut Yuri Romanenko spent a record 326 days in the Mir space station during 1987. The international space station currently under construction (see Figure 2a, p. x) will provide better facilities for crews spending extended periods in space.

Exploration of other planets in the Solar System has been achieved by unmanned space probes and vehicles. In particular the robot Sojourner, sent to Mars on the Pathfinder mission, Figure 63.10, roved the surface for several weeks during 1997 sending back extensive data on Martian rocks, and more recently the Mars Global Surveyor satellite has mapped the surface of that planet in extraordinary detail (see Figure 63.8a).

Figure 63.10 Sojourner on the surface of Mars

64 Stars and the Universe

The night sky

The night sky has been an object of wonder and study since the earliest times. On a practical level it provided our ancestors with a calendar, a clock and a compass. On a theoretical level it raised questions about the origin and nature of the Universe and its future. It is only in the last 100 years or so that some progress has been made in finding answers and making sense of what we see.

Although the stars may seem close-by on a dark night, in fact the distances involved are mind-boggling. So much so that we need a new unit of length, the **light-year** (l.y.). This is the distance travelled by light in one year and equals about 10 million million kilometres (10^{13} km) or 6 million million miles. The star nearest to the Solar System is Alpha Centauri, 4.3 l.y. away, which means the light arriving from it at the Earth today left 4.3 years ago. The Pole Star is 142 l.y. from Earth. The whole of the night sky we see is past history and we will have to wait a long time to find out what is happening out there right now.

Galaxies

A **galaxy** is a large collection of stars which, like our Sun, all produce their own light. Millions of galaxies make up the whole **Universe**.

As well as containing stars, galaxies consist of clouds of gas, mostly hydrogen, and dust. They move in space, many rotating as spiral discs like huge Catherine wheels with a dense central bulge.

The **Milky Way**, part of the spiral galaxy (see Figure 1a, p. vi) to which our Solar System belongs, can be seen on dark nights as a narrow band of light spread across the sky, Figure 64.1a. We are near the outer edge of the galaxy, so what we see when we look at the Milky Way is the galaxy's central bulge. Figure 64.1b is an infrared photograph of the galaxy, taken from beyond the Earth's atmosphere, clearly showing the central bulge.

Figure 64.1a The Milky Way from Earth

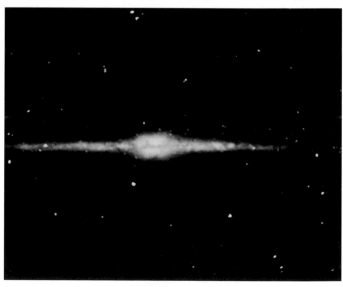

Figure 64.1b An image of the Milky Way from a space probe (the Cosmic Background Explorer)

Galaxies travel in groups and the nearest spiral in our **Local Cluster** is the **Andromeda** galaxy. It is 2.2 million light-years away and visible to the naked eye.

The stars in a galaxy are often millions of times further apart than the planets in the Solar System but millions of times closer than are neighbouring galaxies.

Constellations

On a good night when there is no Moon and the air is clear, although the Earth seems to have an enormous star-spangled dome overhead, only up to 4000 or so stars can be seen by the unaided eye.

From ancient times star-gazers have tried to bring some order into the sky by fitting the stars into patterns, the **constellations**, to which names were given. Some of the major constellations and stars that are seen from the northern hemisphere in winter, looking north and south, are shown in Figures 64.2a and b respectively, The pale bands indicate the Milky Way.

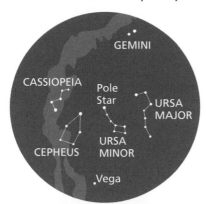

a Looking north

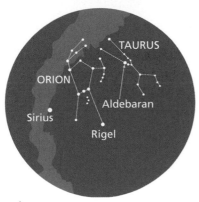

b Looking south

Figure 64.2 Major constellations seen from the northern hemisphere in winter

Ursa Major (or the **Plough**) is one of the most well-known constellations in the northern sky. It consists of seven stars in the shape of a plough, two of these pointing towards the Pole Star which is used in navigation to find north. Opposite the Plough is the W-shaped constellation **Cassiopeia**.

Orion (or the **Hunter**) is one of the most easily recognized constellations in the southern sky on winter nights. He is holding a club (a distorted Y of five stars) aloft in one hand and a shield (a curved line of five stars) in the other hand towards **Taurus** (the **Bull**), situated above and to his right. Orion's belt of three bright stars (from which hangs his dagger of three less bright stars) points towards **Sirius** (the **Dog Star**), the brightest star in the sky. Taurus represents the head of a bull with the large star **Aldebaran** forming one of its eyes.

Motion of the stars

Observation of the night sky at intervals of a few hours shows that in the northern hemisphere the stars appear to revolve anticlockwise about a point near the Pole Star, Figure 64.3a. As they revolve the stars keep the same positions in relation to each other, and in fact the appearance of the constellations has changed very little over the centuries.

Like the Sun and Moon, the stars have a daily journey across the sky, rising upwards in the east and setting in the west – due to the Earth's daily spin about its axis. The **circumpolar** stars, i.e. those near the Pole Star, are an exception to this; they are so close to the pole that they never disappear below the horizon. The trails of the circumpolar stars during a 3-hour period are shown in Figure 64.3b; the brightest trail near the centre is that of the Pole Star.

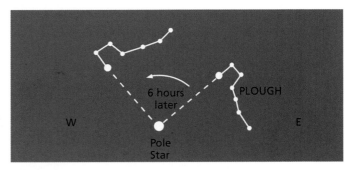

Figure 64.3a Stars appear to revolve about a point close to the Pole Star

Figure 64.3b Trails of the circumpolar stars in a 3-hour period

Although the stars always rise in the same *places* they do not do so at the same *times* every day. They rise about 4 minutes earlier each day since the Earth is progressing in its orbit around the Sun; in almost but not exactly a year they again rise and set at the same time. If a certain star is seen to rise in the east exactly when the Sun sets on a particular day, a few weeks later it will be seen well above the eastern horizon at sunset. Thus the stars in Orion rise in midwinter just after sunset and set before sunrise; by March, they are high in the heavens at sunset and set about midnight. Or again, a circumpolar constellation such as Cassiopeia seen at midnight directly above the Pole Star, will appear directly below it at the *same time* six months later.

331

■ *Life cycle of stars*

A star is a ball of hot hydrogen and helium gas in which nuclear fusion results in the production and emission of electromagnetic radiation.

Stars vary in age, size, mass, surface temperature, colour and brightness. Colour and brightness both depend on surface temperature which in turn increases with the mass of the star. Stars that are blue or white are hotter and brighter (surface temperatures of 6000 to 25 000 °C) than those that are yellow or red (3000 to 6000 °C).

a) Origin of stars

Stars are thought to form when gravitational attraction pulls together clouds of hydrogen gas and dust (called **nebulae**) in regions of space where their density is greater. As the mass of the star increases its core temperature rises as the gravitational p.e. of the gas molecules is converted to k.e. When the hydrogen is hot enough nuclear fusion occurs. If the 'young' star has a very large mass (more than twenty times that of the Sun) it forms a blue or white star. If it has a smaller mass (about equal to that of the Sun) if forms a yellow or red star and this is more common.

b) Yellow dwarf star

When a star is of this type, as our Sun is at present, the very strong forces of gravity pulling it together are balanced by opposing forces trying to make it expand due to its extremely high temperature (this 'thermal pressure' arises from the k.e. of the nuclei in the core). The star is then in a stable state which may last for about 10 000 million years, during which time the hydrogen in its core is changed to helium.

When there is no more hydrogen left, the star becomes unstable, expands, cools and turns into a **red giant**. (In the case of our Sun, which is halfway through its stable state, this stage would be reached in about 5000 million years from now. The inner planets Mercury, Venus and Earth would be devoured and all forms of life destroyed.) Later in its life it contracts under its own gravity to become a **white dwarf** in which matter may be millions of times denser than any on Earth. White dwarfs cool to **red dwarfs** and finally, after a very long time, to cold **black dwarfs**, Figure 64.4a.

c) Blue star

This type has a much shorter stable stage of about 100 million years. When its hydrogen is used up it expands to a **blue supergiant** and then cools down to a **red supergiant**. It may then become unstable and explode forming a **supernova**, Figure 64.4b. During the explosion there is a huge increase in its brightness, and gas and dust from its outer layers are thrown into space.

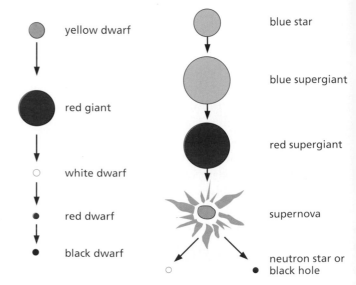

a Small mass   **b** Large mass

Figure 64.4 Life cycle of stars

The Crab Nebula is the remains of the supernova seen by Chinese astronomers on Earth in 1054. It is visible through a telescope as a hazy glow in the constellation Taurus. Figure 64.5 is an image of the Crab Nebula taken by the Hubble Space Telescope.

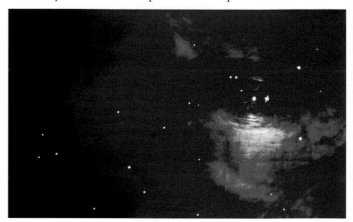

Figure 64.5 The Crab Nebula

The centre of the supernova collapses to a very dense **neutron star**, which spins rapidly and acts as a **pulsar**, sending out pulses of radio waves. If the supergiant is very massive, the remnant at the centre of the supernova has such a large density that its gravitational force stops anything escaping from its surface, even light. This is a **black hole**. In a black hole matter is packed so densely that the mass of the Earth would only occupy the volume of one cubic centimetre! Since neither matter nor radiation can escape from a black hole we cannot see it directly. However if nearby material, such as gas from a neighbouring star, falls towards a black hole, intense X-ray radiation may be emitted which alerts us to its presence. Objects believed to be massive black holes have also been identified at the centre of many spiral galaxies; in some they appear to power very luminous objects, while in others they appear to be 'dormant'.

Origin of the Solar System

While the Sun was probably formed like other stars, as described previously, the formation of the planets is not so clear. One theory is that the Solar System was all formed at the same time, about 5000 million years ago, and the planets were created when a disc of matter was pushed out from the centre of the cloud of hydrogen gas that produced the Sun. Evidence for this as the approximate age of the Earth comes from the radioactive dating of minerals in rocks (Chapter 58). Further confirmation was provided by rocks brought back from the Moon, found to be about 4500 million years old.

This view that the whole Solar System was formed at the same time is supported by the fact that the orbits of the planets are more or less in the same plane and all revolve round the Sun in the same direction. Neither of these are likely to have happened by chance.

To account for the existence of heavier chemical elements in the Sun and inner planets, it is thought these might have come from an exploding supernova. During the lifetime of a star, atoms of hydrogen, helium and other light elements are changed into atoms of heavier elements. It is possible that matter from previous stellar explosions could have mixed with interstellar hydrogen before our Sun and the planets formed, making our Sun a 'second generation' star.

Theories of the Universe

a) The expanding Universe

In developing a theory about the origin of the Universe two discoveries about galaxies have to be taken into account. The first is that light from other galaxies is 'shifted' to the red end of the spectrum. The second is that the farther away a galaxy is from us, the greater is this **red-shift**. The shift is detected by noting the positions of certain wavelengths due to a known element in the star's spectrum compared with their positions in a laboratory-produced spectrum of the element.

Red-shift can be explained if other galaxies are moving away from us very rapidly, and the farther away they are, the faster their speed of recession. That is, if the Universe is expanding. This explanation is based on the Doppler effect, which occurs when a source of waves is moving. If the source approaches us the waves are crowded into a smaller space and their wavelength appears smaller and their frequency greater; if the source moves away, the wavelength seems larger, Figure 64.6. It occurs with sound waves and explains the rise and fall of pitch of a siren as the vehicle approaches and passes us. The same effect is shown by light. When the light source is receding the wavelength seems longer, i.e. the light is redder. From the size of the red-shift, the speed of recession of a galaxy can be calculated; the most distant ones visible are receding with speeds up to one-third that of light.

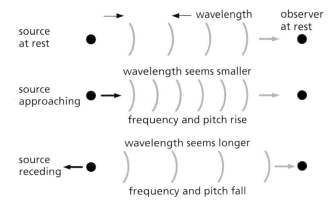

Figure 64.6 The Doppler effect

b) The Big Bang theory

If the galaxies are receding from each other, it follows that in the past they must have been closer together. It is therefore possible that initially all the matter in the Universe was packed together in an extremely dense state. The Big Bang theory proposes that this was the case and that the Universe started about 14 billion years ago from one place with a huge explosion – the Big Bang.

The resulting expansion continues today but predictions vary as to what will happen in the future. One is that the expansion will go on. Another is that the expansion will be succeeded by contraction to a dense state, followed by expansion again and so on, leading to a pulsating Universe. The rate of expansion appears to be decreasing due to the gravitational attraction between galaxies. If the mass (or more strictly the density) of the Universe equals a critical value then gravitational forces will be sufficient to eventually halt the expansion; if the mass exceeds the critical value the Universe will then contract, but if the mass is less than the critical value the Universe will continue to expand for ever. Scientists are not sure how much matter the Universe contains; much of it is invisible in the form of 'dark matter' (such as black holes) which does not emit radiation but does exert a gravitational force and it may be that this hidden mass is sufficient to lead the Universe to collapse. The gravitational force between masses not only determines the motion and the evolution of planets, stars and galaxies but will also control the ultimate fate of the Universe.

c) Other theories

The Steady State or Continuous Creation theory was a popular alternative theory in the 1960s but became less favoured as evidence for the Big Bang theory emerged. It proposed that new matter (hydrogen) is created all the time to fill up the empty space arising from the expansion of the Universe. In this case the Universe would not change and would always 'look' the same.

The whole question of how the Universe originated and what happened before it formed is one that may remain an impenetrable mystery, perhaps beyond our understanding. Nevertheless it never fails to fascinate.

Age of the Universe

Hubble, an American astronomer, discovered that the speed of recession v of a galaxy is directly proportional to its distance away s. This is called **Hubble's law** and can be written

$$v \propto s \quad \text{or} \quad v = H_0 \times s$$

where H_0 is the **Hubble constant**. Its value is found by measuring the speeds of recession of galaxies (from red-shifts) whose distances away are known from other astronomical data. Considerable difficulties are involved, but H_0 is estimated to be 23 km per second per million light-years. This means that a galaxy is receding at 23 km/s for every million light-years of distance from us. One that is 2 million light-years away will be receding at 46 km/s. The further away a galaxy is from us the faster it is receding.

The age of the Universe can be shown to equal $1/H_0$ and can be calculated very roughly. We assume that the time for which two galaxies were close together at the Big Bang when they were formed is negligible compared with their present ages, i.e. most of their lifetimes have been spent apart. Their ages are thus approximately the age of the Universe, which equals the time since expansion began.

For the two galaxies, assuming the speed of recession has not changed, we have

$$\text{age} = \frac{\text{distance apart}}{\text{speed of recession}} = \frac{s}{v}$$

From Hubble's law, it follows that

$$\text{age} = \frac{1}{H_0} = \frac{1}{23 \, \text{km/s/million light-years}}$$

$$= \frac{10^{19} \, \text{km}}{23 \, \text{km/s}} \quad (\text{since } 1 \, \text{l.y.} = 10^{13} \, \text{km})$$

$$= 4.3 \times 10^{17} \, \text{s}$$

But $1 \, \text{year} \approx 3.2 \times 10^7 \, \text{s}$

$$\therefore \quad \text{age of Universe} \approx \frac{4.3 \times 10^{17} \, \text{s}}{3.2 \times 10^7}$$

$$\approx 1.4 \times 10^{10} \, \text{y}$$

$$\approx 14\,000 \, \text{million years}$$

This result has an uncertainty of ± 4000 million years; it could be between 10 000 and 18 000 million years.

Extra-terrestrial life

Our exploration of space has not yet led to the discovery of life on any of the other planets or moons in our Solar System. The most promising conditions for living species to develop are thought to be those where a source of energy and water are available; there are indications that there was once water on Mars (see p. 325) and water may also be present beneath the icy surface of Jupiter's moon Europa, Figure 64.7. Astronomers are currently trying to locate other stars that have planets circling them; several distant solar systems have now been identified so perhaps an environment as friendly to life as the Earth's exists elsewhere in the galaxy.

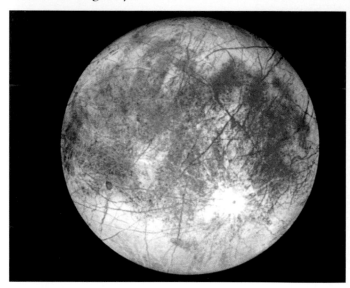

Figure 64.7 Could life exist in a subterranean ocean on Europa?

The search for living organisms, micro-organisms, or their fossilized remains, can be tackled in a number of different ways.

(i) Manned space craft could be landed on the planet or moon in our Solar System and a search made for signs of life.

(ii) Unmanned probes could be sent further into space and robots used to transmit photographs or collect specimens to bring back to Earth.

(iii) Chemical analysis of the gases in the atmosphere of a planet could reveal the presence of life, since living organisms can change the chemical composition of an atmosphere (there is much more oxygen in the Earth's atmosphere than is derived from geological and chemical processes).

(iv) The SETI project (Search for Extra-Terrestrial Intelligence) has recruited people with home computers around the world to help in the analysis of radio signals from distant galaxies; they are hoping to find non-random signals in a narrow band of wavelengths which could be coming from intelligent life elsewhere in the Universe.

Questions

1 Which of the following statements are *true*?

A A galaxy is a large collection of stars.
B The Universe is a collection of galaxies.
C The Solar System belongs to the Andromeda galaxy.
D A light-year is the distance travelled by light in 1 year.
E The Pole Star is the brightest star in the sky.

2 a Why is the Pole Star useful for finding direction?
b How does the Plough help in finding the Pole Star?
c If the Plough is in the position shown in Figure 64.8 at 6 p.m. on a certain evening, draw two diagrams to show where it will be
  (i) at midnight on the same evening,
  (ii) at 6 p.m. 6 months later.

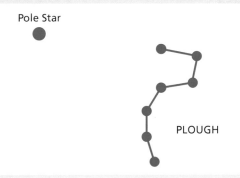

Figure 64.8

3 a The Sun is at the stable stage of its life. Explain, in terms of the forces acting on the Sun, what this means.
b At the end of the stable stage of its life a star will change. Describe and explain the changes that could take place.

(AQA (NEAB) Higher, June 99)

4 Figure 64.9 shows the orbit time and average distance from the Sun of some planets and some asteroids. Asteroids are sometimes called minor planets.

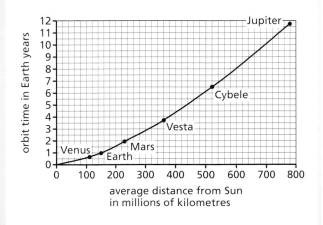

Figure 64.9

a Suggest why Jupiter takes longer than Mars to orbit the Sun.
b The asteroid Ida is in orbit at an average distance of 430 million km from the Sun. Use the graph to find out how long it takes to orbit the Sun.
c The orbit of Ida about the Sun is, in fact, elliptical. This means its speed varies during the orbit, like a comet.

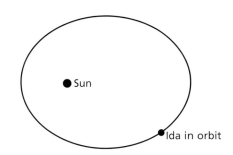

Figure 64.10

  (i) Copy Figure 64.10 and write an X to mark the place where Ida will be travelling at its highest speed.
  (ii) Explain why it will be travelling at its highest speed at this place.
d The light from stars in distant galaxies is observed to be **red shifted**. Measurements of red shift allow astronomers to calculate the speed that the galaxies are travelling **away** from us. The graph in Figure 64.11 shows how the speed that galaxies move away from us varies with their distance from us.

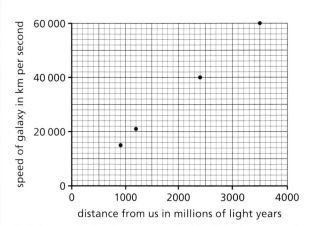

(a light year is the distance travelled by light in one year)

Figure 64.11

  (i) What is meant by **red shift**?
  (ii) What type of shift would be observed in the light from stars travelling **towards** us?

b The diagram below shows the way a building vibrates when an earth tremor is first felt. The vibrations are caused by P and S waves (seismic waves) which travel through the mantle.

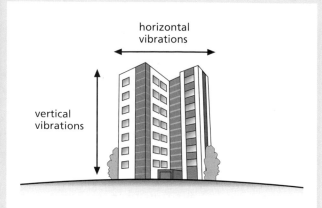

horizontal vibrations

vertical vibrations

The building vibrates vertically at first. A very short time later it begins to vibrate horizontally as well. Explain why.

c Information about the Earth's structure has been obtained by studying the shock (seismic) waves produced by earthquakes. Diagrams A, B and C below show the paths of seismic waves through the mantle and the core.

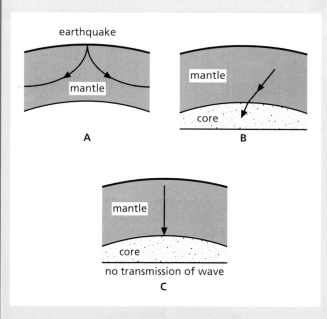

earthquake

mantle

A

mantle

core

B

mantle

core

no transmission of wave

C

(i) Describe and explain what happens to both types of wave travelling through the mantle (diagram A).

(ii) Describe and explain what happens to the wave travelling from the mantle to the core (diagram B). State which type of wave is shown.

(iii) Explain why the wave shown in diagram C does not travel through the core. State which type of wave is shown.

(*AQA (NEAB) Higher, June 98*)

The Solar System; stars and the Universe

4 Explain the following facts.
  a A year on Mercury is less than one on Earth.
  b The only planets to have phases like the Moon are Venus and Mercury.
  c The nights on Mars are very cold.
  d The four giant outer planets have atmospheres containing hydrogen gas but the Earth does not.

5 Some students observed the sky at night for several weeks. Apart from the Moon, several bright objects were seen. Some of these are shown in the diagram.

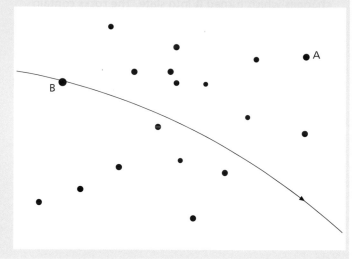

A is one of many bright objects which stay in a fixed pattern.

B is an object which moves slowly across the fixed pattern.

C is an object, not shown in the diagram, which moves quickly across the sky several times in one night.

a Choose words from the list below to name A, B and C.

planet   satellite   star   the Sun

b What is seen in the night sky which
  (i) gives out its own light,
  (ii) reflects light from the Sun,
  (iii) moves round (orbits) the Earth?
(*NEAB Foundation, June 98*)

6 a Complete each sentence by choosing the correct word or phrase from the box. Each word or phrase should be used once or not at all.

milky way	moon	planet
solar system	star	universe

The Sun is the nearest ...... to the Earth.

The Sun is in the galaxy called the ....

Within the ...... there are millions of galaxies.

Pluto is orbited by one ....

b The diagram shows the path taken by the Voyager 2 spacecraft.

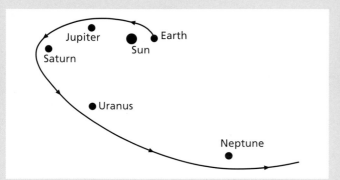

Choose the force from the box which caused the spacecraft to change direction each time it got close to a planet.

air resistance	friction	gravity

(*AQA (SEG) Foundation, Summer 99*)

7 Outline the evidence for:
a the view that all members of the Solar System were formed about the same time,
b the expanding Universe.

8 Three space vehicles, P (mass m), Q (mass m) and R (mass $2m$) are launched into circular orbits above the Earth's atmosphere as shown in the diagram.

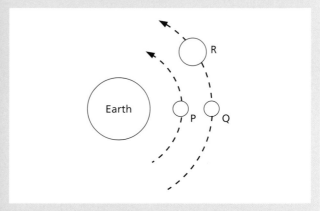

a Why must Q be moving more slowly than P, although it has the same mass?
b Why must Q and R have the same speed, although they have different masses?

9 a State Hubble's law.
b Write down the equation connecting the age of the Universe to the Hubble constant H_0.
c Some astronomers believe that the true value of H_0 is 30 km/s/million light-years. Using this value, calculate the approximate age of the Universe. (1 light-year $= 10^{19}$ km; 1 year $= 3.2 \times 10^7$ s)

Revision questions

Light and sight

1 The image of a window on the screen of a pinhole camera is

A virtual, inverted and larger
B virtual, upright and smaller
C real, upright and larger
D real, inverted and larger
E real, inverted and smaller.

2 In the diagram a ray of light is shown reflected at a plane mirror. What is
a the angle of incidence,
b the angle the reflected ray makes *with the mirror*?

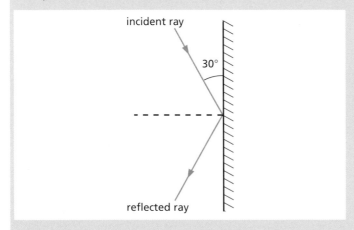
incident ray
30°
reflected ray

3 In the diagram below a ray of light IO changes direction as it enters glass from air.
a What name is given to this effect?
b Which line is the normal?
c Is the ray bent towards or away from the normal in the glass?

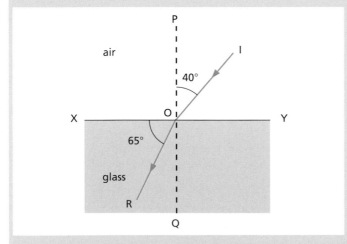
P
air
40°
X ——— O ——— Y
65°
glass
R
Q

d What is the value of the angle of incidence in air?
e What is the value of the angle of refraction in glass?

4 In the diagram below which of the rays A to E is most likely to represent the ray emerging from the parallel-sided sheet of glass?

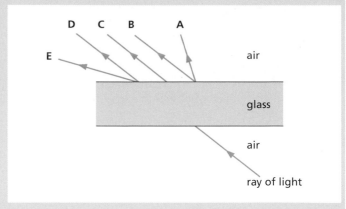
D C B A
E
air
glass
air
ray of light

5 If the critical angle for glass is 42°, which diagram A to E below does *not* represent the behaviour of a ray of light falling on a glass–air boundary?

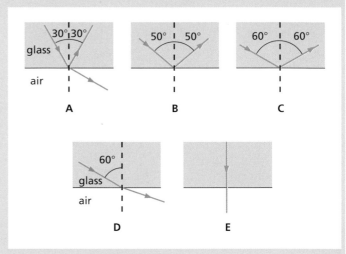
30° 30°
glass
air
A
50° 50°
B
60° 60°
C
60°
glass
air
D
E

6 A ray of light is shown passing through the upper prism of a prismatic periscope. In which position A to E must the lower prism be placed for someone at X to use it?

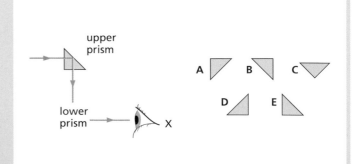
upper prism
lower prism
X
A B C
D E

7 A convex lens produces a magnified, inverted image of an object if the distance of the object from the lens is

A greater than two focal lengths
B equal to two focal lengths
C between one and two focal lengths
D equal to one focal length
E less than one focal length.

8 The eye forms an image on the retina which

1 is inverted
2 is focused by the lens changing shape
3 has its brightness controlled by the size of the pupil.

Which statement(s) is (are) correct?

A 1, 2, 3   **B** 1, 2   **C** 2, 3   **D** 1   **E** 3

9 A narrow beam of white light is shown passing through a glass prism and forming a spectrum on a screen.
   a What is the effect called?
   b Which colour of light appears at (i) A, (ii) B?

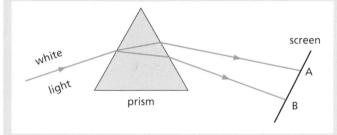

10 When using a magnifying glass to see a small object

1 an upright image is seen
2 the object should be less than one focal length away
3 a real image is seen.

Which statement(s) is (are) correct?

A 1, 2, 3   **B** 1, 2   **C** 2, 3   **D** 1   **E** 3

11 a Why is a concave mirror used as a make-up mirror?
   b Why is a convex mirror used as a driving mirror?
   c Why is a concave parabolic mirror used as
      (i) a car headlamp reflector,
      (ii) a dish aerial?

12 The diagram below shows a cross-section of an eye.

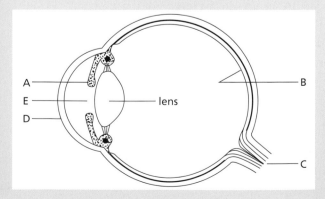

a Label the parts marked A, B, C, D and E.
b An object OB is placed 12 cm in front of a converging lens of focal length 9 cm. The diagram below is drawn to scale.

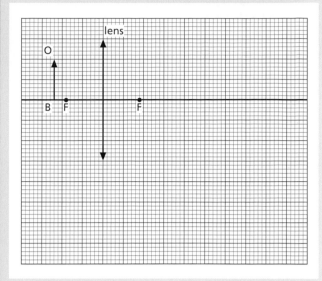

(i) Copy the diagram on graph paper and draw the ray diagram to show the position and size of the image. Draw and label the image.
(ii) Write down two ways in which the image is different from the object.

c Projectors and cameras use converging lenses to produce an image of an object. Write down two ways in which the image produced differs from the object when a converging lens is used in
(i) a projector,
(ii) a camera.

(NEAB Higher, June 98)

Waves and sound

13 In the transverse wave shown below distances are in centimetres. Which pair of entries **A** to **E** is correct?

	A	**B**	**C**	**D**	**E**
Amplitude	2	4	4	8	8
Wavelength	4	4	8	8	12

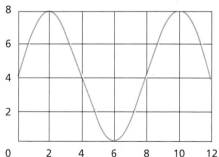

14 When water waves go from deep to shallow water, the changes (if any) in its speed, wavelength and frequency are

	Speed	Wavelength	Frequency
A	greater	greater	the same
B	greater	less	less
C	the same	less	greater
D	less	the same	less
E	less	less	the same

341

15 When the straight water waves in the diagram pass through the narrow gap in the barrier they are diffracted. What changes (if any) occur in
a the shape of the waves,
b the speed of the waves,
c the wavelength?

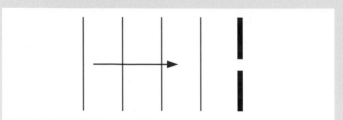

16 a When two sets of waves arrive at the same point at the same time, under what conditions do they
(i) completely cancel out,
(ii) produce a larger wave?
b If A and B below are two dippers vibrating together (in phase) in a ripple tank, what happens
(i) at P if PA = PB,
(ii) at Q if QB − QA = half a wavelength,
(iii) at R if RB − RA = one wavelength?

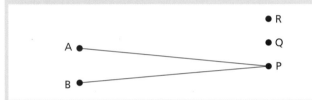

17 Sometimes 'light + light = darkness'. This behaviour of light is called

A reflection   B refraction   C dispersion
D interference   E deviation.

18 Which one of the following is not electromagnetic waves?

A infrared   B gamma rays   C ultraviolet
D X-rays   E sound

19 Compared to radio waves, the wavelength, frequency and speed of light are

	Wavelength	Frequency	Speed
A	greater	smaller	the same
B	greater	smaller	greater
C	greater	greater	the same
D	smaller	greater	smaller
E	smaller	greater	the same

20 The wave travelling along the spring in the diagram is produced by someone moving end X of the spring to and fro in the directions shown by the arrows.
a Is the wave longitudinal or transverse?
b What is the region called where the coils of the spring are (i) closer together, (ii) farther apart, than normal?

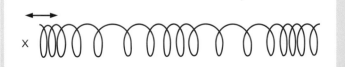

21 Sound and light waves

A travel through a vacuum
B travel as longitudinal waves
C travel with the same speed in air
D can be diffracted
E have similar wavelengths.

22 The signal sent out by a sonar echo sounder in a ship is received back from the sea bed directly below the ship (see diagram) 2 seconds later. The speed of sound in sea water, in m/s, is

A 750   B 1500   C 2250   D 3000   E 3750

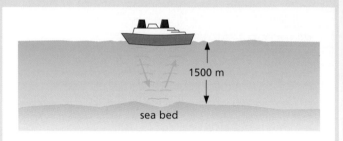

sea bed

23 If a note played on a piano has the same pitch as one played on a guitar, they have the same

A frequency   B amplitude   C quality
D loudness   E harmonics.

24 The waveforms of two notes P and Q are shown below. Which one of the statements **A** to **E** is true?

A P has a higher pitch than Q and is not so loud.
B P has a higher pitch than Q and is louder.
C P and Q have the same pitch and loudness.
D P has a lower pitch than Q and is not so loud.
E P has a lower pitch than Q and is louder.

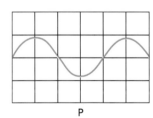
P

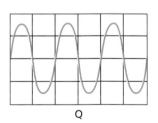
Q

25 The loudness of the note produced by a vibrating wire can be increased by

A increasing the length of the wire
B decreasing the length of the wire
C increasing the amplitude of vibration
D increasing the tension of the wire
E decreasing the tension of the wire.

26 Examples of transverse waves are

1 water waves in a ripple tank
2 all electromagnetic waves
3 sound waves.

Which statement(s) is (are) correct?

A 1, 2, 3   B 1, 2   C 2, 3   D 1   E 3

27 Notes of the same pitch played on the guitar and the clarinet sound different because

1 different overtones are present in each case
2 the fundamental frequencies are different
3 a guitar string vibrates transversely and the air column in a clarinet vibrates longitudinally.

Which reason(s) is (are) correct?

A 1, 2, 3   **B** 1, 2   **C** 2, 3   **D** 1   **E** 3

28 **a** An endoscope is an instrument used by doctors for looking inside patients. A bundle of thin optical fibres pass light into the patient's body, a second bundle of fibres carry reflected light back to the doctor.

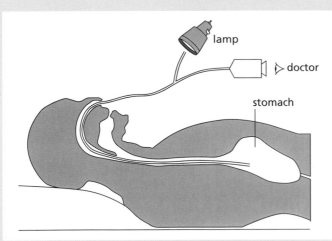

(i) Copy and complete the diagram below to show how an optical fibre is able to pass light into a patient's body.

(ii) Give *one* advantage of using lots of thin fibres to make the bundles, rather than a few thick fibres.
(iii) Give *one* further example of the practical use of an optical fibre.
b The diagram shows a wave travelling through a stretched spring.

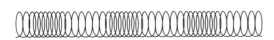

In what way is this wave the same as a sound wave?
c Sound waves travel faster in liquids than in gases. Why?
d A bat uses ultrasound to find its way around. Explain how.

(AQA (SEG) Foundation, Summer 99)

29 When prospecting for oil geologists may cause a small explosion at the surface of the Earth. The distance to possible oil-bearing rocks can then be found by recording how long the sound wave reflected from the rocks takes to return to the surface.

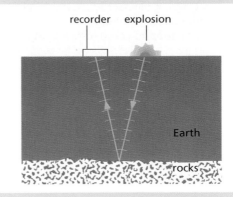

In an exploration 3.6 s is recorded between the explosion and the return of the reflected wave. Taking the average speed of sound through the Earth as 1800 m/s, the rock causing the reflection is at a depth in metres of

A 1800×3.6

B $1800/3.6$

C $\dfrac{1800 \times 3.6}{2}$

D $\dfrac{1800 \times 2}{3.6}$

E $1800 \times 3.6 \times 2$

Matter and molecules

30 The basic SI units of mass, length and time are

	Mass	Length	Time
A	kilogram	kilometre	second
B	gram	centimetre	minute
C	kilogram	centimetre	second
D	gram	centimetre	second
E	kilogram	metre	second

31 Density can be calculated from the expression

A mass/volume
B mass × volume
C volume/mass
D weight/area
E area × weight

32 A cube of density 1000 kg/m³ and side of length 2 m has a mass in kg of

A 2000   **B** 4000   **C** 8000   **D** 16 000
E 32 000

33 Which of the following properties are the same for an object on Earth and on the Moon?

1 weight   **2** mass   **3** density

Use the answer code:

A 1, 2, 3   **B** 1, 2   **C** 2, 3   **D** 1   **E** 3

34 Diffusion occurs more quickly in a gas than in a liquid because

A molecules in a gas have more frequent collisions than molecules in a liquid
B gas molecules are larger
C gas molecules move randomly
D on average, molecules in a gas are further apart than molecules in a liquid
E heat is needed to cause diffusion in a liquid.

35 Jerry makes a spring balance (newton-meter) for weighing fish. She uses a stiff spring and parts of a bicycle pump. She puts a scale in newtons on the rod.

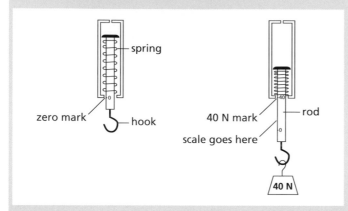

a (i) What happens to the spring when the 40 N weight is put on the hook?
(ii) Suggest why the spring balance would not be suitable for weighing up to 80 N.

b Jerry loads the spring balance. Each time she adds another 5 newtons to the weight, she makes another scale mark on the rod.

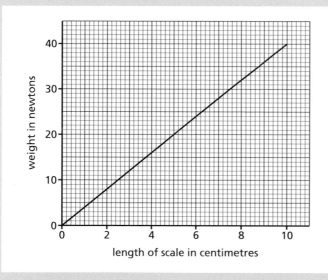

The graph shows how adding weights increases the length of the scale.
(i) How can you tell from the graph that the gaps between the marks on the scale are equal?
(ii) Use the graph to find how far along the scale the 30 N mark is.
(iii) Which word shows the correct link between the **weight** and the **length of the scale**?

compressed   constant   parallel   proportional

c (i) Use the graph to find the value of the spring constant in N/cm. You *must* show how you work out your answer.
(ii) What is the value of the spring constant in N/m?

d Jerry catches a really big fish. It is too heavy for the balance. Her friend Tom also has a balance. They weigh the fish using both balances.

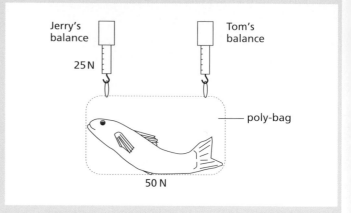

(i) What is the reading on Tom's balance?
(ii) Explain your answer.

e Jerry and Tom drive along a track with regular bumps. This makes their van bounce up and down. At a certain speed the van bounces 8 times in 10 seconds.
(i) What is the frequency of the bounces?
(ii) If the van slows down slightly the bounces become more violent. Suggest why this happens. Use your ideas about oscillations.
(iii) What could they do to make the bounces less violent?

(*OCR Foundation, June 99*)

36 a The smallest division marked on a metre rule is 1 mm. A student measures a length with the ruler and records it as 0.835 m. Is he justified in giving three significant figures?
b The SI unit of density is

A kg m   B kg/m^2 C kg m^3 D kg/m
E kg/m^3

Forces and pressure

37 The work done by a force is

1 calculated by multiplying the force by the distance moved in the direction of the force
2 measured in joules
3 the amount of the energy changed.

Which statement(s) is (are) correct?

A 1, 2, 3   B 1, 2   C 2, 3   D 1   E 3

38 The main energy change occurring in the device named is

	Device	Main energy change
1	electric lamp	electrical to heat and light
2	battery	chemical to electrical
3	pile driver	k.e. to p.e.

Which statement(s) is (are) correct?

A 1, 2, 3   B 1, 2   C 2, 3   D 1   E 3

39 If a lift of mass 200 kg is raised by an electric motor through a height of 15 m in 20 s,

1 the weight of the lift is 2000 N
2 the useful work done is 30 000 J
3 the power output of the motor is 1.5 kW.

Which statement(s) is (are) correct?

A 1, 2, 3   **B** 1, 2   **C** 2, 3   **D** 1   **E** 3

40 The efficiency of a machine which raises a load of 200 N through 2 m when an effort of 100 N moves 8 m is

A 0.5%   **B** 5%   **C** 50%   **D** 60%   **E** 80%

41 Which one of the following statements is *not* true?

A Pressure is the force acting on unit area.
B Pressure is calculated from force/area.
C The SI unit of pressure is the pascal (Pa) which equals 1 newton per square metre (1 N/m²).
D The greater the area over which a force acts the greater is the pressure.
E Force = pressure × area.

42 A mercury manometer is connected to a gas supply as shown. The pressure of the gas supply exceeds atmospheric pressure by the pressure exerted by a column of mercury of length

A 15 mm   **B** 20 mm   **C** 25 mm   **D** 30 mm
E 45 mm

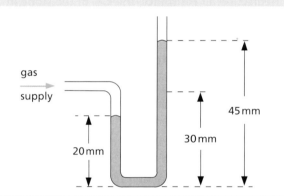

43 If the piston in the diagram below is pulled out of the cylinder from position X to position Y, without changing the temperature of the air enclosed, the air pressure in the cylinder is

A reduced to a quarter   **B** reduced to a third
C the same   **D** trebled
E quadrupled.

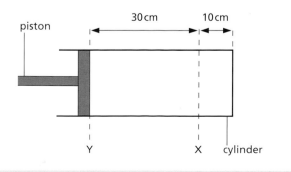

44 a When water flows through a tube, the flow can be streamline or turbulent. How is streamline flow different from turbulent flow? You may wish to draw a diagram to help your answer.
b The diagram shows a horizontal tube through which a streamlined flow of water passes. Some water will rise up each vertical tube. The higher the water goes, the higher the pressure at the bottom of that tube.

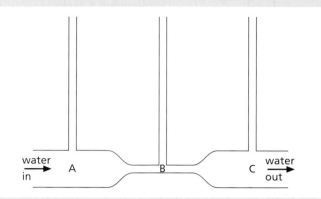

(i) Copy the diagram and draw the probable water level in each of the vertical tubes.
(ii) At which point A, B or C, in the horizontal tube is the water flowing fastest?
(iii) Write down the connection between the speed at which the water flows through the horizontal pipe and the pressure exerted by the water.
c The diagram shows an airbrush paint spray.

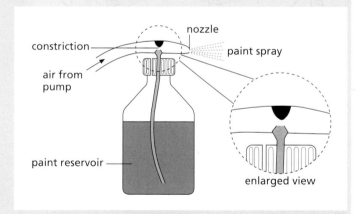

Using the ideas from part **b**, explain why the paint inside the reservoir moves towards the nozzle.
(*SEG Foundation, Summer 98*)

Motion and energy

45 The equally spaced dots on the ticker tape below are made by a vibrator of frequency 50 Hz. The speed of the tape in m/s is

A 0.04   **B** 0.1   **C** 0.2   **D** 0.4   **E** 1

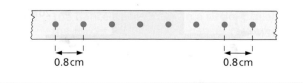

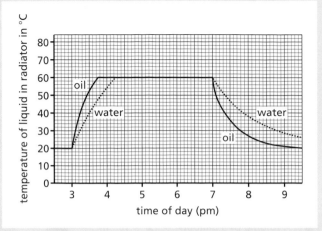

a The radiators are switched on at the mains at 3 pm and left for 4 hours. The graph shows how the temperature of the liquid in each radiator changes. The oil-filled radiator takes 45 minutes to reach 60 °C.
(i) How much longer does it take the water-filled radiator to reach 60 °C?
(ii) The oil-filled radiator heats up faster than the water-filled radiator. What is the **biggest** difference between the temperatures of the two radiators during the heating-up time?

b Explain why it is a good idea for radiators to have grooved surfaces.

c A radiator is heated by a 400 W electrical element. Calculate how much energy is given out by the electrical element in 120 s. Use the equation below. You *must* show how you work out your answer.

$$\text{energy} = \text{power} \times \text{time}$$

d Calculate the energy that needs to be transferred to raise the temperature of the water in the radiator by 20 °C. The specific heat capacity of the water is 4000 J/kg °C. Use the equation below. You *must* show how you work out your answer.

energy transfer = mass × specific heat capacity
× temperature change

e The amount of energy needed to raise the temperature of the water-filled steel radiator is actually **more** than the correct answer to part **d**. Suggest why.

f A student writes this.

> The oil gives out more heat energy than the water between 7 and 8 o'clock because the slope of the line for the oil is steeper.

Discuss whether the student is correct. Use your ideas about specific heat capacity and the rate of energy transfer.

(OCR Higher, June 99)

Electricity and electromagnetic effects

72 If the two uncharged metal spheres R and S on insulating stands are separated while the negatively charged polythene strip is held near R, the charges on R and S are

	A	B	C	D	E
R	+	−	−	+	zero
S	+	−	+	−	zero

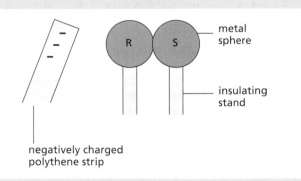

negatively charged
polythene strip

73 For the circuit below calculate
a the total resistance,
b the current in each resistor,
c the p.d. across each resistor.

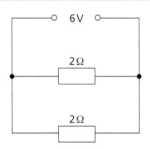

74 Repeat question 73 for the circuit below.

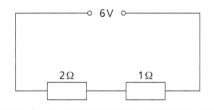

75 An electric kettle for use on a 230 V supply is rated at 3000 W. For safe working, the cable supplying it should be able to carry at least

A 2 A   B 5 A   C 10 A   D 15 A   E 30 A

76 Which one of the following statements is *not* true?

A In a house circuit lamps are wired in parallel.
B Switches, fuses and circuit breakers should be placed in the neutral wire.
C An electric fire has its earth wire connected to the metal case to prevent the user receiving a shock.
D When connecting a three-core cable to a 13 A three-pin plug the *brown* wire goes to the *live* pin.
E The cost of operating three 100 W lamps for 10 hours at 10p per unit is 30p.

77 The diagrams **A** to **E** represent magnetic fields caused by current-carrying conductors. Which one is due to
  a a long straight wire,
  b a circular coil,
  c a solenoid?

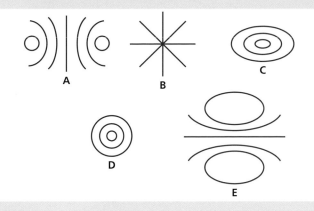

78 Which of the following statements is/are true?

  1 An electromagnet consists of a coil of wire wound on a soft iron core.
  2 The strength of the magnetic field produced by an electromagnet increases if the strength of the current and/or the number of turns of wire is increased.
  3 In the diagram below when the switch is closed the gap G increases.

  A 1, 2, 3   **B** 1, 2   **C** 2, 3   **D** 1   **E** 3

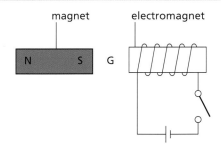

79 Which one of the following statements about the diagram below is *not* true?

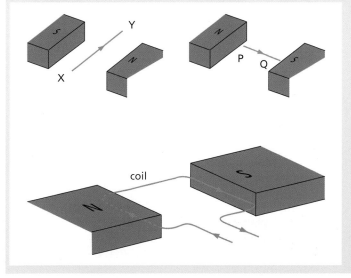

A If a current is passed through the wire XY, a vertically upwards force acts on it.
B If a current is passed through the wire PQ, it does not experience a force.
C If a current is passed through the coil, it rotates clockwise.
D If the coil had more turns and carried a larger current, the turning effect would be greater.
E In a moving-coil loudspeaker a coil moves between the poles of a strong magnet.

80 Which statement(s) is (are) correct?

  1 An ammeter is connected in series in a circuit and a voltmeter in parallel.
  2 An ammeter has a high resistance.
  3 A voltmeter has a low resistance.

  A 1, 2, 3   **B** 1, 2   **C** 2, 3   **D** 1   **E** 3

81 Which one of the following statements is *not* true when a magnet is pushed N pole first into a coil as in the diagram below?

A A p.d. is induced in the coil and causes a current through the galvanometer.
B The induced p.d. increases if the magnet is pushed in faster and/or the coil has more turns.
C Mechanical energy is changed to electrical energy.
D The coil tends to move to the right because the induced current makes face X a N pole which is repelled by the N pole of the magnet.
E The effect produced is called electrostatic induction.

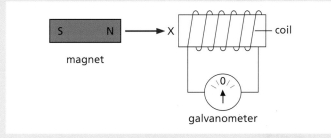

82 Which of the following statements is/are true?

  1 A transformer changes an alternating p.d. from one value to another according to the equation $V_s/V_p = N_s/N_p$.
  2 Transformers are used to send electrical energy by cable over long distances at high voltage and low current to cut heat loss.
  3 The number of turns on the primary of a transformer which has 200 turns on the secondary and is designed to deliver 10 V from the 230 V mains supply, see below, is 5000.

  A 1, 2, 3   **B** 1, 2   **C** 2, 3   **D** 1   **E** 3

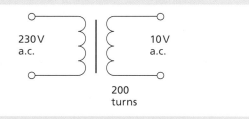

83 Which of the units **A** to **E** could be used to measure
a electric charge,
b electric current,
c p.d.,
d energy,
e power?

A ampere **B** joule **C** volt **D** watt
E coulomb

84 In the circuit shown below all four lamps are identical. All four switches are closed (ON). All four lamps are lit (ON).

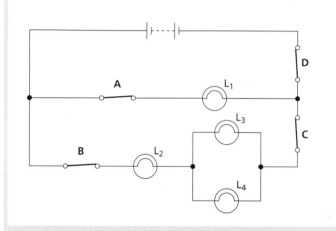

a Which *single* switch, **A** to **D**, should be opened in order to
 (i) turn OFF *all four* lamps?
 (ii) turn OFF one lamp only?
b When *all four* switches are closed (ON), state which lamp L_1 to L_4 will be the brightest. Give a reason for your answer.
c Lamps are sometimes used in electronic systems as output devices. Other devices are used as input sensors. Below there is a list of output devices and input sensors. Write down the *three* input sensors.

buzzer heater LDR

motor switch thermistor

(AQA (NEAB) Foundation, June 99)

85 a Link the picture to the correct circuit symbol. The lamp has been done for you.

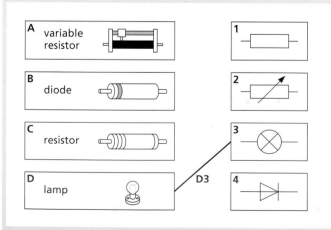

b A family tent is to be fitted with a simple lighting circuit.

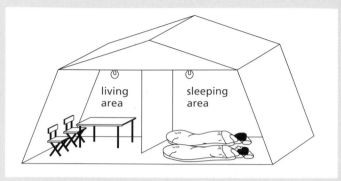

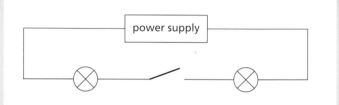

The diagram shows the first circuit used.

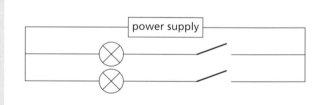

(i) Are the lamps connected in series or in parallel?
(ii) This is not a good circuit for using in the tent. Why?
The diagram below shows the second circuit used.

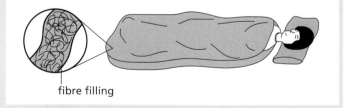

(iii) Give *two* reasons why this circuit is better than the first circuit.
c Many people use a sleeping bag when they sleep in a tent. Sleeping bags, designed to keep a person warm, have a fibre filling.

fibre filling

(i) Complete the sentence by choosing the correct words from the box.

conduction   convection   radiation

The fibre is designed to reduce heat transfer by ...... and ....
(ii) Explain why the fibre is good at reducing heat loss from a person sleeping in the bag.
(AQA (SEG) Foundation, Summer 99)

86 The diagram shows a washing machine.

a Complete the following sentence.

The washing machine is designed to transfer electrical energy as ...... energy and ...... energy.

b The washing machine is connected to the mains electricity supply by a three-core cable and plug.
(i) What colour wire must be connected to the earth terminal in the plug?
(ii) To which colour wire is the fuse in the plug connected?

c The power of the washing machine is 2.75 kW. The washing machine is used for 3 hours. Use the equation below to calculate how much energy is transferred to the washing machine.

energy transferred = **power** × **time**
(kilowatt hour, kW) (kilowatt, kW) (hour, h)

(NEAB Foundation, June 98)

87 The circuit below is for a transformer supplying a lamp. If the transformer is 100% efficient the resistance of the lamp is

A 1.5 Ω   **B** 3 Ω   **C** 6 Ω   **D** 12 Ω   **E** 24 Ω

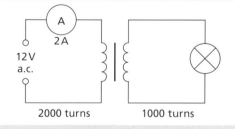

88 a The diagram shows a simple electricity generator. Rotating the loop of wire causes a current which lights the lamp.

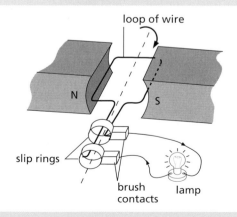

State *three* ways to increase the current produced by the generator.

b Most electricity in Britain is generated by coal fired power stations.
(i) Complete the sequence of useful energy transfers which take place in the power station.

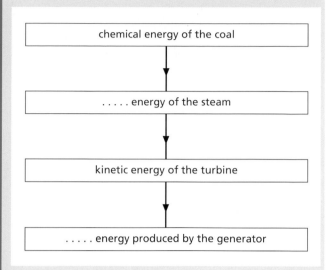

chemical energy of the coal

. . . . . energy of the steam

kinetic energy of the turbine

. . . . . energy produced by the generator

(ii) The diagram shows a cross-section of a power station generator.

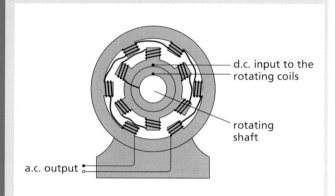

Starting with the current going to the rotating coils, explain how the generator produces electricity.

c The diagram shows a wind turbine which is used to produce electricity using energy from the wind.

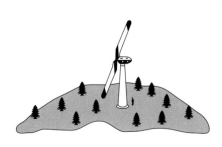

(i) What is the source of energy which creates winds?
(ii) Explain the advantage of using a wind turbine to produce electricity.

(AQA (SEG) Higher, Summer 99)

Electrons and atoms

89 Which of the following statements about the deflection of the beam of electrons by the p.d. between the plates P and Q below is/are true?

1 Plate P is negative.
2 The deflection would be greater if the p.d. was greater.
3 The deflection would be greater if the electrons were moving slower.

A 1, 2, 3   **B** 1, 2   **C** 2, 3   **D** 1   **E** 3

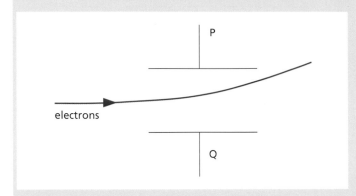

90 Which of the patterns **A** to **E** could appear on the screen of a CRO if an alternating p.d. is applied to the Y-plates with the time base
a off,
b on?

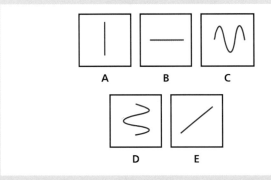

91 Identify the radiations X, Y and Z in the diagram below.

	A	**B**	**C**	**D**	**E**
X	alpha	beta	gamma	gamma	beta
Y	beta	alpha	alpha	beta	gamma
Z	gamma	gamma	beta	alpha	alpha

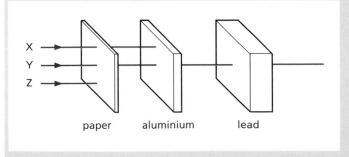

92 The radioactive substance whose decay curve is given below has a half-life in minutes of

A 1   **B** 2   **C** 3   **D** 4   **E** 5

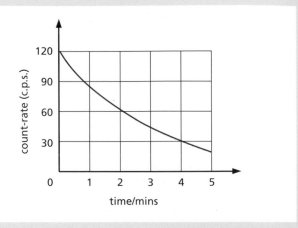

93 A radioactive source which has a half-life of 1 hour gives a count-rate of 100 c.p.s. at the start of an experiment and 25 c.p.s. at the end. The time taken by the experiment was, in hours,

A 1   **B** 2   **C** 3   **D** 4   **E** 5

94 Which one of the following statements is *not* true?

A An atom consists of a tiny nucleus surrounded by orbiting electrons.
B The nucleus contains protons and neutrons, called nucleons, in equal numbers.
C A proton has a positive charge, a neutron is uncharged and their mass is about the same.
D An electron has a negative charge of the same size as the charge on a proton but it has a much smaller mass.
E The number of electrons equals the number of protons in a normal atom.

95 A lithium atom has a nucleon (mass) number of 7 and a proton (atomic) number of 3.

1 Its symbol is $^{7}_{4}$Li.
2 It contains 3 protons, 4 neutrons and 3 electrons.
3 One of its isotopes has 3 protons, 3 neutrons and 3 electrons.

Which statement(s) is (are) correct?

A 1, 2, 3   **B** 1, 2   **C** 2, 3   **D** 1   **E** 3

96 Which symbol **A** to **E** is used in equations for nuclear reactions to represent
a an alpha particle,
b a beta particle,
c a neutron,
d an electron?

A $^{0}_{-1}$e   **B** $^{1}_{0}$n   **C** $^{4}_{2}$He   **D** $^{0}_{-1}$e   **E** $^{1}_{1}$n

97 **a** Radon $^{220}_{86}$Rn decays by emitting an alpha particle to form an element whose symbol is

A $^{216}_{85}$At   **B** $^{216}_{86}$Rn   **C** $^{218}_{84}$Po   **D** $^{216}_{84}$Po   **E** $^{217}_{85}$At

b Thorium $^{234}_{90}$Th decays by emitting a beta particle to form an element whose symbol is

A $^{235}_{90}$Th   **B** $^{230}_{89}$Ac   **C** $^{234}_{89}$Ac   **D** $^{232}_{88}$Ra   **E** $^{234}_{91}$Pa

98 Which one of the following statements about the transistor circuit shown below is *not* true?

A The collector current I_C is zero until base current I_B flows.

B I_B is zero until the base–emitter p.d. V_{BE} is +0.6V.

C A small I_B can switch on and control a large I_C.

D When used as an amplifier the input is connected across B and E.

E X must be connected to supply − and Y to the +.

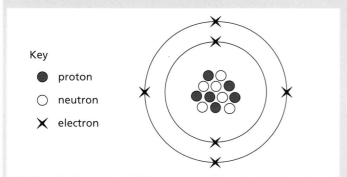

99 a The diagram below represents an atom of carbon-12 ($^{12}_{6}C$).

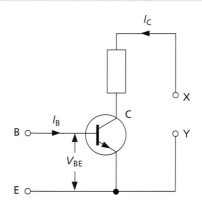

Key

● proton

○ neutron

✕ electron

Copy and complete the table of data about an atom of carbon-12.

Particle	Number in atom	Type of charge (neutral, negative or positive)
proton		
neutron		
electron		

b Carbon-14 ($^{14}_{6}C$) is a radioactive isotope of carbon. Describe the difference between an atom of carbon-14 and an atom of carbon-12.

c Carbon-14 decays by emitting a beta particle.

(i) Which part of the atom emits the beta particle?

(ii) Which statement is the correct description of a beta particle?

A a fast-moving proton

B a fast-moving neutron

C a fast-moving electron

(London Foundation, June 99)

100 In a nuclear power station, energy from nuclear fission is used to generate electricity.

a The diagram below illustrates the fission of uranium-235.

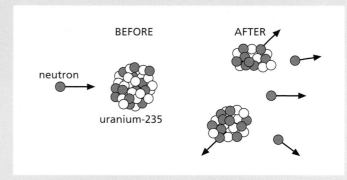

(i) Describe the process shown in this diagram.

(ii) In the fission of uranium-235, energy is released from the nucleus. What happens to this energy?

b The diagram below shows the main parts of a nuclear power station.

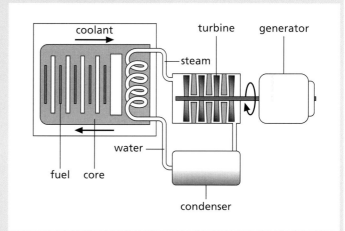

Explain how energy from the fuel is used to generate electricity.

(London Foundation, June 98)

101 The waveform (i) is displayed on a CRO. A student then alters *two* controls and obtains the waveform (ii). The two controls adjusted were

A time base and Y-shift

B X-shift and Y-gain

C Y-gain and time base

D focus and X-gain

E X-gain and Y-gain.

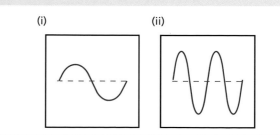

102 The diagram below shows the paths of two alpha particles, A and B, into and out of a thin piece of metal foil.

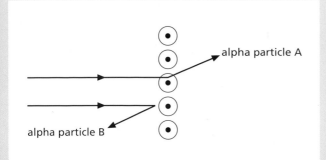

The paths of the alpha particles depend on the forces on them in the metal.

a Describe the model of the atom which was used to explain the paths of alpha particles aimed at thin sheets of metal foil.

b Scientists used to believe that atoms were made up of negative charges embedded in a positive 'dough'. This is called the 'plum pudding' model of the atom. The diagram below shows a model of such an atom.

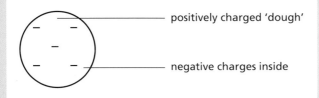

(i) Explain how the 'plum pudding' model of the atom can explain why alpha particle A is deflected through a very small angle.
(ii) Explain why the 'plum pudding' model of the atom can *not* explain the large deflection of alpha particle B.

(NEAB Higher, June 98)

103 The diagram shows a simple AM (amplitude modulation) radio system for generating a radio signal.

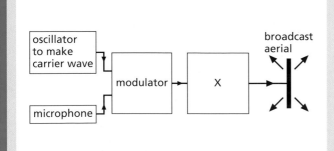

a What does the microphone do?
b What is required in box X to make the signal strong enough to be received over a wide area?

c The wave trace for the carrier wave is shown below. Sketch a possible wave trace for the audio wave (Box 1).
Now show how the carrier wave (Box 2) is reshaped by the modulator.

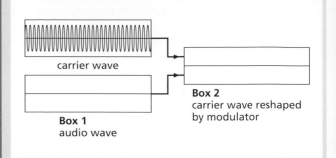

d Each radio station sends out a modulated carrier wave in this way. How is it possible to listen to the signal from just one radio station at a time?

(OCR Higher, June 99)

104 a A moving coil microphone is a transducer. Explain what is meant by a **transducer**.

b (i) A moving coil microphone produces an analogue signal. Explain the difference between an analogue and a digital signal. You may use diagrams if you wish.
(ii) State and explain *one* advantage of using digital signals for the transmission of information.

c The diagram below shows the construction of a moving coil microphone.

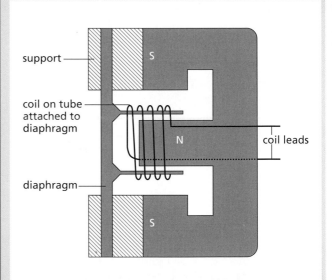

When a variable sound wave signal arrives at the diaphragm, an analogue voltage is produced across the coil.
(i) Explain how an analogue voltage is produced.
(ii) State *one* way to increase the sensitivity of the microphone.

(London Higher, June 99)

Earth and space physics

105 **a** Draw a diagram to show the Earth's structure.
  b On the diagram you have drawn, sketch in lines showing the paths taken by P, S and L waves from an earthquake.
  c Why are seismic waves refracted in a *curved* path when they pass through the Earth's mantle?

106 **a** The diagram shows the Earth at four positions in its orbit around the Sun.

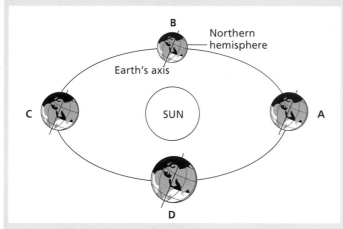

  (i) How long does the Earth take to orbit once around the Sun?
  (ii) In which *one* of the positions, **A**, **B**, **C** or **D** would it be summer in the Northern hemisphere? Give a reason for your answer.
  b Explain how the movement of the Earth gives us day and night. You may wish to draw a diagram to help explain your answer.
  c Complete the following sentences.

  The solar system has one ...... called the Sun. There are ...... planets in orbit around the Sun. Some planets have one or more ...... in orbit around them.

  (SEG Foundation, Summer 98)

107 The table below contains data about the Earth and Saturn.

	Earth	Saturn
distance from Sun	150 million km	1427 million km
average temperature	20 °C	−180 °C
surface gravitational field strength	9.8 N/kg	8.9 N/kg
time to rotate on axis	24 hours	11 hours
time to orbit Sun	1 year	29.5 years
density	5.5 g/cm³	0.7 g/cm³

  Use the data in the table to answer the questions.
  a How long is a day on Saturn?
  b Suggest why Saturn is colder than the Earth.
  c Give *one* reason why astronomers think that Saturn is composed mainly of gases.
  d A space probe lands on Saturn's surface. How will the weight of the probe on Saturn compare with its weight on Earth?

  (London Foundation, June 99)

108 A space shuttle is in orbit round the Earth at a certain height.
  a What keeps it in orbit?
  b If the shuttle reduces its mass by launching a communications satellite, how does this affect the centripetal force?
  c If the shuttle moves to a higher orbit does there have to be an increase or a decrease in
  (i) the centripetal force,
  (ii) the orbital speed?

109 The Sun radiates energy E at a rate of 4.0×10^{26} watts by converting some of its mass m according to the equation $E = mc^2$. If $c = 3 \times 10^8$ m/s, calculate how much matter the Sun loses per second.

110 **a** The Cassini spacecraft launched in 1997 will take seven years to reach Saturn. The journey will take the spacecraft close to several other planets.

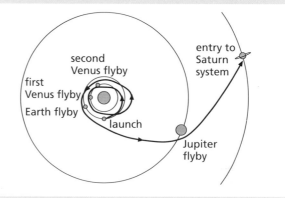

  Each time the spacecraft approaches a planet it changes direction and gains kinetic energy. Explain why.
  b Cassini carries a probe which will travel to Titan, Saturn's largest moon. The 320 kg probe will enter Titan's atmosphere at 6 km/s. After plunging 100 km, parachutes will open to reduce the speed of the probe before it lands on Titan's surface.
  (i) Use the equation below to calculate in joules, the kinetic energy of the probe as it enters Titan's atmosphere. Show clearly how you work out your answer.

$$\text{kinetic energy} = \tfrac{1}{2}mv^2$$

  (ii) The outside of the probe is fitted with a heat shield designed to withstand very high temperatures. Explain why.
  (iii) Why do parachutes reduce the speed of falling objects?
  c The Big Bang theory attempts to explain the origin of the Universe.
  (i) What is the Big Bang theory?
  (ii) What can be predicted from the Big Bang theory about the size of the Universe?
  d (i) Explain how stars like the Sun were formed.
  (ii) The Sun is made mostly of hydrogen. Eventually the hydrogen will be used up and the Sun will 'die'. Describe what will happen to the Sun from the time the hydrogen is used up until the Sun 'dies'.

  (AQA (SEG) Higher, Summer 99)

Mathematics for physics

USE THIS SECTION AS THE NEED ARISES

■ Solving physics problems

When tackling physics problems using mathematical equations it is suggested that you **do not substitute numerical values until you have obtained the expression in symbols which gives the answer**. That is, work in symbols until you have solved the problem and only then insert the numbers in the expression to get the final result.

This has two advantages. First, it reduces the chance of errors in the arithmetic (and in copying down). Second, you write less since a symbol is usually a single letter whereas a numerical value is often a string of figures.

Adopting this 'symbolic' procedure frequently requires you to change round an equation first. The next two sections and the questions that follow them are intended to give you practice in doing this and then substituting numerical values to get the answer.

■ Equations – type 1

In the equation $x = a/b$, the subject is x. To change it we **multiply or divide both sides** of the equation by the same quantity.

To change the subject to a

We have

$$x = \frac{a}{b}$$

If we multiply both sides by b the equation will still be true.

$$\therefore \qquad x \times b = \frac{a}{b} \times b$$

The b's on the right-hand side cancel

$$\therefore \qquad b \times x = \frac{a}{\not b} \times \not b = a$$

$$\therefore \qquad a = b \times x$$

To change the subject to b

We have

$$x = \frac{a}{b}$$

Multiplying both sides by b as before, we get

$$a = b \times x$$

Dividing both sides by x:

$$\frac{a}{x} = \frac{b \times x}{x} = \frac{b \times \not x}{\not x} = b$$

$$\therefore \qquad b = \frac{a}{x}$$

Now try the following questions using these ideas.

Questions

1 What is the value of x if
  a $2x = 6$ **b** $3x = 15$ **c** $3x = 8$

  d $\frac{x}{2} = 10$ **e** $\frac{x}{3} = 4$ **f** $\frac{2x}{3} = 4$

  g $\frac{4}{x} = 2$ **h** $\frac{9}{x} = 3$ **i** $\frac{x}{6} = \frac{4}{3}$

2 Change the subject to
  a f in $v = f\lambda$ **b** λ in $v = f\lambda$
  c I in $V = IR$ **d** R in $V = IR$

  e m in $d = \frac{m}{V}$ **f** V in $d = \frac{m}{V}$

  g s in $v = \frac{s}{t}$ **h** t in $v = \frac{s}{t}$

3 Change the subject to
  a I^2 in $P = I^2R$ **b** I in $P = I^2R$
  c a in $s = \frac{1}{2}at^2$ **d** t^2 in $s = \frac{1}{2}at^2$
  e t in $s = \frac{1}{2}at^2$ **f** v in $\frac{1}{2}mv^2 = mgh$

  g y in $\lambda = \frac{ay}{D}$ **h** ρ in $R = \frac{\rho l}{A}$

4 By replacing (substituting) find the value of $v = f\lambda$ if
  a $f = 5$ and $\lambda = 2$ **b** $f = 3.4$ and $\lambda = 10$
  c $f = 1/4$ and $\lambda = 8/3$ **d** $f = 3/5$ and $\lambda = 1/6$
  e $f = 100$ and $\lambda = 0.1$ **f** $f = 3 \times 10^5$ and $\lambda = 10^3$

5 By changing the subject and replacing find
  a f in $v = f\lambda$ if $v = 3.0 \times 10^8$ and $\lambda = 1.5 \times 10^3$
  b h in $p = 10hd$ if $p = 10^5$ and $d = 10^3$
  c a in $n = a/b$ if $n = 4/3$ and $b = 6$
  d b in $n = a/b$ if $n = 1.5$ and $a = 3.0 \times 10^8$
  e F in $p = F/A$ if $p = 100$ and $A = 0.2$
  f s in $v = s/t$ if $v = 1500$ and $t = 0.2$

■ Equations – type 2

To change the subject in the equation $x = a + by$ we **add or subtract the same quantity from each side**. We may also have to divide or multiply as in type 1. Suppose we wish to change the subject to y in

$$x = a + by$$

Subtracting a from both sides,

$$x - a = a + by - a = by$$

Dividing both sides by b,

$$\frac{x-a}{b} = \frac{by}{b} = y$$

$$\therefore \qquad y = \frac{x-a}{b}$$

Questions

6 What is the value of x if
  a $x+1=5$ **b** $2x+3=7$ **c** $x-2=3$

  d $2(x-3)=10$ **e** $\frac{x}{2}-\frac{1}{3}=0$ **f** $\frac{x}{3}+\frac{1}{4}=0$

  g $2x+\frac{5}{3}=6$ **h** $7-\frac{x}{4}=11$ **i** $\frac{3}{x}+2=5$

7 By changing the subject and replacing, find the value of a in $v=u+at$ if
  a $v=20$, $u=10$ and $t=2$
  b $v=50$, $u=20$ and $t=0.5$
  c $v=5/0.2$, $u=2/0.2$ and $t=0.2$

8 Change the subject in $v^2=u^2+2as$ to a.

◼ Proportion (or variation)

One of the most important mathematical operations in physics is finding the relation between two sets of measurements.

a) Direct proportion

Suppose that in an experiment two sets of readings are obtained for the quantities x and y as in Table 1 (units omitted).

Table 1

x	1	2	3	4
y	2	4	6	8

We see that when x is doubled, y doubles; when x is trebled, y trebles; when x is halved, y halves; and so on. There is a one-to-one correspondence between each value of x and the corresponding value of y.

We say that y is **directly proportional** to x, or y **varies directly** as x. In symbols

$$y \propto x$$

Also, the **ratio** of one to the other, e.g. y to x, is always the same, i.e. it has a constant value which in this case is 2. Hence

$$\frac{y}{x} = \text{a constant} = 2$$

The constant, called the **constant of proportionality** or **variation**, is given a symbol, e.g. k, and the relation (or law) between y and x is then summed up by the equation

$$\frac{y}{x} = k \quad \text{or} \quad y = kx$$

Notes

1 In practice, because of inevitable experimental errors, the readings seldom show the relation so clearly as here.

2 If instead of using numerical values for x and y we use letters, e.g. x_1, x_2, x_3, etc., and y_1, y_2, y_3, etc., then we can also say

$$\frac{y_1}{x_1} = \frac{y_2}{x_2} = \frac{y_3}{x_3} = \ldots\ldots = k$$

or $\qquad y_1 = kx_1,\ y_2 = kx_2,\ y_3 = kx_3, \ldots\ldots$

b) Inverse proportion

Two sets of readings for the quantities p and V are given in Table 2.

Table 2

p	3	4	6	12
V	4	3	2	1

There is again a one-to-one correspondence between each value of p and the corresponding value of V but when p is doubled V is halved; when p is trebled, V has one-third its previous value; and so on.

We say that V is **inversely proportional** to p, or V **varies inversely** as p, i.e.

$$V \propto \frac{1}{p}$$

Also, the **product** $p \times V$ is always the same ($=12$) and we write

$$V = \frac{k}{p} \quad \text{or} \quad pV = k$$

where k is the constant of proportionality or variation and equals 12 in this case.

Using letters for values of p and V we can also say

$$p_1 V_1 = p_2 V_2 = p_3 V_3 = \ldots\ldots = k$$

Graphs

Another useful way of finding the relation between two quantities is by a graph.

a) Straight line graphs

When the readings in Table 1 are used to plot a graph of y against x, a **continuous** line joining the points is **a straight line passing through the origin**, Figure M1. Such a graph shows there is direct proportionality between the quantities plotted, i.e. $y \propto x$. But note that the line must go through the origin.

A graph of p against V using the readings in Table 2 is a curve, Figure M2. However if we plot p against $1/V$ (or V against $1/p$) we get a straight line through the origin, showing that $p \propto 1/V$, Figure M3 (or $V \propto 1/p$).

Table 3

p	V	1/V
3	4	0.25
4	3	0.33
6	2	0.50
12	1	1.00

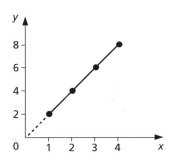

Figure M1

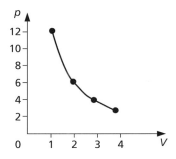

Figure M2

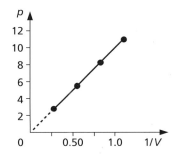

Figure M3

b) Slope or gradient

The slope or gradient of a straight line graph equals the constant of proportionality. In Figure M1, the slope is $y/x = 2$; in Figure M3 it is $p/(1/V) = 12$.

In practice points plotted from actual measurements may not lie exactly on a straight line due to experimental errors. The 'best straight line' is then drawn 'through' them so that they are equally distributed about it. This automatically averages the results. Any points that are well off the line stand out and may be investigated further.

c) Practical points

(i) The axes should be labelled giving the quantities being plotted and their units, e.g. I/A meaning current in amperes.

(ii) If possible the origin of both scales should be on the paper and the scales chosen so that the points are spread out along the graph.

(iii) Mark the points $\odot$ or $\times$.

Questions

9 In an experiment different masses were hung from the end of a spring held in a stand and the extensions produced were as shown below.

Mass/g	100	150	200	300	350	500	600
Extension/cm	1.9	3.1	4.0	6.1	6.9	10.0	12.2

    a Plot a graph of **extension** along the vertical (*y*) axis against **mass** along the horizontal (*x*) axis.
    b What is the relation between **extension** and **mass**? Give a reason for your answer.

10 Pairs of readings of the quantities m and v are given below.

m	0.25	1.5	2.5	3.5
v	20	40	56	72

    a Plot a graph of m along the vertical axis and v along the horizontal axis.
    b Is m directly proportional to v? Explain your answer.
    c Use the graph to find v when $m = 1$.

11 The distance s (in metres) travelled by a car at various times t (in seconds) are shown below.

s	0	2	8	18	32	50
t	0	1	2	3	4	5

    Draw graphs of
    a s against t,
    b s against t^2.
    What can you conclude?

Answers

Higher level questions are marked with *.

Light and sight

1 Light rays
1 Larger, less bright
2 a 4 images b Brighter but blurred
3 C
4 Before; sound travels slower than light

2 Reflection of light
1 a 40° c 40°, 50°, 50° d Parallel
2 A
3 Top half

3 Plane mirrors
1 B
2 D
3 4 m towards mirror

4 Curved mirrors
2 7.2 cm from mirror, 1.6 cm high
3 Concave, parabolic
4 Gives a wide field of view

5 Refraction of light
3 250 000 km/s
4 C

6 Total internal reflection
1 a Angle of incidence = 0
  b Angle of incidence > critical angle
3 Periscope, binoculars

7 Lenses
1 Parallel
2 a Convex
  c Image 9 cm from lens, 3 cm high
3 Distance from lens:
  a > 2F b 2F c between F and 2F
  d < F

8 The eye
1 a Iris b Retina c Ciliary muscles
2 E
3 a (i) no (ii) yes (iii) no
  b Concave
4 a (i) no (ii) yes b Convex

9 Colour
2 a The plastic absorbs all colours except green
  b Green
3 a Blue b White

10 Simple optical instruments
1 Towards
2 a Larger, blurred, less bright
  b Moved closer to the slide
3 a 4 cm
  b 8 cm behind lens, virtual, m = 2

11 Telescopes
1 a Objective, 100 cm
  b 100 cm

  c Eyepiece
  d Real image formed by A; 5 cm
  e 105 cm
2 a E b B
3 a To collect as much light as possible and give greater detail
  b Optical telescope collects light, radio telescope collects radio waves; both have a concave reflector
  c Less distortion to image as light does not have to pass through the Earth's atmosphere

Additional questions
1 B
2 A
3 b (ii) refracted
  d The image is as far behind the mirror as the belt is in front, it is the same size as the belt, it is virtual
4 a Ray passes into air and is refracted away from the normal
  b Total internal reflection occurs in water
5 E
6 A
8 a (i) Moon blocks out light from Sun
  b Move the card nearer the lens
9 C
10 a 3 b 3 cm
11* 14 m
12 a A retina, B cornea, C iris, D optic nerve
  b (i) lens (ii) thinner (iii) fatter (iv) light (v) larger
  c (i) distant (ii) concave lens (iii) near (iv) convex lens
13 a Long sight b Convex lens
14 A
15 A
16* A: convex f = 10 cm
   B: convex f = 5 cm
17* b (i) A larger and real, B larger and virtual
   (ii) A projector, B magnifying glass

Waves and sound

12 Mechanical waves
1 a 1 cm b 1 Hz c 1 cm/s
2 A, C
3 a Speed of ripple depends on depth of water
  b AB since ripples travel more slowly towards it, therefore water shallower in this direction

4 a Trough
  b (i) 3.5 mm (ii) 17 mm/s (iii) 5 Hz

13 Light waves
1 a $S_1O = S_2O$ b $S_1O_1 - S_2O_1 = \lambda/2$
  c $S_1O_2 - S_2O_2 = \lambda$
2 a Wavelength of light is very small
  b Sources not coherent
3 a 0.7 µm b 0.4 µm
4 a Polaroid sheet, reflection from water
  b Sunglasses c No

14 Electromagnetic radiation
1 a C
  b Microwaves, infrared
  c X-rays
  d (i) sunburn, cancer
   (ii) use sunscreen and wear protective clothing
2 a B b D
3 a Ultraviolet
  b Microwaves
  c Gamma rays
  d Infrared
4 a 3 m b 2×10^{-4} s

15 Sound waves
1 1650 m (about 1 mile)
2 a Reflection
  b (i) 2040 m (ii) 340 m/s
3 b decreases
  c (i) 1.6 m (ii) 320 m/s
4 a $2 \times 160 = 320$ m/s
  b $240/(3/4) = 320$ m/s
  c 320 m
5 a Reflection, refraction, diffraction, interference
  b Vibrations are perpendicular to rather than along the direction of travel of the wave; longitudinal

16 Musical notes
1 b (i) 1.0 m (ii) 2.0 m
2 a B
  c (i) higher (ii) quieter
  d (i) ultrasonic
   (ii) medical ultrasound imaging, cleaning jewellery

Additional questions
1 a $30/6 = 5$ cm b 4 Hz c $v = f\lambda$
  d 20 cm/s
3* (i) X down (ii) Y up (iii) Z up
4 E
5 a Light is a wave
  b Longer wavelength c Red
6 a (i) L (ii) N
7 a (i) microwave
   (ii) ultraviolet
   (iii) X-rays

b Infrared, visible, ultraviolet
c Kills cancer cells
d (ii) lead shielding
8 B
9* b (i) 10 m (ii) 30 Hz
(iii) beats of 2 Hz

Matter and molecules

17 Measurements
1 a 10 **b** 40 **c** 5 **d** 67 **e** 1000
2 a 3.00 **b** 5.50 **c** 8.70 **d** 0.43
e 0.1
3 a 1×10^5; 3.5×10^3; 4.28×10^8;
5.04×10^2; 2.7056×10^4
b 1000; 2 000 000; 69 000; 134;
1 000 000 000
4 a 1×10^{-3}; 7×10^{-5}; 1×10^{-7};
5×10^{-5}
b 5×10^{-1}; 8.4×10^{-2}; 3.6×10^{-4};
1.04×10^{-3}
5 10 mm
6 a two **b** three **c** four **d** two
7 24 cm³
8 40 cm³; 5
9 80
10 a 250 cm³ **b** 72 cm³

18 Density
1 a (i) 0.5 g (ii) 1 g (iii) 5 g
b (i) 10 g/cm³ (ii) 3 kg/m³
c (i) 2.0 cm³ (ii) 5.0 cm³
2 a 8.0 g/cm³ **b** 8.0×10^3 kg/m³
3 15 000 kg
4 130 kg
5 1.1 g/cm³

19 Weights and stretching
1 a 1 N **b** 50 N **c** 0.50 N
2 a 120 N **b** 20 N
3 a 2000 N/m **b** 50 N/m
4 A

20 Molecules
1 B
2 a Air is readily compressed
b Steel is not easily compressed
3 Bromine molecules collide with air
molecules
4 The forces between the layers are
weak
5 a Molecules move further apart in
a gas than in a liquid
b Molecules collide with each other
c Rubber is porous, so gas
molecules can diffuse out of the
balloon

Additional questions
1 a Metre. kilogram, second
b Different number of significant
figures
c (i) πr^2 (ii) $\frac{4}{3}\pi r^3$ (iii) $\pi r^2 h$
2 0.88 g/cm³
3 b 8 cm **c** 2 cm
4 b 6 mm
d Spring will be <u>longer than</u> 20 mm

---

5 a (i) compressed (iii) shortest
b Use more or stronger springs
c (ii) downward, Earth
6 a (i) D (ii) B
b Collisions with gas molecules
c (i) faster (ii) increases

Forces and pressure

21 Moments and levers
1 E
2 E
3 (i) C (ii) A (iii) B
4 Distance multiplier

22 Centres of gravity
1 c Moment of force is greater
2 a B **b** A **c** C
3 Tips to right
4 (i) 50 cm (iii) 1.2 N
5* a Narrow track and short wheel
base; passengers raise centre of
mass
b 6000 N m
c Centre of mass raised so that
weight lies outside wheel base at
smaller angles of tilt

23 Adding forces
1 40 N
2 50 N
3 25 N
4 50 N at an angle of 53° to the 30 N
force

24 Energy transfer
1 a Electrical to sound
b Sound to electrical
c k.e. to p.e.
d Electrical to light (and heat)
e Chemical to electrical to light and
heat
2 A chemical; B heat; C k.e.; D
electrical
3 180 J
4 1.5×10^5 J
5 a 150 J **b** 150 J **c** 10 W
6 500 W
7 a $(300/1000) \times 100 = 30\%$
b Heat **c** Warms surroundings
8 a 100 J **b** 150 J **c** 67%

25 Pressure in liquids
1 a Make a footprint, copy outline
and estimate area using graph
paper
b 34 375 N/m²
c Floor will be damaged by spikes
because the area of a spike is
small so the pressure exerted is
high
2 a (i) 25 Pa (ii) 0.50 Pa (iii) 100 Pa
b 30 N
3 a (i) pressure = force/area
(ii) 12 kPa
b Raft, so pressure lower and house
does not sink

---

4 a 100 Pa **b** 200 N
5 a A liquid is nearly incompressible
b A liquid transfers the pressure
applied to it
6 1 150 000 Pa (1.15×10^6 Pa) (ignoring
air pressure)
7* a (i) liquid transfers pressure
(ii) at same height so pressures
equal
b 2.5×10^6 N/m²
c Small force × large
distance = large force × small
distance

26 Floating, sinking and flying
1 a 10 N **b** 8 N **c** 6 N **d** 5 N
2 a 480 kg **b** 4800 N
c 4800 − 1600 = 3200 N
3 25 000 − 20 000 = 5000 N
4 a 30 g **b** 30 g **c** 30 cm³
5 Raft rises (displaces less water),
water level in pool rises (more
water displaced by fully submerged
swimmer)

27 Fluid flow
1 a River flows faster in middle
b River gets wider so flows slower

Additional questions
1 a Moment = force × perpendicular
distance from pivot
b 0.75 N m
c (i) decreased (ii) decreased
2 a Moments about A:
of $P = P \times 0 = 0$;
of 500 N = 500 × 1 N m
(clockwise);
of 200 N = 200 × 2 N m
(clockwise);
of $Q = Q \times 3$ (anticlockwise)
b 300 N **c** 700 N **d** 400 N
3 a 7 N **b** 13 N
4 a Electricity transferred to k.e.
and heat
b Electricity transferred to heat
c Electricity transferred to sound
5 a (i) light energy to electrical
energy
(ii) electrical energy to chemical
energy
(iii) electrical energy to kinetic
energy
b Heat
6 3.5 kW
7* a (i) 10 N/kg (see p.132)
(ii) 450 N
b (i) force × distance
(ii) 1800 J
c 720 W
8 E
9 B
10 a 80 N **b** 80 N **c** 8 kg **d** 0.01 m³
e 0.01 m³

Motion and energy

28 Velocity and acceleration
1 a 20 m/s b 6.25 m/s
2 a (i) Ahmed
     (ii) longest time
  b (i) speed = distance/time
     (ii) 8 m/s
3 a 15 m/s b 900 m
4 2 m/s^2
5 50 s
6 a 6 m/s b 14 m/s
7 a Uniform acceleration
  b 75 cm/s^2
8 4 s
9 a 1 s
  b (i) 10 cm/tentick^2
     (ii) 50 cm/s per tentick
     (iii) 250 cm/s^2
  c 0

29 Graphs and equations
1 a 60 km b 5 hours c 12 km/h
  d 2 e 1½ hours
  f 60 km/3½ h = 17 km/h
  g Steepest line: EF
2 a 100 m b 20 m/s c Slows down
3 a 5/4 m/s^2 b (i) 10 m (ii) 45 m
  c 22 s
4 96 m
5 a 4 s b 24 m
6* a (i) 9400 m
     (ii) 26 hours 30 minutes
  b (i) none (ii) falls rapidly
  c (i) F (ii) D (iii) B
7* a 8400 m/s b 2016 km
  c Fuel burnt so total mass reduced

30 Falling bodies
1 a (i) 10 m/s (ii) 20 m/s (iii) 30 m/s
     (iv) 50 m/s
  b (i) 5 m (ii) 20 m (iii) 45 m
     (iv) 125 m
2 3 s; 30 m/s
3 a 10 s b 2000 m
4 B

31 Newton's laws of motion
1 D
2 20 N
3 a 5000 N b 15 m/s^2
4 a 4 m/s^2 b 2 N
5 a 0.5 m/s^2 b 2.5 m/s c 25 m
6 a 1000 N b 160 N
7 a 5000 N b 20 000 N; 40 m/s^2
8 a (i) weight (ii) air resistance
  b Falls at constant velocity
     (terminal velocity)
10* a 350 N upwards b 0.46 m/s^2
  c Acceleration decreases, since air
     resistance increases when
     velocity increases
  d Increases

32 Momentum
1 a 50 kg m/s b 2 kg m/s
  c 100 kg m/s
2 2 m/s

3 4 m/s
4 0.5 m/s
5 2.5 m/s
6 a 40 kg m/s b 80 kg m/s
  c 20 kg m/s^2 d 20 N
7 2.5 m/s
8 a 10 000 N b 10 m/s^2
9* a (i) 0 (ii) kg m/s
  b (i) 80 m/s
     (ii) follow through to extend Δt
     or strike harder
     (iii) higher

33 Kinetic and potential energy
1 a 2 J b 160 J c 100 000 = 10^5 J
2 a 20 m/s b (i) 150 J (ii) 300 J
3 a 1.8 J b 1.8 J c 6 m/s d 1.25 J
  e 5 m/s
4 3.5 × 10^9 W = 3500 MW
5* c s ∝ v^2 d −1.5 m/s^2
6* a 7 m
  b (i) fatigue, poor eyesight
     (ii) A increases; B none; C
     increases
  c Less friction
  d (i) −6.25 m/s^2 (ii) 5625 N
     (iii) 72 m

34 Circular motion
2 a Sideways friction between tyres
     and road
  b (i) larger (ii) smaller (iii) larger
3 Slicks allow greater speed in dry
  conditions but in wet conditions
  treads provide frictional force to
  prevent skidding
4 5000 s (83 min)
5 b (i) less (ii) greater (iii) less
6* a Gravity b Different orbital radii
  c (i) 24 hours (ii) faster
  d (i) communication (ii) weather
  e Distortion due to Earth's
     atmosphere eliminated

Additional questions
1 a OA, BC: accelerating;
     DE: decelerating;
     AB, CD: uniform velocity
  b OA: a = +80 km/h^2;
     AB: v = 80 km/h;
     BC: a = +40 km/h^2;
     CD: v = 100 km/h;
     DE: a = −200 km/h^2
  c OA 40 km; AB 160 km;
     BC (5 + 40) = 45 km;
     CD 100 km; DE 25 km
  d 370 km e 74 km/h
2 a Uniform velocity b 600 m
  c 20 m/s
3 A
4 E
5* a 4.0 s b 80 m c 40 m/s
  d 20 m/s
6* a B and C
  b (i) speed = distance/time
     (ii) 2500 s (iv) 6300 m

7 a A, E
  b speeding up A > E; constant
     speed A = E; slowing down A < E
  c When van at rest and A = 0
  d W to X decelerating; X to Y
     constant velocity; Y to Z
     accelerating
  e Backwards force; forwards;
     backwards
8 a B, less air resistance
  b (i) X caused by gravity
     (ii) Y caused by air resistance
     (iii) speed of sky diver will go
     up
     (iv) X stays the same
  c Y increases
9 10 000 m/s
10 75 km/h; yes
11 a (i) zero; not moving
     (ii) momentum before collision
     is <u>smaller than</u> momentum after
     collision
     (iii) to right
  b (i) gains
     (ii) ball loses; stump gains
     (iii) ball gains; cannon gains
12 a Increases b (i) 24 hours
     (ii) 35.5 × 10^6 m
  c Orbits more often; greater
     image detail
13* a (i) 6 kg m/s (ii) 6 kg m/s
     (iii) 100 kg
14* c 8 m d s ∝ v^2
  e Driver slows down
  f Braking distance would be less
15* a (i) centripetal (ii) 2.2 × 10^{12} J
     (iii) heat
  b (ii) −2 m/s^2 c (ii) 156 kN

Heat and energy

35 Thermometers
1 a 1530 °C b 19 °C c 0 °C
  d −12 °C e 37 °C
2 C
3 a Property must change
     continuously with temperature
  b Volume of a liquid, resistance,
     pressure of a gas
  c (i) platinum resistance
     (ii) thermocouple (iii) alcohol

36 Expansion of solids and liquids
2 Aluminium
3 a Aluminium b 1.009 m
  c Linear expansivities the same
  d Platinum
  e Shrink fit by cooling steel in
     liquid nitrogen
4 a 0.1 m b 0.0004 m c 0.3 m

37 The gas laws
1 a 4 m^3 b 1 m^3
2 600 K (327 °C)
3 a (i) 26 °C (ii) add 273 °C
  b (i) absolute zero (ii) zero
     (iii) 0 K

4 a 15 cm³ **b** 6 cm³

5* b (i) P proportional to $1/V$ **c** 196

38 Specific heat capacity

1 15 000 J

2 A = 2000 J/(kg °C); B = 200 J/(kg °C); C = 1000 J/(kg °C)

3 a Steel **b** 56 000 J

4 Specific heat capacity of jam is higher than that of pastry so it cools more slowly

39 Latent heat

1 a 3400 J **b** 6800 J

2 a $5 \times 340 + 5 \times 4.2 \times 50 = 2750$ J **b** 1700 J

3 680 s

4 a 0 °C **b** 45 g

5 a 9200 J **b** 25 100 J

6 157 g

7 a Ice has a high latent heat of fusion

  b Water has a high latent heat of vaporization

8 Heat drawn from the water when it evaporates

9 Heat drawn from the milk when the water evaporates

10* a (i) to measure heat supplied from the surroundings
    (ii) 440 J/g
    (iii) not all the heat goes to melting the ice

  b (i) no change
    (ii) increase p.e. of molecules

11* a (i) liquid (ii) liquid

  b (i) 440 °C

  c (i) constant (ii) p.e. increases

40 Conduction and convection

3 a If small amounts of hot water are to be drawn off frequently it may not be necessary to heat the whole tank

  b If large amounts of hot water are needed it will be necessary to heat the whole tank

4 a (i) 2.2 MJ (ii) 1.1 MJ **b** 300 W

41 Radiation

1 a (i) radiation (ii) convection
    (iii) conduction

  b (i) black
    (ii) black surfaces are good absorbers of radiation; shiny surfaces are poor absorbers of radiation

4* a Hot air rises at A so cooler air is drawn in at C

  b (i) radiation
    (ii) black surfaces are good emitters of radiation

5* a (i) vacuum
    (ii) no medium is present to transfer heat by conduction or convection

  b (i) silvered inner surfaces
    (ii) radiation reflected

42 Energy sources

1 a 2% **b** Water **c** Cannot be used up **d** Solar, wind

  e All energy ends up as heat which is difficult to use and there is only a limited supply of non-renewable sources

2 Renewable, non-polluting (i.e. no CO_2, SO_2 or dangerous waste), low initial building cost of station to house energy converters, low running costs, high energy density, reliable, allows output to be readily adjusted to varying energy demands

4 b 0.25

Additional questions

2 B

3 A

4 B

5 a Pressure arises from the collision of gas molecules with the sides of the container

  b (i) speed increases
    (ii) pressure increases as collisions with the walls are more frequent
    (iii) they will absorb radiation, heat up and may explode

6* a (i) −260 °C
    (ii) absolute zero
    (iii) molecules have zero k.e.

  b Weight of liquid balanced by excess pressure of trapped air. The air pressure would increase if volume reduced (collisions with walls would occur more frequently if volume smaller)

7 1200 J

8* a (i) 1 800 000 J (ii) 0.75 kg

  b 2.1 °C

  c Less heat is lost by conduction, convection and radiation as the air temperature gets closer to body temperature; more heat has then to be lost by sweating to prevent body overheating

9 Metal is a better conductor of heat than rubber

11* a 0.1 m³ **b** 76 kg

  c Higher specific heat capacity and more mass allows more energy to be stored

  d 850 kJ

  e (i) 1.3 °C/minute
    (ii) temperature rises more slowly at 1.2 °C/minute

12* 6 kW

Electricity and electromagnetic effects

43 Static electricity

1 D

2 Electrons are transferred from the cloth to the polythene

3 a (i) so particles separate
    (ii) to attract paint and help it stick

  b To discharge aircraft

44 Electric current

1 a 5 C **b** 50 C **c** 1500 C

2 a 5 A **b** 0.5 A **c** 2 A

4 B

5 C

6 All read 0.25 A

45 Potential difference

1 a 12 J **b** 60 J **c** 240 J

2 a 6 V **b** (i) 2 J (ii) 6 J

3 B

4 b Very bright
  c Normal brightness
  d No light
  e Brighter than normal
  f Normal brightness

5 a 6 V **b** 360 J

6 $x = 18$, $y = 2$, $z = 8$

46 Resistance

1 3 Ω

2 20 V

3 C

4 A = 3 V; B = 3 V; C = 6 V

5 2 Ω

6 a 15 Ω **b** 1.5 Ω

7 D

8* b (ii) 2 A (iii) 5 Ω
  c (i) increases

47 Capacitors

2 a (i) maximum (ii) zero
  b (i) maximum (ii) zero

48 Electrolysis and cells

1 a Ions
  b (i) faster deposition of copper
    (ii) greater mass of copper deposited

49 Electric power

1 a 100 J **b** 500 J **c** 6000 J

2 a 24 W **b** 3 J/s

3 C

4 2.99 kW

5 a (i) 0.2 A (ii) 0.6 A **b** (ii) 4 J/s

6* b (i) 15 A (ii) 60 J/s

50 Electricity in the home

1 Fuse is in live wire in (a) but not in (b)

3 a 3 A **b** 13 A **c** 13 A

4 40p

5 a (i) live, neutral, earth (ii) brown
  b (i) 0.2 kWh (ii) 2p

6* a (i) lamp
    (ii) large current drawn by kettle; earth connection needed

  b (i) 2040 W (ii) 4.25 A
    (iii) longer time needed

7* c (ii) 500 ms

Additional questions: Electricity

1 C
2 b (i) loses
  (ii) negative; positive
  (iii) attract
3 C
4 B
5 a (ii) $2\,\Omega$
6 B
7 a $2\,\Omega$ **b** $\frac{2}{3}$A **c** $\frac{1}{3}$A
8 $\frac{4}{3}\,\Omega$; 2A
9* b (i) 1.6A (ii) $2.5\,\Omega$
  c Slope increases
10* $30\,\Omega$
11* $86\,\Omega$
12 a (i) 2kW (ii) 60W (iii) 850W
  b 4A
13 b (i) metal cover
  (ii) electrical; heat
  (iii) convection
14 a 800W
  b (i) $P = IV$ (ii) 3.5A (iii) 5A
  c 0.95 **d** 0.8p

51 Magnetic fields

1 C
3 A N, B N, C S, D S, E N, F S,
4 a (i) S–N; N–S (ii) magnet 2
  b (i) becomes magnetized
  (ii) easily magnetized
  (iii) to slide easily
  (iv) steel paper clip (becomes magnetized and is attracted to fixed iron tip; aluminium is not magnetized)

52 Electromagnets

1 a N **b** E
2 S
3 a (i) 4.5N
  (ii) increase number of turns of wire
  b (i) 1A (ii) no

53 Electric motors

1 E
2 Clockwise
4 E

55 Generators

1 a (i) zero (ii) negative

56 Transformers

2 B
3 a 24 **b** 1.9A
4* a (i) increase number of turns on the secondary coil
  (ii) core is divided into sheets which are insulated from each other
  b (i) voltage can be easily changed
  (ii) used to step voltages up or down
  (iii) to avoid contact with high voltage
  c (i) $N_s/N_p = V_s/V_p$ (ii) 4000
  d (i) resistance of cable decreases
  (ii) radiation

Additional questions: Electromagnetic effects

1 c (i) L, N, O
  (ii) proportional
2 a To complete the circuits to the battery negative
  b One contains the starter switch and relay coil; the other contains the relay contacts and starter motor
  c Carries much larger current to starter motor
  d Allows wires to starter switch to be thin since they only carry the small current needed to energize the relay
4 B
7* a 5A **b** 400W **c** 80V
8* a (i) electromagnetic induction
  (ii) deflected to left
  (iii) larger deflection to right occurs more quickly

Electrons and atoms

57 Electrons

1 a A −ve, B +ve
  b Down
2 a B
  b (iii) change value of high voltage
  (iv) no beam, electrons repelled by anode
  c (i) +12V (ii) −12V
3 a Shorter wavelength
  b Reduced intensity
5* a (i) electrons emitted from filament by thermionic emission and attracted to +ve M
  (ii) less electrons emitted by filament
  (iii) electrons collide with gas atoms and do not reach M
  b (i) 3.2×10^{-17}J (ii) 3.2mA

58 Radioactivity

1 a α **b** γ **c** β **d** γ
  e α **f** α **g** β^- **h** γ
2 a Cosmic rays, radon gas
  b Lift with forceps, keep away from eyes
  c 120, 60, 30, 15, 0, 0
  e (i) absorbed (ii) γ
3 25 minutes
4 D
5* a (i) β, γ (ii) γ
  d (i) 1/4 (ii) 9000 million years
6* a Electron (β^--particle)
  b Approx. 6000 years
  c Cloth about 800 years old; possible

59 Atomic structure

1 d Reactor; heat exchanger; steam; generator
2* a (i) B (ii) more electrons

b (i) A and C
  (ii) same number of electrons and protons
c (i) ● (ii) ○ (iii) ×
d ●●
  ○○
3* a 15
  c Lies above the stability line
  d (i) A neutron; B proton; C beta (electron)
  (ii) composed of quarks
  (iii) $N = Z$; lies close to the stability line
4* c (ii) electron, $_{-1}^{0}e$
  (iii) positron, $_{+1}^{0}e$

60 Electronics and control

2 a $V_1 = V_2 = 3$V
  b $V_1 = 1$V, $V_2 = 5$V
  c $V_1 = 4$V, $V_2 = 2$V
4 A: AND; B: OR; C: NAND; D: NOR
5 A: OR; B: NOT; C: NAND; D: NOR; E: AND
6 a (i) NOT (ii) 0 1
        1 0
  b Switch 0; gate output high
8 a (i) OR (ii) 0, 1, 1, 1
  b (i) AND gates
  (ii) power off or door open or water level low
9 a (i) thermistor
  b (ii) analogue, smooth
  c (i) $700\,\Omega$ (ii) $2100\,\Omega$
  (iii) 1V
  (iv) output remains high as input to NOT is <1V
10 a Reset **b** Q = 1 **c** Q = 1
  d R = 0
11* a P diode; Q transistor; R_1 LDR
  b (i) 4.8V
  (ii) output NOT gate is low
  c 0.06V
12* a (iii) stabilizes system
  b (i) 1 0 1 0 (ii) goes off
        0 0 1 0

61 Telecommunications

2 a (i) tuner
  (ii) audio frequency amplifier
  (iii) loudspeaker
  b (i) amplitude modulated
4 b Decoder – brain; receiver – outer ear; transmitter – mouth
6* a B sky; C space
  c (i) B (ii) fades
  (iii) no satellite close enough
7* a 10^9Hz
  c (i) loss of signal
  (ii) size of opening; wavelength
  (iii) frequency too low

Additional questions

1 a Adjust focus and brightness; adjust X and Y shifts
  b Calculate (displacement in cm) × (volts/cm)

c (i) turn on time base
(ii) vertical displacement doubles
2 a Electrons attracted to positive plate
3 a (i) 6 protons, 6 neutrons
(ii) 6 protons, 8 neutrons, $N>Z$
(iii) 6 protons
b (ii) $N=Z$
(iii) $^{12}_{6}C$, $N=Z$
c (i) number of protons increases to 7; number of neutrons decreases to 7
(ii) electron
4* a (iii) 12 seconds
b (i) 22 800 years
5* b (i) α
(ii) $^{116}_{84}Po + ^{4}_{2}He$
(iii) radon decays
(iv) radioactive radon gas can be breathed in and may lodge in the lungs
6 a (i) diode (ii) capacitor
(iii) transistor
(iv) multimeter
(v) microphone
b (i) A input sensor; B control circuit; C output sensor
(ii) AND
(iii) 1 1 1 1
     1 0 0 0
     0 1 0 0
     0 0 0 0
(iv) key on and saddle pressed
7* a F input; G processor; H output
b (i) 0 0 1
     0 1 1
     1 0 1
     1 1 0
(ii) AND + NOT
(iii) 0 0 1
     0 1 0
     1 0 0
     1 1 0
Alarm sounds only if both pressure switches released
8* a (i) B (ii) C **b** OR
c (i) LDR or photodiode
(ii) thermistor
d $3\,k\Omega$
e (i) $5.5\,k\Omega$ (ii) $1.09\,mA$
f (i) $2.7\,V$ (ii) yes (iii) $0.75\,V$
(iv) no
10 a Sound; electrical; light
b Vibrate; increases; magnetic field; vibrate
11 a (i) Y gain, (ii) time base
b $7\,V$ **c** Half-wave rectified
e So the current from dynamo does not by-pass the lamp by flowing through the battery
12* **a** $1.6 \times 10^{-16}\,J$ **b** $1.9 \times 10^{7}\,m/s$

Earth and space physics

62 Structure of the Earth
1 d is incorrect
3* a A crust; B core; C mantle
b Both refracted, and both transmitted through solids; P travel faster than S; P are longitudinal and S are transverse waves
c (i) P and S (ii) P

63 The Solar System
1 A, B, E
2 B, E
4 a Lower
b More time
5 a Composed of gas not liquid or solids
b (i) magnetic field strength
(ii) further from the Sun
(iii) further from the Sun
c Density of surface rocks is lower (mainly composed of sulphur); no impact craters (due to active volcanoes)
d (i) burnt up
(ii) fell slower so heated more slowly

64 Stars and the Universe
1 A, B, D
4* a Jupiter is further from the Sun
b 5 years
c (i) X is the point nearest to the Sun
d (ii) blue
(iii) proportional
(iv) 1750 million light years
(v) 1750 million years
(vi) 17 500 million years
(vii) age of the universe

Additional questions
1 A: originally sedimentary, later parts became metamorphic;
B: originally igneous, later parts became metamorphic;
C: originally sedimentary, later parts became metamorphic
2* d (i) 10 (ii) 1000
3* a A crust; B mantle; C outer core; D inner core
b P (longitudinal) waves arrive first followed by S (transverse) waves
c (i) total internal reflection
(ii) refraction (P)
(iii) S wave not transmitted in liquid
5 a A star; B planet; C satellite
b (i) star (ii) planet or moon
(iii) moon
6 a Star; Milky Way; Universe; moon
b Gravity
9* b Age = $1/H_0$
c 10 000 million years

Revision questions
1 E
2 a 60° **b** 30°
3 a Refraction **b** POQ
c Towards **d** 40°
e $90 - 65 = 25°$
4 C
5 D
6 B
7 C
8 A
9 a Dispersion
b (i) red (ii) violet
10 B
11 a Magnifies
b Wide field of view
c (i) produces a parallel beam from a small light at focus
(ii) focuses light from distant objects
12* a A iris; B retina; C optic nerve; D cornea; E pupil
b (ii) inverted, magnified
c (i) inverted, larger
(ii) inverted, smaller
13 C
14 E
15 a Circular **b, c** No change
16 a (i) troughs from one arrive at same time as crests from other
(ii) crests from one arrive at same time as crests from other
b (i) larger wave
(ii) cancel out
(iii) larger wave
17 D
18 E
19 E
20 a Longitudinal
b (i) compression
(ii) rarefraction
21 D
22 B
23 A
24 D
25 C
26 B
27 D
28 a (ii) greater detail obtained in image
(iii) communications
b Longitudinal
c Liquids are denser so molecules closer together
d Echo technique
29* C
30 E
31 A
32 C
33 C
34 D
35 a (i) compressed
(ii) not long enough
b (i) linear plot (ii) 7.5 cm
(iii) proportional

c (i) 4 N/cm (ii) 400 N/m
d (i) 25 N
e (i) 0.8 Hz
  (ii) bumps occur at the resonant frequency of spring system
  (iii) add extra weight to the springs or drive faster
36* **a** Yes, 1 mm = 0.001 m **b** E
37 A
38 B
39 A
40 C
41 D
42 C
43 A
44 **b** (i) B lowest (ii) B
  (iii) higher speed, lower pressure
  c Air pressure is lower at constriction than in the reservoir and air flow is in the direction of the nozzle
45 D
46 **a** 480 m **b** 6 m/s
  c 8/20 = 0.4 m/s²
47 E
48 C
49 B
50 D (F = 20 − 12 = 8 N; m = 2 kg; a = F/m = 8/2 = 4 m/s²)
51 D
52 E
53 D
54 **a** B **b** A
55 A
56 E
57 **a** (i) B
  b (i) reaction time of driver, tyre tread
  (ii) 8 m
58* **a** (iii) 3.55 m/s²
  b (i) 195 N (ii) 550 N
  (iii) 1386 J (iv) 693 W
  (v) energy lost as heat
  c Heat used to evaporate water of sweat
59* **a** (i) equal in size
  (ii) opposite in direction
  b 0.83 m/s
  c Friction exerted by ice on skates provides centripetal force
60 C
61 **a** (i) 280 K (ii) 298 K (iii) 250 K
  b (i) 27 °C (ii) 300 °C (iii) −73 °C
62 B
63 C
64 A
65 D
66 C
67 C
68 **a** Solid M, liquid O, gas L
  c (i) bimetallic strip bends and contact is broken
  (ii) digital (on or off)

69 **a** (i) coal, oil, gas
  (ii) occurs without using up any of the Earth's energy reserves
  b (i) no sulphur dioxide emitted and more efficient
  (ii) less reserves
  c Not many suitable sites where it is always windy; large ground area needed for windmill farm
  d Reserves of fossil fuels finite
70* **a** Particles collide with walls of cylinder
  b 8.33 m³ **c** 833 balloons
71* **a** (i) 30 minutes (ii) 12 °C
  b Larger surface area to radiate heat
  c 48 kJ **d** 800 kJ
  e Radiates as heated
  f Student incorrect; calculate for each case *specific heat capacity* × *temperature drop*
72 D
73 **a** 1 Ω **b** 3 A **c** 6 V
74 **a** 3 Ω **b** 2 A
  c 4 V across 2 Ω; 2 V across 1 Ω
75 D
76 B
77 **a** D **b** A **c** E
78 B
79 C
80 D
81 E
82 B
83 **a** E **b** A **c** C **d** B **e** D
84 **a** (i) D (ii) A
  b L₁; takes largest current
  c LDR, switch, thermistor
85 **a** A2, B4, C1, D3
  b (i) series
  (ii) if lamp is on in the living area it must also be on in the sleeping area
  (iii) lamps can be switched on independently; lamps brighter
  c (i) conduction and convection
86 **a** Kinetic and heat
  b (i) yellow/green (ii) brown
  c 8.25 kWh
87* A
88* **a** Increase speed of rotation, area of coil, strength of magnet or number of turns on coil
  b (i) heat; electrical
  c (i) Sun (ii) renewable
89 C
90 **a** A **b** C
91 E
92 B
93 B
94 B
95 C (symbol is ₃⁷Li)
96 **a** C **b** A **c** B **d** A
97 **a** D **b** E

98 E
99 **a** 6 +ve protons; 6 neutral neutrons; 6 −ve electrons
  b 2 extra neutrons in carbon-14
  c (i) nucleus (ii) C
100 **a** (ii) becomes k.e. of the particles
101* C
103* **a** Convert sound to electrical signal
  b Amplifier
  d Frequency selected in tuner
105 **c** Because the density of the mantle changes gradually not abruptly
106 **a** (i) 365 days
  (ii) C (northern hemisphere tilted towards the Sun)
  b Rotation of the Earth on its axis
  c Star; nine; moons
107 **a** 11 hours **b** Further from Sun
  c Low density **d** Less
108 **a** Gravity **b** Reduces
  c (i) decreases (ii) decreases
109 4.4 × 10⁹ kg/s
110* **a** Gravitational field accelerates probe
  b (i) 5.8 × 10⁹ J
  (ii) friction generates heat when probe enters Titan's atmosphere
  (iii) large area increases air resistance opposing motion
  c (ii) Universe is expanding

Mathematics for physics

1 **a** 3 **b** 5 **c** 8/3 **d** 20 **e** 12 **f** 6
  g 2 **h** 3 **i** 8
2 **a** f = v/λ **b** λ = v/f **c** I = V/R
  d R = V/I **e** m = d × V
  f V = m/d **g** s = vt **h** t = s/v
3 **a** I² = P/R **b** I = √(P/R) **c** a = 2s/t²
  d t² = 2s/a
  e t = √(2s/a) **f** v = √(2gh)
  g y = Dλ/a **h** ρ = AR/l
4 **a** 10 **b** 34 **c** 2/3 **d** 1/10 **e** 10
  f 3 × 10⁸
5 **a** 2.0 × 10⁵ **b** 10 **c** 8 **d** 2.0 × 10⁸
  e 20 **f** 300
6 **a** 4 **b** 2 **c** 5 **d** 8 **e** 2/3 **f** −3/4
  g 13/6 **h** −16 **i** 1
7 a = (v − u)/t **a** 5 **b** 60 **c** 75
8 a = (v² − u²)/2s
9 **b** Extension ∝ mass since the graph is a straight line through the origin
10 **b** No: graph is a straight line but does not pass through the origin
  c 32
11 **a** is a curve
  b is a straight line through the origin, therefore s ∝ t² or s/t² = a constant = 2

367

Index